AF358796

Exercise Testing and Prescription Strategies to Improve Quality of Life

Exercise Testing and Prescription Strategies to Improve Quality of Life

Editors

Rafael Oliveira
João Brito

Basel • Beijing • Wuhan • Barcelona • Belgrade • Novi Sad • Cluj • Manchester

Editors

Rafael Oliveira
School of Sport
Santarém Polytechnic
University
Rio Maior
Portugal

João Brito
School of Sport
Santarém Polytechnic
University
Rio Maior
Portugal

Editorial Office
MDPI AG
Grosspeteranlage 5
4052 Basel, Switzerland

This is a reprint of articles from the Special Issue published online in the open access journal *Healthcare* (ISSN 2227-9032) (available at: www.mdpi.com/journal/healthcare/special_issues/CNEH7D1A4N).

For citation purposes, cite each article independently as indicated on the article page online and as indicated below:

Lastname, A.A.; Lastname, B.B. Article Title. *Journal Name* **Year**, *Volume Number*, Page Range.

ISBN 978-3-7258-1632-3 (Hbk)
ISBN 978-3-7258-1631-6 (PDF)
doi.org/10.3390/books978-3-7258-1631-6

Contents

About the Editors ix

Preface xi

Rafael Oliveira and João Paulo Brito
Updating Exercise Testing Strategies and Exercise Prescription Protocols
Reprinted from: *Healthcare* **2024**, *12*, 901, doi:10.3390/healthcare12090901 1

Billy C. L. So, Manny M. Y. Kwok, Nakita W. L. Lee, Andy W. C. Lam, Anson L. M. Lau, Allen S. L. Lam, et al.
Lower Limb Muscles' Activation during Ascending and Descending a Single Step-Up Movement: Comparison between In water and On land Exercise at Different Step Cadences in Young Injury-Free Adults
Reprinted from: *Healthcare* **2023**, *11*, 441, doi:10.3390/healthcare11030441 5

Francisco Javier Pérez-Quero, Antonio Granero-Gallegos, Antonio Baena-Extremera and Raúl Baños
Goal Orientations of Secondary School Students and Their Intention to Practise Physical Activity in Their Leisure Time: Mediation of Physical Education Importance and Satisfaction
Reprinted from: *Healthcare* **2023**, *11*, 568, doi:10.3390/healthcare11040568 20

Bruno Lombardi Amado, Claudio Andre Barbosa De Lira, Rodrigo Luiz Vancini, Pedro Forte, Taline Costa, Katja Weiss, et al.
Comparison of Knee Muscular Strength Balance among Pre- and Post-Puberty Adolescent Swimmers: A Cross-Sectional Pilot Study
Reprinted from: *Healthcare* **2023**, *11*, 744, doi:10.3390/healthcare11050744 31

Mohamed Salih Mahfouz, Ahmad Y. Alqassim, Nasser Hadi Sobaikhi, Abdulaziz Salman Jathmi, Fahad Omar Alsadi, Abdullah Mohammed Alqahtani, et al.
Physical Activity, Mental Health, and Quality of Life among School Students in the Jazan Region of Saudi Arabia: A Cross-Sectional Survey When Returning to School after the COVID-19 Pandemic
Reprinted from: *Healthcare* **2023**, *11*, 974, doi:10.3390/healthcare11070974 40

Lorena Rodríguez-García, Halil Ibrahim Ceylan, Rui Miguel Silva, Ana Filipa Silva, Amelia Guadalupe-Grau and Antonio Liñán-González
Effects of 10-Week Online Moderate- to High-Intensity Interval Training on Body Composition, and Aerobic and Anaerobic Performance during the COVID-19 Lockdown
Reprinted from: *Healthcare* **2024**, *12*, 37, doi:10.3390/healthcare12010037 50

Carla Brites-Lagos, Liliana Ramos, Anna Szumilewicz and Rita Santos-Rocha
Feasibility of a Supervised Postpartum Exercise Program and Effects on Maternal Health and Fitness Parameters—Pilot Study
Reprinted from: *Healthcare* **2023**, *11*, 2801, doi:10.3390/healthcare11202801 61

Peng-Shuai Ma, Wi-Young So and Hyongjun Choi
Using the Health Belief Model to Assess the Physical Exercise Behaviors of International Students in South Korea during the Pandemic
Reprinted from: *Healthcare* **2023**, *11*, 469, doi:10.3390/healthcare11040469 77

Holger Stephan, Udo Frank Wehmeier, Tim Förster, Fabian Tomschi and Thomas Hilberg
Additional Active Movements Are Not Required for Strength Gains in the Untrained during
Short-Term Whole-Body Electromyostimulation Training
Reprinted from: *Healthcare* **2023**, *11*, 741, doi:10.3390/healthcare11050741 87

**Jerrican Tan, Oleksandr Krasilshchikov, Garry Kuan, Hairul Anuar Hashim, Monira I.
Aldhahi, Sameer Badri Al-Mhanna, et al.**
The Effects of Combining Aerobic and Heavy Resistance Training on Body Composition, Muscle
Hypertrophy, and Exercise Satisfaction in Physically Active Adults
Reprinted from: *Healthcare* **2023**, *11*, 2443, doi:10.3390/healthcare11172443 99

**Guilherme Corrêa De Araújo Moury Fernandes, José G. G. Barbosa Junior, Aldo Seffrin,
Lavínia Vivan, Claudio A. B. de Lira, Rodrigo L. Vancini, et al.**
Amateur Female Athletes Perform the Running Split of a Triathlon Race at Higher Relative
Intensity than the Male Athletes: A Cross-Sectional Study
Reprinted from: *Healthcare* **2023**, *11*, 418, doi:10.3390/healthcare11030418 111

**Ana Myriam Lavín-Pérez, Juan Luis León-Llamas, Francisco José Salas Costilla, Daniel
Collado-Mateo, Raúl López de las Heras, Pablo Gasque Celma, et al.**
Validity of On-Line Supervised Fitness Tests in People with Low Back Pain
Reprinted from: *Healthcare* **2023**, *11*, 1019, doi:10.3390/healthcare11071019 118

Xiaozhu Ma, Jiangtao Yan and Wanjun Liu
An Early Indicator in Evaluating Cardiac Dysfunction Related to Premature Ventricular
Complexes: Cardiorespiratory Capacity
Reprinted from: *Healthcare* **2023**, *11*, 2940, doi:10.3390/healthcare11222940 133

**Dejan Reljic, Fabienne Frenk, Hans Joachim Herrmann, Markus Friedrich Neurath and
Yurdagül Zopf**
Maximum Heart Rate- and Lactate Threshold-Based Low-Volume High-Intensity Interval
Training Prescriptions Provide Similar Health Benefits in Metabolic Syndrome Patients
Reprinted from: *Healthcare* **2023**, *11*, 711, doi:10.3390/healthcare11050711 144

**Mónica Sousa, Rafael Oliveira, João Paulo Brito, Alexandre Duarte Martins, João Moutão
and Susana Alves**
Effects of Combined Training Programs in Individuals with Fibromyalgia: A Systematic Review
Reprinted from: *Healthcare* **2023**, *11*, 1708, doi:10.3390/healthcare11121708 162

**Wei-Hsiu Hsu, Wei-Bin Hsu, Zin-Rong Lin, Shr-Hsin Chang, Chun-Hao Fan,
Liang-Tseng Kuo, et al.**
Effects of 24 Weeks of a Supervised Walk Training on Knee Muscle Strength and Quality of Life
in Older Female Total Knee Arthroplasty: A Retrospective Cohort Study
Reprinted from: *Healthcare* **2023**, *11*, 356, doi:10.3390/healthcare11030356 184

**Inés Muñoz-Paredes, Azael J. Herrero, Natalia Román-Nieto, Alba M. Peña-Gomez and
Jesús Seco-Calvo**
Influence of Transcranial Direct Current Stimulation and Exercise on Fatigue and Quality of
Life in Multiple Sclerosis
Reprinted from: *Healthcare* **2023**, *11*, 84, doi:10.3390/healthcare11010084 195

**Ángel Denche-Zamorano, David Manuel Mendoza-Muñoz, Sabina Barrios-Fernandez,
Carolina Perez-Corraliza, Juan Manuel Franco-García, Jorge Carlos-Vivas, et al.**
Physical Activity Reduces the Risk of Developing Diabetes and Diabetes Medication Use
Reprinted from: *Healthcare* **2022**, *10*, 2479, doi:10.3390/healthcare10122479 210

**Irene Ferrando-Terradez, Lirios Dueñas, Ivana Parčina, Nemanja Ćopić,
Svetlana Petronijević, Gianfranco Beltrami, et al.**
Women's Involvement in Steady Exercise (WISE): Study Protocol for a Randomized Controlled
Trial
Reprinted from: *Healthcare* **2023**, *11*, 1279, doi:10.3390/healthcare11091279 . . . . **224**

**Carolina Alexandra Cabo, Orlando Fernandes, María Mendoza-Muñoz,
Sabina Barrios-Fernandez, Laura Muñoz-Bermejo, Rafael Gómez-Galán, et al.**
An Active Retirement Programme, a Randomized Controlled Trial of a Sensorimotor Training
Programme for Older Adults: A Study Protocol
Reprinted from: *Healthcare* **2023**, *11*, 86, doi:10.3390/healthcare11010086 . . . . **238**

About the Editors

Rafael Oliveira

Rafael Franco Soares Oliveira is an adjunct professor at the Escola Superior de Desporto de Rio Maior, Instituto Politécnico de Santarém (Santarém Polytechnic University, School of Sport), an integrated member of the Research Center in Sports Sciences, Health, and Human Development, and a collaborator of the Life Quality Research Centre. He completed his Ph.D. in Sports Sciences at the University of Beira Interior, , Covilhã, Portugal.

His research activity has been focusing on exercise physiology, testing, control and prescription, load monitoring/quantification, strength and conditioning, and sports training methodology. Currently, he is an author of more than 130 published scientific articles (indexing in *Journal of Citation Reports or Scientific Journal Ranking*) and about 20 books/book chapters.

Furthermore, he was awarded first place in research and development by the Polytechnic Institute of Santarem in 2022, 2023, and 2024. He was also awarded with an honorable mention as the researcher with the best quotation in scientific production in 2022 and first place as the researcher with the best quotation in scientific production in 2023 by the Life Quality Research Centre.

João Brito

João Paulo Moreira de Brito is a full professor at the Escola Superior de Desporto de Rio Maior, Instituto Politécnico de Santarém (Santarém Polytechnic University, School of Sport), , an integrated member of the Research Center in Sports Sciences, Health, and Human Development, and a collaborator of the Life Quality Research Centre. He completed his Ph.D. in Sports Sciences at the University of Lleida, Lleida, Spain.

His research activity has been focusing on exercise physiology, physiological assessment of athletes from different sports, determination of physical and physiological profiles of elite athletes, subjective and objective load monitoring/quantification, and strength training to improve performance in male and female athletes. Currently, he is an author of more than 80 published scientific articles in indexed journals and about 17 books or book chapters.

Preface

The connections between exercise and improved physical and psychological health are well established. The benefits apply to both healthy individuals and those with chronic illnesses. The purpose of this reprint is to explore in greater detail the links between exercise and quality of life (QOL) in both healthy individuals and those with chronic diseases, as well as to examine issues that impact the development, implementation, and evaluation of exercise programs and exercise testing. Topics to be addressed include the following: characterization of QOL outcomes in exercise studies, development of exercise goals, exercise prescriptions, exercise testing, exercise adherence, methods to evaluate the efficacy of exercise in relation to QOL, and the translation of research into practice.

This reprint contributes to the field with 19 articles and 1 editorial, and it was derived from a Special Issue of *Healthcare* with the focus on lifestyle and physical activity as a reference for improving quality of life and increasing life expectancy. Despite updated information, such as the book from the American College of Sports Medicine about exercise testing and prescriptions for general populations in different stages of life (e.g., children, pregnancy, and older) or with some diseases (e.g., type 2 diabetes, hypertension, obesity, dyslipidemia, or chronic pulmonary obstructive diseases), research still focuses on different types of testing and exercise prescriptions that can be described in details in this type of book. Furthermore, not all strength and conditioning programs can explore all fitness components, as well as not all studies can apply all types of testing. In this sense, the knowledge about more training programs for a specific target population and how to manage them are still relevant and needed. In the same way, different testing and monitoring tools can provide different information and interpretations.

Therefore, the aim of this Special Issue, which now constitutes a reprint, was to compile and provide updated information on exercise testing and prescription to provide new and effective strategies to improve quality of life. Several authors, each notable in their respective fields, have contributed to this reprint through their studies. We are grateful for their collaboration and the confidence they placed in this Special Issue of the *Healthcare* journal.

Any fitness professional, physiologist, or practitioner who is interested in the topics of fibromyalgia, multiple sclerosis, knee arthroplasty, obesity with metabolic syndrome, low back pain, postpartum women, and people with premature ventricular beats are welcome to read the present document.

Rafael Oliveira and João Brito
Editors

MDPI

Updating Exercise Testing Strategies and Exercise Prescription Protocols

Rafael Oliveira [1,2,*] **and João Paulo Brito** [1,2]

[1]   Sports Science School of Rio Maior, Polytechnic Institute of Santarém, 2040-413 Rio Maior, Portugal; jbrito@esdrm.ipsantarem.pt

[2]   Research Centre in Sport Sciences, Health Sciences and Human Development, 5001-801 Vila Real, Portugal

*   Correspondence: rafaeloliveira@esdrm.ipsantarem.pt

Citation: Oliveira, R.; Brito, J.P. Updating Exercise Testing Strategies and Exercise Prescription Protocols. *Healthcare* **2024**, *12*, 901. https://doi.org/10.3390/healthcare12090901

Received: 10 April 2024
Revised: 19 April 2024
Accepted: 25 April 2024
Published: 26 April 2024

Exercise testing and prescription is still a hot topic. It is evidenced by the constant updating of the guidelines for exercise testing and prescription provided by the American College of Sports Medicine (ACSM). While the guidelines describe the most relevant test for general population with or without specific conditions, it also describes Frequency, Intensity, Time, and Type (FITT) of exercise prescription [1,2].

Another example of this hot topic is the "physical activity on prescription" model adopted by Sweden in which all licensed healthcare professionals may prescribe it. In 2019, a systematic review about the main elements of this model was published. Specifically, the study included studies that compared adults who received this models versus those who did not and the main findings showed that there was a tendency for higher levels of physical activity in those who followed the Swedish model [3]. Additionally, this model was considered the best by the European Commission and it was then used by 10 European countries in a project to implement this approach [4].

Even so, there are several questions that are still not well addressed such as other specific tests in the different contexts, the specific exercises or even questions about training periodization.

Therefore, this Special Issue updated information on exercise testing and prescription strategies to improve quality of life. The present Special Issue contributed to the field with 19 articles:

1. Cabo, C.A.; Fernandes, O.; Mendoza-Muñoz, M.; Barrios-Fernandez, S.; Muñoz-Bermejo, L.; Gómez-Galán, R.; Parraca, J.A. An Active Retirement Programme, a Randomized Controlled Trial of a Sensorimotor Training Programme for Older Adults: A Study Protocol. *Healthcare* **2023**, *11*, 86. https://doi.org/10.3390/healthcare1101008 6

2. Ferrando-Terradez, I.; Dueñas, L.; Parčina, I.; Ćopić, N.; Petronijević, S.; Beltrami, G.; Pezzoni, F.; San Martín-Valenzuela, C.; Gijssel, M.; Moliterni, S.; et al. Women's Involvement in Steady Exercise (WISE): Study Protocol for a Randomized Controlled Trial. *Healthcare* **2023**, *11*, 1279. https://doi.org/10.3390/healthcare11091279

3. Sousa, M.; Oliveira, R.; Brito, J.P.; Martins, A.D.; Moutão, J.; Alves, S. Effects of Combined Training Programs in Individuals with Fibromyalgia: A Systematic Review. *Healthcare* **2023**, *11*, 1708. https://doi.org/10.3390/healthcare11121708

4. Denche-Zamorano, Á.; Mendoza-Muñoz, D.M.; Barrios-Fernandez, S.; Perez-Corraliza, C.; Franco-García, J.M.; Carlos-Vivas, J.; Pastor-Cisneros, R.; Mendoza-Muñoz, M. Physical Activity Reduces the Risk of Developing Diabetes and Diabetes Medication Use. *Healthcare* **2022**, *10*, 2479. https://doi.org/10.3390/healthcare10122479

5. Muñoz-Paredes, I.; Herrero, A.J.; Román-Nieto, N.; Peña-Gomez, A.M.; Seco-Calvo, J. Influence of Transcranial Direct Current Stimulation and Exercise on Fatigue and Quality of Life in Multiple Sclerosis. *Healthcare* **2023**, *11*, 84. https://doi.org/10.3390/healthcare11010084

6. Hsu, W.-H.; Hsu, W.-B.; Lin, Z.-R.; Chang, S.-H.; Fan, C.-H.; Kuo, L.-T.; Hsu, W.-W.R. Effects of 24 Weeks of a Supervised Walk Training on Knee Muscle Strength and Quality of Life in Older Female Total Knee Arthroplasty: A Retrospective Cohort Study. *Healthcare* **2023**, *11*, 356. https://doi.org/10.3390/healthcare11030356

7. De Araújo Moury Fernandes, G.C.; Barbosa Junior, J.G.G.; Seffrin, A.; Vivan, L.; de Lira, C.A.B.; Vancini, R.L.; Weiss, K.; Knechtle, B.; Andrade, M.S. Amateur Female Athletes Perform the Running Split of a Triathlon Race at Higher Relative Intensity than the Male Athletes: A Cross-Sectional Study. *Healthcare* **2023**, *11*, 418. https://doi.org/10.3390/healthcare11030418

8. So, B.C.L.; Kwok, M.M.Y.; Lee, N.W.L.; Lam, A.W.C.; Lau, A.L.M.; Lam, A.S.L.; Chan, P.W.Y.; Ng, S.S.M. Lower Limb Muscles' Activation during Ascending and Descending a Single Step-Up Movement: Comparison between In water and On land Exercise at Different Step Cadences in Young Injury-Free Adults. *Healthcare* **2023**, *11*, 441. https://doi.org/10.3390/healthcare11030441

9. Ma, P.-S.; So, W.-Y.; Choi, H. Using the Health Belief Model to Assess the Physical Exercise Behaviors of International Students in South Korea during the Pandemic. *Healthcare* **2023**, *11*, 469. https://doi.org/10.3390/healthcare11040469

10. Pérez-Quero, F.J.; Granero-Gallegos, A.; Baena-Extremera, A.; Baños, R. Goal Orientations of Secondary School Students and Their Intention to Practise Physical Activity in Their Leisure Time: Mediation of Physical Education Importance and Satisfaction. *Healthcare* **2023**, *11*, 568. https://doi.org/10.3390/healthcare11040568

11. Reljic, D.; Frenk, F.; Herrmann, H.J.; Neurath, M.F.; Zopf, Y. Maximum Heart Rate- and Lactate Threshold-Based Low-Volume High-Intensity Interval Training Prescriptions Provide Similar Health Benefits in Metabolic Syndrome Patients. *Healthcare* **2023**, *11*, 711. https://doi.org/10.3390/healthcare11050711

12. Stephan, H.; Wehmeier, U.F.; Förster, T.; Tomschi, F.; Hilberg, T. Additional Active Movements Are Not Required for Strength Gains in the Untrained during Short-Term Whole-Body Electromyostimulation Training. *Healthcare* **2023**, *11*, 741. https://doi.org/10.3390/healthcare11050741

13. Amado, B.L.; De Lira, C.A.B.; Vancini, R.L.; Forte, P.; Costa, T.; Weiss, K.; Knechtle, B.; Andrade, M.S. Comparison of Knee Muscular Strength Balance among Pre- and Post-Puberty Adolescent Swimmers: A Cross-Sectional Pilot Study. *Healthcare* **2023**, *11*, 744. https://doi.org/10.3390/healthcare11050744

14. Mahfouz, M.S.; Alqassim, A.Y.; Sobaikhi, N.H.; Jathmi, A.S.; Alsadi, F.O.; Alqahtani, A.M.; Shajri, M.M.; Sabi, I.D.; Wafi, A.M.; Sinclair, J. Physical Activity, Mental Health, and Quality of Life among School Students in the Jazan Region of Saudi Arabia: A Cross-Sectional Survey When Returning to School after the COVID-19 Pandemic. *Healthcare* **2023**, *11*, 974. https://doi.org/10.3390/healthcare11070974

15. Lavín-Pérez, A.M.; León-Llamas, J.L.; Salas Costilla, F.J.; Collado-Mateo, D.; López de las Heras, R.; Gasque Celma, P.; Villafaina, S. Validity of On-Line Supervised Fitness Tests in People with Low Back Pain. *Healthcare* **2023**, *11*, 1019. https://doi.org/10.3390/healthcare11071019

16. Tan, J.; Krasilshchikov, O.; Kuan, G.; Hashim, H.A.; Aldhahi, M.I.; Al-Mhanna, S.B.; Badicu, G. The Effects of Combining Aerobic and Heavy Resistance Training on Body Composition, Muscle Hypertrophy, and Exercise Satisfaction in Physically Active Adults. *Healthcare* **2023**, *11*, 2443. https://doi.org/10.3390/healthcare11172443

17. Brites-Lagos, C.; Ramos, L.; Szumilewicz, A.; Santos-Rocha, R. Feasibility of a Supervised Postpartum Exercise Program and Effects on Maternal Health and Fitness Parameters—Pilot Study. *Healthcare* **2023**, *11*, 2801. https://doi.org/10.3390/healthcare11202801

18. Ma, X.; Yan, J.; Liu, W. An Early Indicator in Evaluating Cardiac Dysfunction Related to Premature Ventricular Complexes: Cardiorespiratory Capacity. *Healthcare* **2023**, *11*, 2940. https://doi.org/10.3390/healthcare11222940

19. Rodríguez-García, L.; Ceylan, H.I.; Silva, R.M.; Silva, A.F.; Guadalupe-Grau, A.; Liñán-González, A. Effects of 10-Week Online Moderate- to High-Intensity Interval Training on Body Composition, and Aerobic and Anaerobic Performance during the COVID-19 Lockdown. *Healthcare* **2024**, *12*, 37. https://doi.org/10.3390/healthcare12010037

In resume, the articles addressed at least one dimension of physical fitness (cardiorespiratory endurance, body composition, muscular strength, muscular endurance, and flexibility) as well as quality of life, mainly measured by questionnaires (see Table 1). Moreover, it is relevant to highlight that this special issue also included studies that addressed specific conditions such as fibromyalgia, multiple sclerosis, knee arthroplasty, obesity with metabolic syndrome, low back pain, postpartum women, and people with premature ventricular beats.

Table 1. Analysis of the published contributions in the Special Issue.

Contributor Number	Target Population	Study Type	Exercise Testing	Exercise Type
1	Older adults	Protocol	Body composition, physical fitness, and questionnaires: health related quality of life; physical activity level	Cardiorespiratory and strength
2	Women adults	Protocol	Daily steps, program adherence, anthropometry, body composition, plank test, 6-min walking test; questionnaires: International Physical Activity; healthy lifestyle and personal control; Pittsburgh Sleep Quality Index; visual analogue scale; physical activity enjoyment scale	Cardiorespiratory (HIIT) training and behavior change theories and techniques.
3	People with fibromyalgia	Systematic Review	Several instruments/tests were used to assess pain, sleep quality, health status and strength gains in the upper and lower limbs	Combined training, HIIT, Tai Chi, aerobic exercise, body balance and strength
4	Participants from 15 to 69 years	Cross-sectional	Diabetes prevalence and Diabetes medication use in diabetics, Physical Activity Level, Body Mass Index, Spanish National Health Survey	NA
5	People with multiple sclerosis	Cross-over	Questionnaires: Multiple Sclerosis International Quality of Life; International Physical Activity; Kurtzke Expanded Disability Status Scale; Spanish version of the Modified Fatigue Impact Scale	Strength and aerobic
6	Older women with knee arthroplasty	Cohort	Isokinetic strength, 6-min walk test, the 8-foot up-and-go test, and the 30-s chair stand test, Knee Injury and Osteoarthritis Outcome Score (KOOS) Questionnaire	Cardiorespiratory exercise
7	Amateur male and female adult athletes	Cross-sectional	Cardiopulmonary exercise test	NA
8	Young adults	Cross-sectional	Muscle activation in water and land step exercise	NA
9	Adults	Cross-sectional	Health belief model (questionnaire)	NA
10	Young (from 12 to 19 years)	Cross-sectional	Questionnaires: Perception of Success; Importance of Physical Education; Satisfaction with Physical Education Intention to partake in leisure time physical activity.	NA
11	Obese people with metabolic syndrome	Cohort	Hydration, blood, anthropometric, cardiopulmonary exercise test, quality of life (questionnaire), daily nutrition.	Cardiorespiratory (HIIT)
12	Adults	Cohort		Strength with electromyostimulation
13	Adolescent swimmers	Cross-sectional	Sexual maturity, body composition, isokinetic strength test	NA
14	Young (from 12 to 18 years)	Cross-sectional	Questionnaires: Fels PAQ; Depression Anxiety Stress Scales; Pediatric Quality of Life Inventory	NA
15	Adult and older (45–72 years) with low back pain	Cross-sectional	Online evaluations: 30-s chair stand-up test; arm curl test; 2-min step test in place; chair-stand and reach test; back scratch; 8-foot up-and-go test; Sharpened Romberg; one-legged stance test	NA
16	Adults	Cohort	body composition, muscle hypertrophy, and exercise satisfaction	Combined aerobic and strength
17	Postpartum women	Cohort pilot study	Blood pressure; anthropometry, body composition; chair stand test; cardiopulmonary exercise test, push-up test; V-sit and reach. Questionnaires: Physical Activity Readiness Questionnaire for Everyone; International Physical Activity; World Health Organization Quality of Life Questionnaire; Pelvic Girdle; Roland-Morris Disability; Fatigue Assessment Scale; Edinburgh Postpartum Depression Scale	Combined training: cardiorespiratory, postural, functional/resistance training, neuromotor training, stretching, breathing, and relaxation exercises
18	People with premature ventricular beats	Cross-sectional	Cardiopulmonary exercise test	NA
19	Women adults	Cohort	Cardiopulmonary exercise test	Cardiorespiratory (HIIT)

HIIT, high-intensity interval training; NA, non-applicable.

The current special issue provided and constitutes relevant information for fitness professional and exercise physiologists. At the same time, it showed meaningful findings about the online exercise testing procedures (see contributors 15 and 17). In addition, it was observed that there are still several research that uses only one type of exercise, despite the general guidelines of the ACSM recommend more that one type [1]. Furthermore, there is a recommendation for future research include behavior change theories in exercise intervention as suggested by the protocol of the contributor 2. Finally, the present special issue also reinforces more research on specific populations, with different ways to control intensity, more specialized tests, while including training periodization practices as well as behavior changes.

Author Contributions: Conceptualization: R.O. and J.P.B.; Writing—Original Draft: R.O.; Writing—Review and Editing: R.O. and J.P.B.; Project administration: R.O. and J.P.B. All authors have read and agreed to the published version of the manuscript.

Funding: R.O. and J.P.B. are research members of the Research Centre in Sports Sciences, Health and Human Development which was funded by National Funds by FCT—Foundation for Science and Technology under the following project UIDB/04045/2020 (https://doi.org/10.54499/UIDB/04045/2020). The funders had no role in the design of the study; in the collection, analyses, or interpretation of data; in the writing of the manuscript, or in the decision to publish the results.

Conflicts of Interest: The authors declare no conflict of interest.

References

1. ACSM—American College of Sports Medicine. *Guidelines for Exercise Testing and Prescription*, 10th ed.; Kluwer, W., Ed.; ACSM: Philadelphia, PA, USA, 2017.
2. ACSM—American College of Sports Medicine. *Guidelines for Exercise Testing and Prescription*, 11th ed.; Liguori, G., Feito, Y., Fountaine, C., Roy, B.A., Eds.; Wolters Kluwer Health: Philadelphia, PA, USA, 2021.
3. Onerup, A.; Arvidsson, D.; Blomqvist, A.; Daxberg, E.L.; Jivegard, L.; Jonsdottir, I.H.; Lundqvist, S.; Mellén, A.; Persson, J.; Sjögren, P.; et al. Physical Activity on Prescription in Accordance with the Swedish Model Increases Physical Activity: A Systematic Review. *Br. J. Sports Med.* **2019**, *53*, 383–388. [CrossRef] [PubMed]
4. EUPAP—European Physical Activity Prescription. Available online: https://www.eupap.org/about (accessed on 15 April 2024).

Article

Lower Limb Muscles' Activation during Ascending and Descending a Single Step-Up Movement: Comparison between In water and On land Exercise at Different Step Cadences in Young Injury-Free Adults

Billy C. L. So *, Manny M. Y. Kwok, Nakita W. L. Lee, Andy W. C. Lam, Anson L. M. Lau, Allen S. L. Lam, Phoebe W. Y. Chan and Shamay S. M. Ng

Gait and Motion Analysis Laboratory, Department of Rehabilitation Sciences, The Hong Kong Polytechnic University, Hong Kong, China
* Correspondence: billy.so@polyu.edu.hk; Tel.: +852-27664377

Abstract: (1) Background: Forward step-up (FSU) simulates the stance phase in stair ascension. With the benefits of physical properties of water, aquatic FSU exercise may be more suitable for patients with lower limb weakness or pain. The purpose of this study is to investigate the effect of progressive steps per min on the surface electromyography (sEMG) of gluteus maximus (GM), biceps femoris (BF), rectus femoris (RF), and gastrocnemius (GA), when performing FSU exercise with different steps per min in water and on land. (2) Methods: Participants (N = 20) were instructed to perform FSU exercises at different steps per min (35, 60, and 95 bpm) in water and on land. The sEMG of the tested muscles were collected. The percentage maximum voluntary isometric contraction (%MVIC) of GM, RF, GA and BF at different environments and steps per min was compared. (3) Result: There was a statistically significant difference of %MVIC of RF at all steps per min comparisons regardless of the movement phases and environments ($p < 0.01$, except for descending phases of 35 bpm vs. 60 bpm). All tested muscles showed a statistically significant lower muscle activation in water ($p < 0.05$) (4) Conclusion: This study found that the %MVIC of the tested muscle in both investigated environments increase as steps per minute increases. It is also found that the movement pattern of FSU exercise activates RF the most among all the tested muscles. Muscle activation of all tested muscles is also found to be smaller in water due to buoyancy property of water. Aquatic FSU exercise might be applicable to patients with lower limb weakness or knee osteoarthritis to improve their lower limb strength.

Keywords: muscles activation; stepping exercises; water immersion

Citation: So, B.C.L.; Kwok, M.M.Y.; Lee, N.W.L.; Lam, A.W.C.; Lau, A.L.M.; Lam, A.S.L.; Chan, P.W.Y.; Ng, S.S.M. Lower Limb Muscles' Activation during Ascending and Descending a Single Step-Up Movement: Comparison between In water and On land Exercise at Different Step Cadences in Young Injury-Free Adults. *Healthcare* **2023**, *11*, 441. https://doi.org/10.3390/healthcare11030441

Academic Editors: Rafael Oliveira and João Paulo Brito

Received: 14 December 2022
Revised: 28 January 2023
Accepted: 29 January 2023
Published: 3 February 2023

1. Introduction

Aquatic exercises are gaining popularity in musculoskeletal rehabilitation, such as knee osteoarthritis, and are recommended to patients who are unable to exercise on land due to physical limitations, such as pain and swelling [1]. The physical properties of water, such as buoyancy, hydrostatic pressure, and drag force, may account for the recommendations made by clinicians [2]. Buoyancy is proposed to be able to reduce compressive force on joint since it can reduce body weight with different immersion depth, for example, immersed to xiphisternum can offload 50–76% of body weight [3,4]. Therefore, buoyancy may reduce joint pain when exercising in water. Hydrostatic pressure, which is directly proportional to immersion depth, could act as a force that aids the resolution of swelling in body parts [5]. Drag force, which is affected by the viscosity and the speed of movement, can be utilized as resistance when a body segment moves relative to water [2]. Drag force is commonly progressed by altering the speed or surface area of limbs [6].

During aquatic exercise, higher speed of moving body parts significantly increases the training load, such that doubling the speed will increase drag force by four times [7].

Therefore, clinicians usually manipulate the parameter of speed, or steps per min to alter the resistance exerted on the body parts to optimize rehabilitation outcomes [1]. Compared to land-based exercise, speed and drag force are unique factors in precise aquatic exercise prescription since they both change the load and movement tasks. The velocity of particular muscle contractions is a crucial factor in specific muscle trainings and performance adaptations which warrant greater attention in aquatic rehabilitation. Previous studies] have investigated the effect of speed on different muscle groups in water and land [8–11]. Speed can be viewed as one of the progression components in aquatic resistance exercise. However, relatively limited studies have investigated the effect of speed on lower limb strengthening exercise, especially unilateral weight-bearing exercise.

Unilateral weight-bearing exercise is commonly prescribed by clinicians in order to stimulate functional muscle recruitment patterns required for daily living and sports, with forward step-up (FSU) exercise being the most common form [12,13]. FSU exercise can be a progression of bilateral weight-bearing exercise as it requires higher muscle activation to achieve [14]. Moreover, FSU exercise is an functional exercise mimicking stair-climbing, which requires simultaneous coordination of hip, knee, and ankle musculature [15]. Aquatic FSU exercise may aid individuals with lower limbs pain or weakness to perform more challenging functional activities with less difficulty.

The potential benefits of FSU exercise are not yet been confirmed and the associated perceptive responses (i.e., the influences of speed and environment) are yet to be known. Therefore, the aim of this study was to investigate the effect of progressive steps per min on the muscle activation of gluteus maximus (GM), rectus femoris (RF), biceps femoris (BF), and gastrocnemius (GA) performed in water and on land. According to Zimmermann et al. [16], three steps per min were selected: 35, 60, and 95 beats per minute (bpm). We hypothesized that muscle activation of the target muscles was lower in water than on land, and they increased as the steps per min increased. The result of this study will aid physiotherapists to customize parameters of FSU exercise to patients in musculoskeletal rehabilitation.

2. Materials and Methods

2.1. Study Design

This study was a cross-sectional study to compare surface electromyography (sEMG) activity of performing FSU exercise in water or on land at three different steps per min. As shown in Figure 1, 26 participants were assessed for eligibility. The included participants (n = 20) were randomly allocated to Group A and Group B in which each group consisted of five males and five females. To eliminate directionality problem, group A participants were asked to perform land trial followed by water trial while Group B participants were asked to perform water trial followed by land trial. The two trials were completed on separate days with sufficient rest in between.

2.2. Sample Size Planning

The sample size was calculated based on the primary outcome of a previous study [17] comparing sEMG of lower extremities between land and water step-up exercise. Using the G* Power software version 3.0.10 and based on the effect size of f = 0.79 between the exercise groups obtained, the primary outcome sEMG of lower extremities assumed a 5% type I error and 80% power. The sample size computed was 15 or more participants. Considering an estimated 30% attrition rate, the total enrolled sample size required to ensure adequate statistical power was 20.

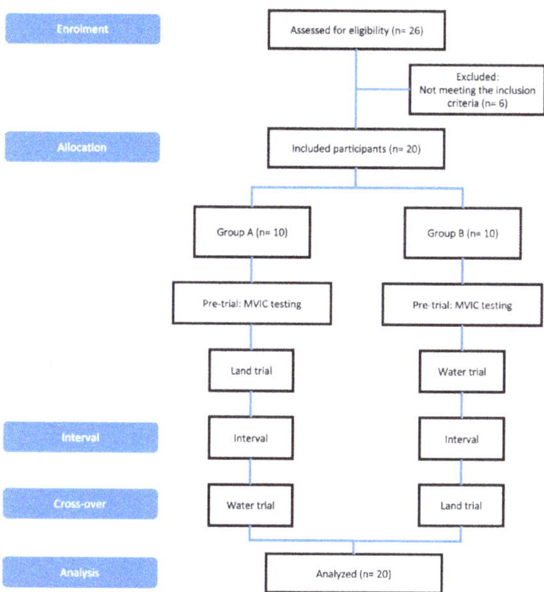

Figure 1. Study flow chart.

2.3. Participants

Participants were recruited by convenient sampling in Hong Kong Polytechnic University. Twenty healthy young adults (10 females and 10 males) aged 18 to 30 participated in this study. Individuals were excluded from this study if they had (1) any musculoskeletal, bone, joint, cardiac, and pulmonary diseases, any infectious diseases, skin conditions, and any known hip or knee injuries (included previous hip or knee surgeries) in recent two years, and (2) any contraindications to aquatic exercises, or prior exposure to aquatic-based exercises. Prior to participation, participants' demographic information including resting heart rate, blood pressure, age, height, weight, BMI, and leg dominance (i.e., the foot used to kick a ball) were obtained. Participants were informed of the nature of the study and signed a consent form prior to voluntary participation. This study was approved by the Departmental Research Committee of the Hong Kong Polytechnic University's Department of Rehabilitation Sciences (Reference Number: HSEARS 20220204005).

2.4. Experimental Set-Up

The trials were videotaped using waterproof camera GoProHERO3 at 90 frames/s. The camera was placed 1.5 m away from the participants and positioned at patellofemoral joint level to prevent any angulation of the video. sEMG activities were recorded using a 16-channel sEMG system (Infinity Mini Wave waterproof, Cometa, Milan, Italy), and a customized data logger at 1000 Hz sampling rate. The sEMG signals were then exported using EMGandMotionTools version 8.3.4.0 (Cometa, Milan, Italy).

2.5. Procedures

The standardized procedures of the study were explained to the participants, and they were as follows:

2.5.1. Skin Preparation, Electrode Placement and Joint Markers

The muscle activation of our target muscles was evaluated by sEMG. To minimize the impedance, the required skin areas for electrode placement were shaved, handled with abrasive material (3M Red Dot Trace Prep), and cleaned with alcohol swab (70% isopropyl).

According to previous studies [16,18], the electrodes were applied to the skin of the dominant leg of participants as follows: GM's electrode: at a point half of the distance between the greater trochanter of the femur and the superior end of the gluteal cleft (Figure 2a); RF's electrode: midway between the anterior superior iliac spine and the superior edge of patella (Figure 2b); BF's electrode: midway between ischial tuberosity and medial joint line of the knee (Figure 2c); GA's electrode: at a point one-fourth of the distance from the medial knee joint line to the base of the calcaneus (Figure 2d). In order to record sEMG signals under water, waterproof technique was adopted by using tegaderm (3M™ Tegaderm™ Transparent Film Roll 16002). Three bony landmarks with markers of 3 cm in diameter were attached over greater trochanter of femur, lateral epicondyle of femur, and lateral malleolus for kinematic tracking (Figure 2e).

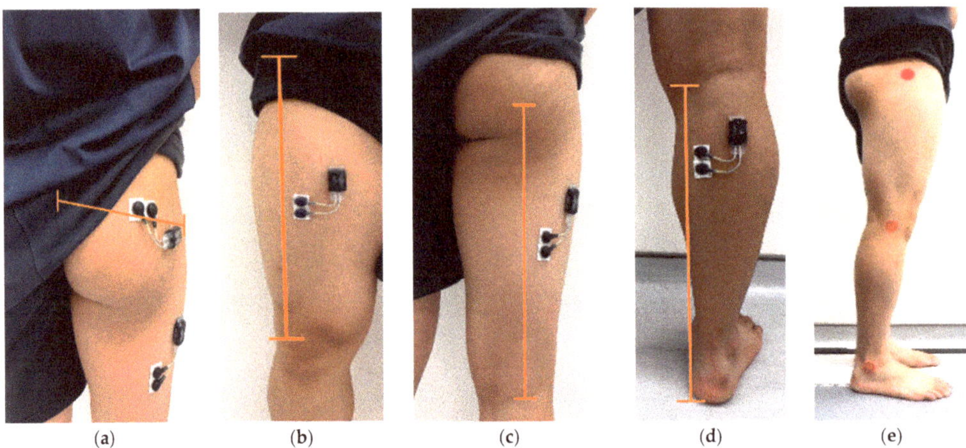

Figure 2. Location of the (**a**) gluteus maximus (GM) electrode placement, (**b**) rectus femoris (RF) electrode placement, (**c**) biceps femoris (BF) electrode placement, (**d**) gastrocnemius (GA) electrode placement, (**e**) three bony landmarks with markers attached.

2.5.2. Pre-Trial: MVIC Tests

Participants performed MVIC tests for each muscle group on land to normalize sEMG data recorded during FSU exercise on land and in water. A 2 min rest was given between each MVIC. Three 5-s MVICs were recorded for each muscle group tested. The order of MVIC tests was GM, RF, BF, and, finally, GA. According to Zimmermann et al. [16] and Yuen et al. [18], MVIC tests were performed as follows: GM MVIC was obtained when the participants extended their hips maximally and maintained 90° of knee flexion in prone; RF MVIC was obtained with the participants seated on a plinth and extending the knee at a secured angle of 45–50° of knee flexion; BF MVIC was obtained when the participants stood on the non-dominant leg and performed isometric knee flexion with the dominant knee flexed at 90°. Participants were allowed to support themselves against the wall using their arms for balance. GA MVIC was obtained with the participants seated with their hips flexed and feet in front of them, then plantar flexed maximally with their knees flexed 20° and feet resting on a stable stool. Consistent verbal encouragement was provided during all MVIC.

2.5.3. FSU Exercise Standard Protocol

Standardized instructions of performing the required movement were given to participants. The participants were instructed to perform FSU exercise with a plyometric box of 21 cm high without using their hands for balance. Each participant performed a 5 min warm-up stepping exercise. After warm-up, 1 min familiarization session was conducted to ensure that the participants accommodated to the test conditions. Participants received researchers' guidance throughout the familiarization session. In the test session, the order of test steps per min (35, 60, and 95 bpm) was randomly assigned to each participant. Three sets of eight repetitions for each step per min (35, 60, and 95 bpm) were performed according to the assigned order. Two min rest was given between each set and 5 min rest was given between each steps per min. Additional verbal cues were given to participants in order to maintain synchronization with steps per min and accuracy of the standard movement. The instructions of the FSU exercise are as follows: The participants started with the foot of the dominant leg placed entirely on the plyometric box with their weight shifted to dominant leg only (Figure 3a). For the ascending phase, they extended the dominant hip and knee to move the body to a standing position (Figure 3b). They were instructed not to use the non-dominant leg to push off. For the descending phase, they descended with the dominant leg and returned to starting position (Figure 3c). They were asked to watch a video clip combining visual and auditory cues while performing FSU exercise.

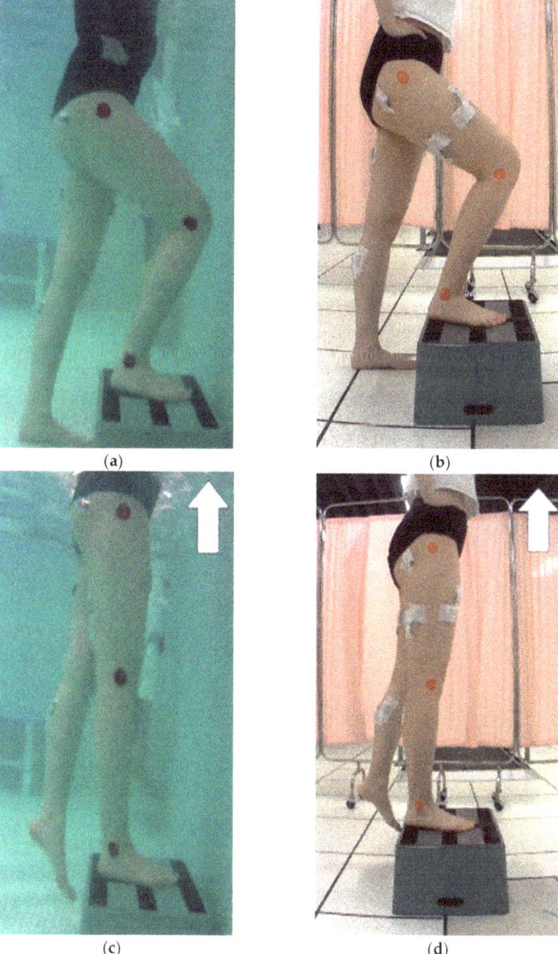

(a)

(b)

(c)

(d)

Figure 3. *Cont.*

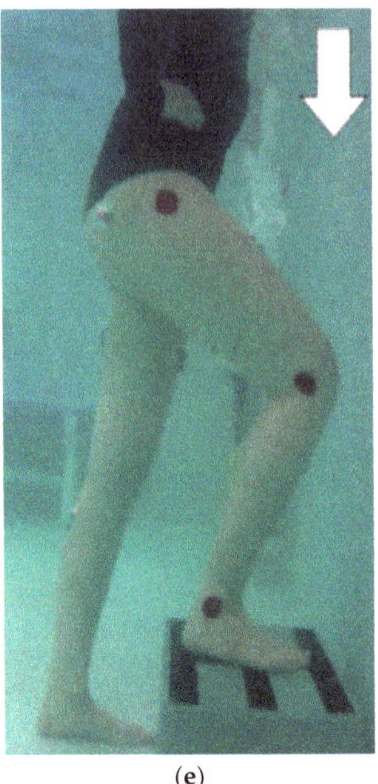 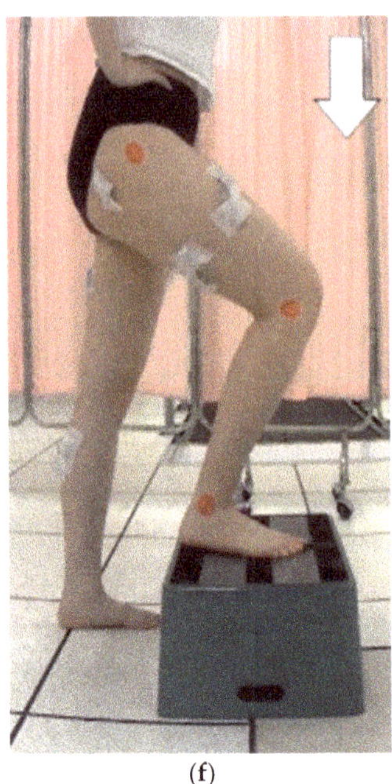

(e)   (f)

Figure 3. FSU exercise: starting position: (**a**) in water and (**b**) on land; ascending phase: (**c**) in water and (**d**) on land; descending phase: (**e**) in water and (**f**) on land.

Regarding the temperature, land trial was maintained at room temperature (25 °C) while water trial was maintained at the indoor swimming pool temperature (28 °C). The water level was set at chest level in starting position. Participants were required to stand on either a 15 cm or a 25 cm tall platform if the water level was above the chest level.

2.6. Outcome Measurements

Normalized muscle activation (%MVIC) was obtained through dividing the recorded muscle activation of the tested muscles individually by the maximum muscle activation values estimated from MVIC tests.

2.7. Data Processing

Raw sEMG signals were processed by bandpass filter (at 20 Hz to 300 Hz) and root-mean-square sliding window (50 ms time constant) (MatLab R2020a; Mathematical computing software, Natick, MA, USA). With reference to Mercer et al. [19], 6 to 12 steps from the 24 steps from each steps per min were selected for analysis. The middle four of the eight steps in each set were selected. For the kinematic data, a customized program was used to determine the period of the middle four FSU movements. Knee angles at the corresponding time were analyzed from markers on participants in the videos taken using motion-tracking software Kinovea (v.0.9.5) (Kinovea, Bordeaux, Nouvelle Aquitaine, France). The initial time for the third step and final time for sixth step will be marked to synchronized with sEMG data for statistical analysis. Amplitudes of EMG signal for the four targeted muscles were calculated and averaged. Raw MVIC data were filtered and smoothed in the same way as raw sEMG signals. MVIC values of the three bursts of contractions were first calculated into three separate means. The greatest mean MVIC value among the three bursts was selected as the MVIC value of the tested muscles. Mean sEMG amplitudes for the ascending and descending phases of the four FSU movements were normalized to these MVIC values and expressed as %MVIC.

2.8. Statistical Analysis

To examine the difference in sEMG activity between aquatic and land environments, cadences and between ascending and descending phases of FSU exercise. The ANOVA three-way was performed for each muscle, using the three main factors (environment, cadence, and phase) and their interactions. The statistical assumptions of normality and sphericity for using the repeated measures ANOVA were tested. In the first place, a descriptive analysis of the main anthropometric variables of the participants and of the maximum activation registered in each of the muscles analyzed in the present study (mean, standard deviation and difference) was conducted. Additionally, each variable was compared (muscle activity of GM, RF, BF, and GA (% MVIC)) between the two environments, at different cadences and various phases and their interactions. For all statistical comparisons, p level was set to ≤ 0.05. Subsequently, an analysis was made regarding the degree of contribution of each of the muscles observed during various cadences and phases. The Bonferroni test was used when there was a statistically significant difference. The effect size was calculated via the partial eta squared with 0.01 indicated a small effect, 0.06 indicated a medium effect, and 0.14 a large effect [20]. All statistical analyses were performed using IBM SPSS Statistics for Windows, Version 26.0 (IBM Corp., Armonk, NY, USA).

3. Results

All participants completed the sessions. There were no adverse effects or safety concerns raised in interventions. Table 1 shows the descriptive characteristics of participants.

Table 1. Descriptive characteristics of the 20 participants.

	n = 20
Gender (Male: Female)	10:10
Age (y)	21.1 ± 1.9
Weight (kg)	61.0 ± 7.7
Height (cm)	169.2 ± 6.0
BMI (kg/m^2)	21.3 ± 1.8
Leg dominance (Left: Right)	0:20

3.1. Comparison of Steps per min during FSU Exercise

Figure 4 shows the change of mean % MVIC of all target muscles. The result indicated a significant difference between the mean %MVIC of all muscles, except for GA, at 35 bpm, 60 bpm, and 95 bpm in both phases on land and in water, respectively.

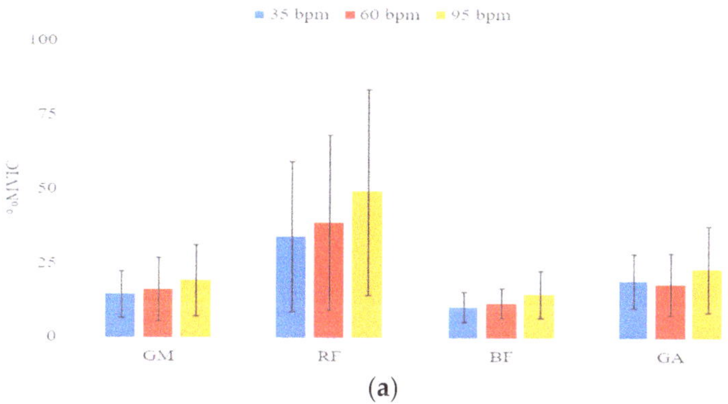

Figure 4. *Cont.*

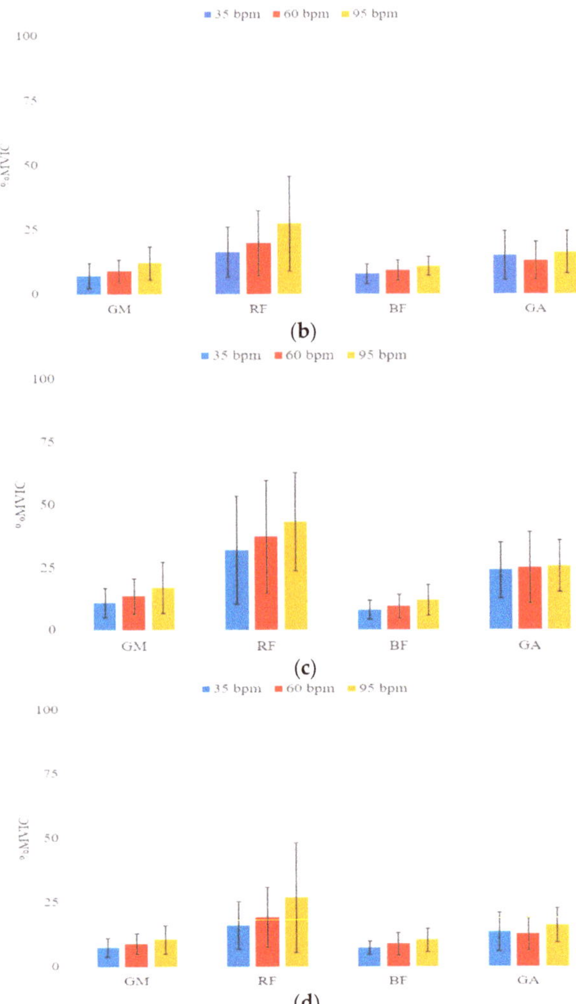

Figure 4. Means of percentage of maximum voluntary isometric contraction (%MVIC) for gluteus maximus (GM), rectus femoris (RF), biceps femoris (BF), and gastrocnemius (GA) performed during (**a**) ascending phase on land, (**b**) ascending phase in water, (**c**) descending phase on land, and (**d**) descending phase in water.

For the comparison between 35 and 60 steps per minute on land, the mean %MVIC of RF at 35 steps per minute was significantly lower than that of 60 steps per minute($p < 0.01$). Regardless of movement phases on land, the mean %MVIC of GM and RF shows no statistically significant difference, except for the descending phase of GM.

For the comparison between 35 and 95 steps per minute on land, the mean %MVIC of GM, RF and BF at 35 steps per minute was significantly lower than that of 95 steps per minute ($p < 0.01$). For the RF, it shows a maximal increase of 44.4% MVIC in the ascending phase.

For the comparison between 60 and 95 steps per minute on land, the mean %MVIC of RF and BF at 60 steps per minute was significantly lower than that of 95 steps per minute

($p < 0.01$). For GM, regardless of movement phases on land, the mean %MVIC of GM shows no statistically significant difference.

For the comparison between 35 and 60 steps per minute in water, mean %MVIC of RF at 35 steps per minute was significantly lower than that of 60 steps per minute ($p < 0.01$). Regardless of movement phases in water, the mean %MVIC of GM and RF shows no statistically significant difference, except for the descending phase of GM.

For the comparison between 35 and 95 steps per minute in water, the mean %MVIC of GM, RF, and BF at 35 steps per minute was significantly lower than that of 95 steps per minute ($p < 0.01$). For RF shows a maximal increase of 68.9% MVIC in the ascending phase at the comparison of 35 and 95 bpm in water.

For the comparison between 60 and 95 steps per minute in water, the mean %MVIC of RF at 60 steps per minute was significantly lower than that of 95 steps per minute ($p < 0.01$). For GM and BF, regardless of movement phases on land, the mean %MVIC of these muscles shows no statistically significant difference.

3.2. Comparison of Environments during FSU Exercise

Table 2 compares the tested muscle activation in different environments. In general, all tested muscles showed lower muscle activation in water when compared to land ($p < 0.05$). RF, GM, and GA showed significant lower muscle activation in water environment regardless movement phases and steps per min ($p < 0.01$, $p < 0.01$ and $p < 0.05$, respectively). RF showed the greatest reduction of 44.8%MVIC in the ascending phase at 95 bpm while GM showed a maximal decrease of 51.7% MVIC in the ascending phase at 35 bpm. The decrease in muscle activation of BF is dependent on movement phases such that a significant lower muscle activation water is only observed in the ascending phase at all investigated steps per min ($p < 0.05$).

Table 2. The comparison of tested muscle activation at different media (GM = gluteus maximus, RF = rectus femoris, BF = biceps femoris, GA = gastrocnemius, A = ascending phase, D = descending phase).

Muscle	Steps per min (bpm)	Phase	On land (%MVIC, Mean ± SD)	In water (%MVIC, Mean ± SD)	p Value	95% Confidence Interval of the Difference		Effect Size
						Lower	Upper	
GM	35	A	14.6 ± 7.9	7.0 ± 4.8	0.000 *	−10.2	−3.9	−0.6
		D	10.7 ± 6.0	7.3 ± 3.7	0.009 *	−5.1	−1.0	−0.4
	60	A	16.2 ± 10.6	8.8 ± 4.5	0.003 *	−12.7	−2.5	−0.5
		D	13.4 ± 7.0	8.8 ± 4.1	0.001 *	2.3	7.0	0.9
	95	A	19.3 ± 11.9	11.9 ± 6.5	0.002 *	−10.8	−3.0	−0.5
		D	16.7 ± 10.2	10.3 ± 5.6	0.009 *	−10.7	−1.9	−0.4
BF	35	A	10.4 ± 5.1	7.6 ± 3.7	0.026 *	−5.0	−0.1	−0.6
		D	7.9 ± 3.9	7.1 ± 2.6	0.363	−2.7	1.0	−0.6
	60	A	11.7 ± 5.0	9.1 ± 3.9	0.011 *	−4.6	−0.5	−0.6
		D	9.3 ± 4.8	8.7 ± 4.3	0.697	−3.1	2.3	−0.6
	95	A	14.6 ± 8.0	10.7 ± 3.6	0.043 *	−7.6	−0.1	−0.5
		D	11.6 ± 6.1	10.2 ± 4.5	0.322	−4.4	1.6	−0.5
RF	35	A	34.0 ± 25.3	16.0 ± 9.7	0.000 *	0.4	5.1	0.5
		D	31.6 ± 21.5	15.9 ± 9.2	0.000 *	−1.0	2.6	0.2
	60	A	38.7 ± 29.4	19.6 ± 12.6	0.000 *	0.7	4.6	0.6
		D	37.0 ± 22.4	19.1 ± 11.8	0.000 *	−2.4	3.5	0.1
	95	A	49.1 ± 34.7	27.1 ± 18.3	0.001 *	0.1	7.7	0.5
		D	43.0 ± 19.7	26.7 ± 21.4	0.002 *	−1.5	4.4	0.2
GA	35	A	19.2 ± 9.2	14.7 ± 9.5	0.067 *	−10.5	0.7	−0.3
		D	23.7 ± 11.1	13.3 ± 7.4	0.003 *	4.1	16.7	0.8
	60	A	18.2 ± 10.5	12.7 ± 7.3	0.028 *	−10.7	−0.4	−0.4
		D	24.6 ± 14.1	12.5 ± 6.3	0.003 *	−18.2	−4.8	−0.5
	95	A	23.3 ± 14.0	16.0 ± 8.2	0.009 *	−11.7	−1.6	−0.4
		D	25.1 ± 10.3	15.9 ± 6.7	0.002 *	4.0	14.5	0.8

* indicates statistically significant difference ($p < 0.05$).

3.3. Comparison of Phases during FSU Exercise

Table 3 compares muscle activation at different movement phases in water and on land, respectively. The muscle activation of all tested muscles was higher in the ascending phase regardless of the exercise environment. However, such difference is not statistically significant in all tested muscles when considered a smaller effect size in water environment. Considering the movement phases and environments, among the target muscles, only BF and GM showed significant difference in muscle activation between movement phases on land ($p < 0.05$), while RF and GA showed no significant difference among movement phases in both environments. BF and GM also showed no significant difference among movement phases in water. On land, BF showed significantly higher muscle activation in ascending phase at all investigated steps per min ($p \leq 0.01$), with the greatest increase of 31.4% MVIC at 35 bpm. GM only showed a significant difference at 35 bpm, with an increase of 35.8% of MVIC. In terms of %MVIC of all target muscles in different movement phases, although RF did not show significant difference in muscle activation among movement phases regardless the exercise environment, it can be seen that the highest %MVIC among the tested muscles, with 27.09% MVIC and 26.68% MVIC in the ascending and descending phase, respectively, in water.

Table 3. The comparison of tested muscle activation at different phases (GM = gluteus maximus, RF = rectus femoris, BF = biceps femoris, GA = gastrocnemius).

	Muscle	Steps per min (bpm)	Ascending Phase (%MVIC, Mean ± SD)	Descending Phase (%MVIC, Mean ± SD)	*p* Value	95% Confidence Interval of the Difference		Effect Size
						Lower	Upper	
On land	GM	35	14.6 ± 7.8	10.7 ± 6.0	0.001 *	−5.5	−2.0	−0.8
		60	16.2 ± 10.6	13.4 ± 7.0	0.079	−7.3	0.8	−0.4
		95	19.3 ± 11.9	16.7 ± 10.2	0.191	−7.1	2.3	−0.3
	RF	35	34.0 ± 25.3	31.6 ± 21.5	0.218	−6.8	0.9	−0.3
		60	38.7 ± 29.4	37.0 ± 22.4	0.351	−7.8	4.4	−0.2
		95	49.1 ± 34.7	43.0 ± 19.7	0.167	−14.3	3.0	−0.3
	BF	35	10.4 ± 5.1	7.9 ± 3.9	0.000 *	1.6	3.4	1.4
		60	11.7 ± 5.0	9.3 ± 4.8	0.003 *	1.0	3.9	0.8
		95	14.6 ± 8.0	11.6 ± 6.1	0.013 *	0.7	5.2	0.6
	GA	35	19.2 ± 9.2	23.7 ± 11.1	0.103	−9.9	1.0	−0.4
		60	18.2 ± 10.5	24.6 ± 14.1	0.167	−2.1	12.6	−0.3
		95	23.3 ± 14.5	25.1 ± 10.3	0.100	−0.9	6.9	−0.4
In water	GM	35	7.0 ± 4.8	7.3 ± 3.7	0.627	−1.7	2.2	−0.1
		60	8.8 ± 4.5	8.8 ± 4.1	0.794	−2.3	2.5	−0.1
		95	11.9 ± 6.5	10.3 ± 5.6	0.391	−3.7	1.1	−0.2
	RF	35	16.0 ± 9.7	15.9 ± 9.2	0.902	−2.1	2.4	0.0
		60	19.6 ± 12.6	19.1 ± 11.8	0.881	−2.0	2.2	0.0
		95	27.1 ± 18.3	26.7 ± 21.4	0.601	−5.9	3.3	−0.2
	BF	35	7.6 ± 3.7	7.1 ± 2.6	0.423	−0.8	1.9	0.2
		60	9.1 ± 3.9	8.7 ± 4.3	0.669	−1.3	2.0	0.1
		95	10.7 ± 3.6	10.2 ± 4.5	0.487	−1.0	2.0	0.2
	GA	35	15.0 ± 10.0	13.5 ± 7.5	0.502	−0.6	9.7	−0.0
		60	12.8 ± 7.3	12.6 ± 6.4	0.881	−2.6	2.0	−0.0
		95	16.1 ± 8.2	16.0 ± 6.6	0.931	−4.2	2.5	−0.1

* indicates statistically significant difference ($p < 0.05$).

3.4. Interaction between Each Muscle Performed Using Three Main Factors (Environment, Cadence, and Phase)

Regarding the various muscle activities examined, the analysis showed that the main effects for environments ($p < 0.001$; ES = 0.51), cadence ($p < 0.001$; ES = 0.657) and phases ($p < 0.001$; ES = 0.315) was significant in GM. Similarly, in RF the main effects for environments ($p < 0.001$; ES = 0.374), cadence ($p < 0.001$; ES = 0.738), and phases ($p < 0.001$; ES= 0.241) was significant. However, in BF, main significant effects could only be found in environments ($p < 0.001$; ES = 0.307) and cadence ($p < 0.001$; ES = 0.659). Among GA, a significant effect was found in environmental factor ($p < 0.05$; ES = 0.258), cadence ($p < 0.001$,

ES = 0.831) and phases ($p < 0.001$; ES = 0.648). The environment* cadence* phases interaction was not significant in RF ($p = 0.685$; ES = 0.02), GA ($p = 0.590$; ES = 0.027) and GM ($p = 0.229$; ES = 0.075) while the interaction was approaching significant in BF ($p = 0.055$; ES = 0.142).

4. Discussion

This study hypothesized that lower limb muscle activation in water is lower than on land, and they increase along steps per min during FSU exercise. The results of this study provide evidence to support both hypotheses. All targeted muscles showed significantly lower levels of activation during the in water movements ($p < 0.05$), with RF showed a maximal reduction of 44.8%MVIC. Moreover, all target muscle, except GA, showed a significant increase in muscle activation as steps per min increased regardless of the environment ($p < 0.05$). The RF showed a more significant increase at all steps per min comparison ($p < 0.01$), with a maximum increase of 44.4% and 68.9% MVIC along the steps per min on land and in water, respectively.

4.1. The Effect of Steps per min on Lower Limb Muscle Activation

Our results showed that all tested muscles indicate an increasing trend of muscle activation, with a maximal increase of 44.4%, regardless of movement phases and environments. This study found that the %MVIC of three of the selected muscles (RF, GM, BF) at 95 bpm was significantly higher than that of 35 bpm, and the %MVIC of RF at 95 bpm was significantly higher than that of itself at all other steps per min comparisons, regardless of movement phases and environments. This result partly echoes with the finding of Zimmermann et al. [16], who investigated the effect of steps per min of stair-stepping exercise on lower limb muscle activation on land and found that the muscle activation of GM, RF, and GA increased significantly with steps per min. However, our results did not show similar trend in GA. Such difference is probably due to the difference of actions in studies. Our study required participants to stabilize their foot on the stool all the time, which GA acts as ankle stabilizer; while the research of Zimmermann et al. [16] involved action of stair climbing, GA acted as the prime mover during ankle plantar flexion. Our result also agrees with the finding of Lee and Lee [21], who found the muscle activation of RF increased as squatting speed increased on land. The similarity of results may be due to the comparable role of RF in both protocols. Lee and Lee [21] proposed that the increase in muscle activation along speed may be due to larger energy output in a shorter time. Clamann [22] found that increase in motor unit recruitment would increase the force output of a muscle, which aligns with the explanation of Lee and Lee [21].

Similar trend of change of muscle activation was also observed in the water environment and this trend may be attributed to the change of drag force. Drag force is the resistance of fluid acts on limbs movement and it increases as the speed of limbs movement in the same direction increases [2]. According to the drag force equation, increasing the speed by twice will increase four times of drag force [7]. Our result echoes with the finding of Chien et al. [6] that the RF muscle activation increased as steps per min increased from 30 bpm to 90 bpm. However, Chien et al. [6] found that the RF muscle activation at 90 bpm in water was comparable with that on land, while our study was unable to reproduce similar result. The discrepancy might be due to the difference of exercise design and angle of limb movement. The study of Chien et al. [6] involved the open chain knee extension movement of participants in a sitting position, which the lower limb movement was mostly contributed by the quadriceps muscle; while our study investigated the muscle activation in close chain FSU exercise, which participants might plantarflexed the ankle and thus reduced the muscle activation and demand on RF. In addition, our study standardized the height of stool instead of the starting position of knee angle at 90 degrees, which is the optimal angle to activate RF and thus, it may affect the muscle activation [23]. Moreover, our results are in line with Miyoshi et al. [24] that the muscle activation of hip extensor increased as the walking speed in water increases. Their study proposed that as hip extensor serves a propulsive function to push the body forward in water, greater water resistance will demand greater muscle activation of the hip extensor [24]. However, it is essential to consider that walking in water is a horizontal displacement while FSU is primarily a vertical displacement, the combination of different role of muscles and direction of water resistance may alter the pattern of muscle activation.

4.2. The Effect of Exercise Environment on Lower Limb Muscle Activation

Our result showed that performing FSU in water will elicit a lower muscle activation of GM, BF, RF and GA than that on land ($p < 0.05$), with a maximal decrease of 44.8% MVIC in the RF at 95 bpm ascending phase with a smaller effect size. This result echoes with finding of Yuen et al. [18], who investigated the muscle activation during squatting. The similar result may be attributed to exercise protocol—both squatting and FSU exercises are close chain. The reduction in muscle activation in water may be explained by the effect of buoyancy. Buoyancy, which equals to weight of the fluid displaced by body, is an upward force exerted by water which causes a reduction in body weight when immersed in water [25]. Participants of this study are immersed to the chest level in water and have 60% of weight reduction when performing FSU exercise in water [2]. Another possible cause could be neuromuscular system deactivation [26]. It is found that immersing participants to the chest level will induce weightlessness to muscle spindles, which decrease the activation of reflex of the pressure receptors within the spindles [26]. Thus, this proprioceptive effect impacts the neuromuscular system in terms of a muscle activation drop. The gravity changes in water may also impact multiple systems besides muscles, such as the vestibular and the visual system, by reducing the stimulation of gravireceptors [26,27]. However, there are uncertainties, such as how and to what extent hydrostatic pressure may affect muscle activation, and whether the kinematics are altered by the reduced ground reaction in water have yet to be discussed in papers [28].

4.3. Comparison of Ascending and Descending Phases of FSU

Our results showed that muscle activation of BF in all the steps per min and GM in 35 bpm were significantly higher during the ascending phase of FSU, with a maximal increase of 25.6% and 35.8%, respectively, while other muscles did not show significantly difference between the movement phases. On land, the range of %MVIC of BF activation found in this study is consistent with that of Ayotte et al. [12] who found that BF activation was about 12% MVIC during FSU. BF acted significantly more actively during the ascending phase than the descending phase on land. One of the possible reasons for higher activation of BF in ascending phase is that BF is a biarticular muscle responsible for hip extension and knee flexion [29]. During the ascending phase, BF acts as a hip extensor which leads to an increase in muscle activation while the descending movement may be facilitated by gravity which may lead to a decrease in muscle activation. In the ascending phase of FSU, the hip is flexed, and participants in our study were instructed to follow the steps per min according to auditory cues, so BF may contract eccentrically to control knee extension. These instructions may also pose greater activity to BF during the ascending phase only. Simenz et al. [13] also found that concentric activations of GM and RF are greater than that of eccentric in FSU. This finding is similar to ours to some extent by which only GM had significant difference between the two phases in 35 bpm. This finding can be explained by the difference in experimental set-up as the step height in their studies was higher (45.72 cm) and their participants were required to perform a loaded step-up by using six RM loads. More importantly, steps per min may be a determining factor of the significant difference as steps per min of FSU exercise was not instructed in their set-up.

In water, no significant difference was found between ascending and descending phases of the four muscles at any steps per min. Compared with the results on land, there is a discrepancy in the results in GM in 35 bpm and BF in all steps per min. The reason for the insignificant results remained unclear but could be a result of the interactions of the water properties with muscle activation. In our study, participants started the FSU exercise with immersion at the chest level and buoyancy can reduce the weight bearing by about 60%

resulting in less muscle activation in the lower limbs [2,30]. The lowered muscle activation in BF was consistent with the findings from Yuen et al. [18] who revealed that BF had about 10% MVIC in both ascending and descending phases during water squatting. Additionally, buoyancy can generate an upward force to assist the upward motion leadings to a decrease in muscle activation of the hip extensors, i.e., BF and GM, during the ascending phase. Drag force may also hinder the knee extension movement because drag force will increase as velocity increases, which may replace BF in controlling the knee extension steps per min [7]. Thus, the muscle activation of GM in 35 bpm and BF in all steps per min in the ascending phase may not be significantly different from that of the descending phase.

The current results revealed insignificant interaction between environment, cadence, and phases among all the muscles despite the interaction was approaching significant in BF. Such insignificant findings might be due to the characteristics of the FSU exercise [31]. In the present study, the FSU exercise was performed with ascending and descending to maintain the exercise cadence with buoyancy supported promotes vertical displacement and resisted to promote resistance trainings. Further research investigating the interaction effects of FSU could be conducted under different factors.

4.4. Clinical Implications

Aquatic FSU exercise may be applicable to patients with lower limb weakness and in the initial phase of musculoskeletal rehabilitation. Our result showed that all tested muscle activation were lower in water, for example, the RF showed a maximal decrease of 44.8% MVIC in water when compared on land. This indicates that exercising in water is less demanding and suitable for patients with lower limb weakness. In addition, the physical properties of water, such as buoyancy and hydrostatic pressure, can reduce pain and swelling for patients in the early phase of post-operative period [32]. Moreover, aquatic FSU exercise can serve as a functional training for stair climbing, especially for patients with knee osteoarthritis, whose stair climbing ability are limited by quadriceps weakness and knee pain [33].

4.5. Limitations of Study

First, this study evaluated healthy population only, which limited the external validity as well as application to patients with musculoskeletal disorders. Second, participants were not restrained from engaging in other lower limb resistance training between land and water trials. It is possible that these influenced the results. Third, MVIC tests were conducted by manual methods. In other words, the maximal force resisted by participants to hold the testing positions is limited by the force of the operator. Nonetheless, in this study, all MVIC tests were conducted by the same instructor in order to minimize deviation.

5. Conclusions

This is the first study investigating the effect of steps per min on lower limb muscle activation during aquatic FSU exercise. This study has shown that GM, RF, BF, and GA muscle activation of healthy individuals increased with steps per min of FSU exercise, which RF showed a maximal increase of 44.4% MVIC. The exercise pattern of FSU exercise activates RF the most as RF acts a prime mover of knee extension and elevating the whole body. In addition, although the %MVIC of BF does not show significant difference in all steps per minute comparison, it serves as the prime mover of hip extension and controlling the speed of knee extension. Nevertheless, these muscle activations were lower in water than on land. The reduction in %MVIC in water is probably due to the buoyancy property of water, the increase in %MVIC along steps per minute is due to the drag force in water. However, the effect of hydrostatic pressure of water is not investigated in this study. Further investigations of factors, such as water immersion depth and angle of limb movement, are needed as these factors may affect the muscle activation pattern. Overall, aquatic FSU exercise can be a functional training exercise to improve stair-climbing ability. Moreover, the muscle activation of all tested muscles is much lower in water, indicating that exercising in water is less demanding and is suitable for patients with lower limb weakness. In future studies, researchers may recruit patients with knee osteoarthritis or knee pain to examine the effect of steps per min on their muscle activation.

Author Contributions: Conceptualization, B.C.L.S. and M.M.Y.K.; methodology, M.M.Y.K., N.W.L.L., A.W.C.L., A.L.M.L., A.S.L.L., and P.W.Y.C.; software, M.M.Y.K., N.W.L.L., A.W.C.L., A.L.M.L., A.S.L.L., and P.W.Y.C.; validation, M.M.Y.K., N.W.L.L., A.W.C.L., A.L.M.L., A.S.L.L., and P.W.Y.C.; formal analysis, M.M.Y.K., N.W.L.L., A.W.C.L., A.L.M.L., A.S.L.L., and P.W.Y.C.; investigation, M.M.Y.K., N.W.L.L., A.W.C.L., A.L.M.L., A.S.L.L., and P.W.Y.C.; resources, B.C.L.S., M.M.Y.K., and S.S.M.N.; data curation, N.W.L.L., A.W.C.L., A.L.M.L., A.S.L.L., and P.W.Y.C.; writing—original draft preparation, N.W.L.L., A.W.C.L., A.L.M.L., A.S.L.L., and P.W.Y.C.; writing—review and editing, B.C.L.S. and M.M.Y.K.; visualization, N.W.L.L., A.W.C.L., A.L.M.L., A.S.L.L., and P.W.Y.C.; supervision, B.C.L.S. and M.M.Y.K.; project administration, B.C.L.S. and M.M.Y.K.; funding acquisition, B.C.L.S., M.M.Y.K., and S.S.M.N. All authors have read and agreed to the published version of the manuscript.

Funding: This research received no external funding.

Institutional Review Board Statement: The study was conducted in accordance with the Declaration of Helsinki and approved by the Institutional Review Board (or Ethics Committee) of THE HONG KONG POLYTECHNIC UNIVERSITY (HSEARS20220311001 and 17 March 2022).

Informed Consent Statement: Informed consent was obtained from all subjects involved in the study.

Data Availability Statement: Not applicable.

Acknowledgments: The authors acknowledge the Club One Spotlight Swimming Pool (Hung Hom) for providing the indoor swimming pool for the study. We thank S. S. Man, ArYu Kwok, and Daniel Tse for statistical advice, and all those who participated in the study.

Conflicts of Interest: The authors have no conflict of interest.

References

1. Heywood, S.M.; McClelland, J.P.; Mentiplay, B.B.; Geigle, P.P.; Rahmann, A.P.; Clark, R.P. Effectiveness of Aquatic Exercise in Improving Lower Limb Strength in Musculoskeletal Conditions: A Systematic Review and Meta-Analysis. *Arch. Phys. Med. Rehabil* **2016**, *98*, 173–186. [CrossRef] [PubMed]
2. Becker, B.E. Aquatic Therapy: Scientific Foundations and Clinical Rehabilitation Applications. *PM R* **2009**, *1*, 859–872. [CrossRef] [PubMed]
3. Harrison, R.A.; Hillman, M.; Bulstrode, S. Loading of the Lower Limb when Walking Partially Immersed: Implications for Clinical Practice. *Physiotherapy* **1992**, *78*, 164–166. [CrossRef]
4. Alberton, C.L.; Zaffari, P.; Pinto, S.S.; Reichert, T.; Bagatini, N.C.; Kanitz, A.C.; Almada, B.P.; Kruel, L.F.M. Water-based exercises in postmenopausal women: Vertical ground reaction force and oxygen uptake responses. *Eur. J. Sport Sci.* **2021**, *21*, 331–340. [CrossRef]
5. Buckthorpe, M.; Pirotti, E.; Della Villa, F. Benefits and use of aquatic therapy during rehabilitation after acl reconstruction—A clinical commentary. *Int. J. Sports Phys. Ther.* **2019**, *14*, 978–993. [CrossRef]
6. Chien, K.-Y.; Kan, N.-W.; Liao, Y.-H.; Lin, Y.-L.; Lin, C.-L.; Chen, W.-C. Neuromuscular Activity and Muscular Oxygenation Through Different Movement Cadences During In-water and On-land Knee Extension Exercise. *J. Strength Cond. Res.* **2017**, *31*, 750–757. [CrossRef] [PubMed]
7. Pöyhönen, T.; Kyröläinen, H.; Keskinen, K.L.; Hautala, A.; Savolainen, J.; Mälkiä, E. Electromyographic and kinematic analysis of therapeutic knee exercises under water. *Clin. Biomech.* **2001**, *16*, 496–504. [CrossRef]
8. Kelly, B.T.; Roskin, L.A.; Kirkendall, D.T.; Speer, K.P. Shoulder Muscle Activation During Aquatic and Dry Land Exercises in Nonimpaired Subjects. *J. Orthop. Sport. Phys. Ther.* **2000**, *30*, 204–210. [CrossRef]
9. Castillo-Lozano, R.; Cuesta-Vargas, A.; Gabel, C.P. Analysis of arm elevation muscle activity through different movement planes and speeds during in-water and dry-land exercise. *J. Shoulder Elbow. Surg.* **2014**, *23*, 159–165. [CrossRef]
10. Kaneda, K.; Sato, D.; Wakabayashi, H.; Nomura, T. EMG activity of hip and trunk muscles during deep-water running. *J. Electromyogr. Kinesiol.* **2009**, *19*, 1064–1070. [CrossRef]
11. Alberton, C.L.; Cadore, E.L.; Pinto, S.S.; Tartaruga, M.P.; da Silva, E.M.; Kruel, L.F. Cardiorespiratory, neuromuscular and kinematic responses to stationary running performed in water and on dry land. *Eur. J. Appl. Physiol.* **2011**, *111*, 1157–1166. [CrossRef] [PubMed]
12. Ayotte, N.W.; Stetts, D.M.; Keenan, G.; Greenway, E.H. Electromyographical analysis of selected lower extremity muscles during 5 unilateral weight-bearing exercises. *J. Orthop. Sports Phys. Ther.* **2007**, *37*, 48–55. [CrossRef] [PubMed]

13. Simenz, C.J.; Garceau, L.R.; Lutsch, B.N.; Suchomel, T.J.; Ebben, W.P. Electromyographical Analysis of Lower Extremity Muscle Activation During Variations of the Loaded Step-Up Exercise. *J. Strength Cond. Res.* **2012**, *26*, 3398–3405. [CrossRef]
14. Khaiyat, O.A.; Norris, J. Electromyographic activity of selected trunk, core, and thigh muscles in commonly used exercises for ACL rehabilitation. *J. Phys. Ther. Sci.* **2018**, *30*, 642–648. [CrossRef] [PubMed]
15. Wang, M.-Y.; Flanagan, S.; Song, J.-E.; Greendale, G.A.; Salem, G.J. Lower-extremity biomechanics during forward and lateral stepping activities in older adults. *Clin. Biomech.* **2003**, *18*, 214–221. [CrossRef]
16. Zimmermann, C.L.; Cook, T.M.; Bravard, M.S.; Hansen, M.M.; Honomichl, R.T.; Karns, S.T.; Lammers, M.A.; Steele, S.A.; Yunker, L.K.; Zebrowski, R.M. Effects of stair-stepping exercise direction and cadence on EMG activity of selected lower extremity muscle groups. *J Orthop Sports Phys Ther* **1994**, *19*, 173–180. [CrossRef]
17. Cuesta-Vargas, Á.; Martín-Martín, J.; Pérez-Cruzado, D.; Cano-Herrera, C.L.; Güeita Rodríguez, J.; Merchán-Baeza, J.A.; González-Sánchez, M. Muscle Activation and Distribution during Four Test/Functional Tasks: A Comparison between Dry-Land and Aquatic Environments for Healthy Older and Young Adults. *Int. J. Environ. Res. Public Health* **2020**, *17*, 4696. [CrossRef]
18. Yuen, C.H.N.; Lam, C.P.Y.; Tong, K.C.T.; Yeung, J.C.Y.; Yip, C.H.Y.; So, B.C.L. Investigation the EMG Activities of Lower Limb Muscles When Doing Squatting Exercise in Water and on Land. *Int. J. Environ. Res. Public Health.* **2019**, *16*, 4562. [CrossRef]
19. Mercer, V.S.; Gross, M.T.; Sharma, S.; Weeks, E. Comparison of gluteus medius muscle electromyographic activity during forward and lateral step-up exercises in older adults. *Phys. Ther.* **2009**, *89*, 1205–1214. [CrossRef]
20. Fritz, C.O.; Morris, P.E.; Richler, J.J. Effect size estimates: Current use, calculations, and interpretation. *J. Exp. Psychol. Gen.* **2012**, *141*, 2–18. [CrossRef]
21. Lee, J.-Y.; Lee, D.-Y. Effect of different speeds and ground environment of squat exercises on lower limb muscle activation and balance ability. *Technol. Health Care* **2018**, *26*, 593–603. [CrossRef] [PubMed]
22. Clamann, H.P. Motor Unit Recruitment and the Gradation of Muscle Force. *Phys. Ther.* **1993**, *73*, 830–843. [CrossRef]
23. Marchetti, P.H.; Jarbas da Silva, J.; Jon Schoenfeld, B.; Nardi, P.S.M.; Pecoraro, S.L.; D'Andréa Greve, J.M.; Hartigan, E. Muscle Activation Differs between Three Different Knee Joint-Angle Positions during a Maximal Isometric Back Squat Exercise. *J. Sport. Med.* **2016**, *2016*, 3846123. [CrossRef] [PubMed]
24. Miyoshi, T.; Shirota, T.; Yamamoto, S.-I.; Nakazawa, K.; Akai, M. Effect of the walking speed to the lower limb joint angular displacements, joint moments and ground reaction forces during walking in water. *Disabil. Rehabil.* **2009**, *26*, 724–732. [CrossRef] [PubMed]
25. Wilcock, I.M.; Cronin, J.B.; Hing, W.A. Physiological Response to Water Immersion: A Method for Sport Recovery? *Sport. Med.* **2006**, *36*, 747–765. [CrossRef]
26. Pöyhönen, T.; Avela, J. Effect of head-out water immersion on neuromuscular function of the plantarflexor muscles. *Aviat. Space Environ. Med.* **2002**, *73*, 1215–1218.
27. Grigoriev, A.; Egorov, A. Nicogossian A, Mohler S, Gazenko O, Grigoriev A ed. *Space Biol. Med.* **1996**, *3*, 475–525.
28. Masumoto, K.; Takasugi, S.-I.; Hotta, N.; Fujishima, K.; Iwamoto, Y. Muscle activity and heart rate response during backward walking in water and on dry land. *Eur. J. Appl. Physiol.* **2005**, *94*, 54–61. [CrossRef]
29. So, B.; Kwok, M.; Fung, V.; Kwok, A.; Lau, C.; Tse, A.; Wong, M.; Mercer, J. A Study Comparing Gait and Lower Limb Muscle Activity During Aquatic Treadmill Running with Different Water Depth and Land Treadmill Running. *J. Human Kinet.* **2022**, *22*, 39–50. [CrossRef]
30. Cuesta-Vargas, A.I.; Cano-Herrera, C.L.; Heywood, S. Analysis of the neuromuscular activity during rising from a chair in water and on dry land. *J. Electromyogr. Kinesiol.* **2013**, *23*, 1446–1450. [CrossRef]
31. Alberton, C.L.; Pinto, S.S.; Cadore, E.L.; Tartaruga, M.P.; Kanitz, A.C.; Antunes, A.H.; Finatto, P.; Kruel, L.F.M. Oxygen Uptake, Muscle Activity and Ground Reaction Force during Water Aerobic Exercises. *Int. J. Sports Med.* **2014**, *35*, 1161–1169. [CrossRef] [PubMed]
32. Cavanaugh, J.T.; Powers, M. ACL Rehabilitation Progression: Where Are We Now? *Curr. Rev. Musculoskelet. Med.* **2017**, *10*, 289–296. [CrossRef] [PubMed]
33. Whitchelo, T.; McClelland, J.A.; Webster, K.E. Factors associated with stair climbing ability in patients with knee osteoarthritis and knee arthroplasty: A systematic review. *Disabil. Rehabil.* **2014**, *36*, 1051–1060. [CrossRef] [PubMed]

healthcare

Article

Goal Orientations of Secondary School Students and Their Intention to Practise Physical Activity in Their Leisure Time: Mediation of Physical Education Importance and Satisfaction

Francisco Javier Pérez-Quero [1], Antonio Granero-Gallegos [1,2,*], Antonio Baena-Extremera [3] and Raúl Baños [4,5]

[1] Department of Education, University of Almeria, 04120 Almeria, Spain
[2] Health Research Centre, University of Almeria, 04120 Almeria, Spain
[3] Department of Musical, Plastic and Corporal Expression, Faculty of Education Sciences, University of Granada, 18071 Granada, Spain
[4] Department of Musical, Plastic and Corporal Expression, Faculty of Social and Human Sciences, University of Zaragoza, Campus de Teruel, 44003 Zaragoza, Spain
[5] Faculty of Sport, Autonomous University of Baja California, Tijuana 22390, Mexico
*   Correspondence: agranero@ual.es

Abstract: The aim of this study was to analyse the mediating role of Physical Education importance and satisfaction/fun between the dispositional goal orientations of secondary school students and their intention to partake in leisure time physical activity. The research design was descriptive, cross-sectional, and non-randomized. In total, 2102 secondary school students participated (M_{age} = 14.87; SD = 1.39) (1024 males; 1078 females). The scales used were the Perception of Success Questionnaire, Importance of Physical Education, Satisfaction with Physical Education, and Intention to Participate in Leisure Time Physical Activity. Structural equation models with the latent variables were also calculated. The results highlight that Physical Education satisfaction/fun has a mediating effect between task orientation and the intention to practice physical activity during leisure time.

Keywords: motivation; satisfaction; task orientation; ego orientation; physical education

Citation: Pérez-Quero, F.J.; Granero-Gallegos, A.; Baena-Extremera, A.; Baños, R. Goal Orientations of Secondary School Students and Their Intention to Practise Physical Activity in Their Leisure Time: Mediation of Physical Education Importance and Satisfaction. *Healthcare* **2023**, *11*, 568. https://doi.org/10.3390/healthcare11040568

Academic Editors: Rafael Oliveira and João Paulo Brito

Received: 22 January 2023
Revised: 9 February 2023
Accepted: 13 February 2023
Published: 14 February 2023

1. Introduction

Physical Education (PE) in Spain has significant importance in Organic Law 3/2020, of 29 December [1]. This law regulates non-university educational teaching in the different age groups and encourages the education system and PE to help people incorporate sports practice into their lives and thus contribute to an active and healthy lifestyle. Moreover, this law adds an additional provision that was not contemplated in the previous one, Organic Law 2/2006, of 3 May [2]. This provision refers to the promotion of physical activity and healthy eating so that they are part of the behaviour of children and young people. This line mentions that the administrations will promote the daily practice of sports and physical exercise by students during the school day, under the terms and conditions that, following the recommendations of the government agencies, guarantee adequate development to promote a healthy and autonomous life, to promote healthy eating habits and active mobility, reducing sedentary lifestyles [1]. Despite this, adolescence is a stage in which the practice of physical activity is still largely neglected; in addition, there is a high rate of physical inactivity in this population [3,4], which is why it is an important issue to address in the scientific field.

The sedentary lifestyle is a global problem, although in Spain, it is particularly associated with various psychological [5] and physical pathologies [6]. In this regard, it should be noted that Spain is the European country with the highest prevalence of mental health issues, with 20.8% of Spanish adolescents suffering from mental disorders such as depression, anxiety, and behavioural disorders, among others, according to the United

Nations Children's Fund [7]. In addition, there is a prevalence of being overweight or obese amongst more than 2.5 million Spanish children and adolescents [8]. According to the National Institute of Statistics [9], the data on physical inactivity are really worrying—31.8% of secondary and upper secondary school students claim to be inactive during their leisure time. The problem is exacerbated by adolescents claiming that they have no intention of practising any physical sports activity in their leisure time, apart from school lessons [10]. Consequently, work needs be done on increasing the students' intention to do physical exercise outside school, and since research on secondary school students has revealed that satisfaction and enjoyment of PE, and the importance given to this subject, can be predictive variables of their intention to practise physical exercise apart from school lessons [11–14], it is advisable to address the mediating role of these variables (satisfaction/fun, importance) between other variables such as the dispositional goal orientations (i.e., task orientation, ego orientation) with which the students approach this subject (i.e., PE) and their intention to practise physical exercise apart from school hours.

The Theory of Planned Behaviour [15] considers "intention" to be the most predictive factor of future behaviour disposing a person towards practising physical activity [16]. According to Ajzen [15], intention can be influenced by three factors: one that reflects the social influence, the subjective norm, understood as the pressure adolescents perceive from their peers or teacher, and two other factors of a personal nature—the attitude towards the behaviour, in which the performance of a certain behaviour is evaluated either positively or negatively, and the perceived behavioural control, which directly predicts the behaviour depending on whether it is under voluntary control or not, and whether there are differences between the control that the person believes they have and that which they actually have. However, it should be noted that there is a gap between intention and behaviour, that is, not every intention to perform a behaviour becomes a behaviour itself [17]. Regarding the intention to be physically active, studies have shown this gap exists between the intention to be physically active and developing a positive behaviour towards physical activity [17,18]. However, the intention is the proximal antecedent of the enactment of a behaviour [19]. In this line, the intention of adolescents to practice physical activity outside school hours has been related both to the importance and usefulness they give to PE [20] and to the satisfaction they experience with the subject [10]. For this reason, we consider it important to investigate how we can act regarding PE, in order to increase the intention of adolescents to be physically active in their free time.

Adolescence is a critical stage for students to develop positive attitudes towards physical activity and sport, with PE playing a fundamental role in generating active lifestyles [12,21]. If students consider PE important and find it useful for their routine life, they are more likely to generate active behaviours in the future, increasing their intention to be physically active in their free time, apart from PE lessons [13,20,22]. In turn, it is important that adolescents have fun and feel satisfied in PE lessons, since satisfaction predicts how they perceive both the importance and usefulness of PE and the intention to be physically active in their leisure time, apart from PE lessons [3,12]. Conversely, if they experience dissatisfaction or boredom, it will detract from the subject and so decrease active behaviours outside of school [23].

The Subjective Well-Being Theory is a theoretical construct that analyses the perception of human beings regarding their satisfaction with life generally, as well as in specific areas of life such as family, friends, school, sports practice, etc. [24]. PE can be considered a specific subject for adolescents in which they might make judgments in terms of satisfaction/fun or dissatisfaction/boredom based on how they learn and experience the subject [10]. In this vein, Balaguer et al. [25] designed an instrument that measured satisfaction with the school, which was later adapted to PE by Baena-Extremera et al. [26], in order to measure the satisfaction and enjoyment experienced by students with PE, in this specific area of adolescent life. This enjoyment with PE acts as an excellent predictor of both the intention to practice physical activity outside PE lessons [9] and their participation in physical sports activities [27]. Thus, generating enjoyable learning environments in PE can be an important variable in approaching the goal of adolescents continuing to practise sports outside of school hours [26]. How adolescents approach their goal perspectives can have a great impact on how satisfied they are with PE [26] and on their intention to be physically active [28].

The Social Cognitive Theory within Achievement Goal Theory describes the two perspectives that predominate in PE—one being task-oriented and the other ego-oriented [29]. According to the author, when an individual judges their own level of ability, an orientation towards the task prevails; in contrast, when the individual compares their ability level to others, an ego orientation predominates. Various studies have linked task orientation with greater fun and enjoyment of PE, whereas students who have an ego orientation suffer more from boredom and do not enjoy practising the subject [26,29,30]. Dispositional goal orientations in PE are so relevant that when adolescent students find themselves in a task-oriented climate, they tend to perceive PE classes as more enjoyable and are more active outside of school [31].

As described above, PE plays an important role in a student's intention to be physically active outside of school hours because it involves self-determined motivation and the importance and usefulness that adolescents find in the subject [13,20,22,32], the enjoyment and fun they experience [10], and the type of dispositional goal orientation that exists in the class [28]. However, to the best of our knowledge, there are no studies of a predictive model that includes all these variables. Therefore, this research aims to analyse the mediating role of satisfaction with PE and the perceived importance and usefulness of PE between the dispositional goal orientations of secondary school students and their intention to practise physical activity in their leisure time. After reviewing the scientific literature, we have considered a hypothesized model (see Figure 1) containing the following hypotheses: (H1) the importance and usefulness that students perceive in PE will have a negative mediating effect between ego orientation and the intention to be active outside of the school environment; (H2) the importance and usefulness that students perceive in PE will have a positive mediating effect between task orientation and the intention to be active outside of the school environment; (H3) satisfaction with PE will have a negative mediating effect between ego orientation and the intention to be active outside of the school environment; (H4) satisfaction with PE will have a positive mediating effect between task orientation and the intention to be active outside of the school environment.

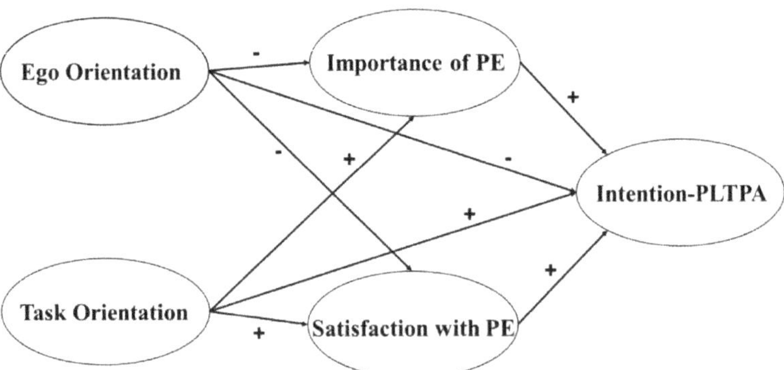

Figure 1. Hypothesized model. Note: dashed lines represent non-significant relationships. Note: PE = Physical Education; Intention-PLTPA = intention to partake in leisure time physical activity.

2. Materials and Methods

2.1. Design and Participants

The study design was cross-sectional. An a priori analysis of the statistical power of the sample size was carried out [28]; it was estimated that a minimum of 1970 students were necessary for effect sizes of $f^2 = 0.126$ with a statistical power of 0.99, and a significance level of $\alpha = 0.05$ in a structural equation model with five latent variables and 23 observable variables [33]. The study involved a total of 2102 secondary school students (1024 male) from 18 secondary schools in Andalusia. Distribution by course was as follows: 34.9% studied in the second grade of Compulsory Secondary Education (CSE); 16.9% in third grade of CSE; 23.6% were fourth graders of CSE; and 24.6% were in their first year of high school (bachillerato). The age ranged from 12 to 19 years ($M = 14.87$; $SD = 1.39$). These students had a medium economic socio-level, with a 3% rate of dropout, and 7% of pupils in the classes were foreign. The classes were mixed (boys and girls), and all pupils had PE as a compulsory subject (two sessions of 60 min/per week). In addition, 28.3% of the students participated in sporting competitions outside of school during the week.

2.2. Procedure

After obtaining authorization from the schools to carry out the research, the students were informed of the study objective and their rights as participants in it, how to answer the questionnaire, that the answers would be kept anonymous, that they would not affect their grades in the subject, and that they could stop participating in the study at any time. The data were collected in person by a researcher during the PE class, after previous agreement with the teacher. The students had 15 min to answer the questionnaire, and the PE teacher was not present in the classroom. Prior consent was obtained from the parents/legal guardians of all the participants included in the study. Approval for the research protocol was obtained from the University Ethics Committee (Ref: 19002018) and was carried out in accordance with the Declaration of Helsinki.

2.3. Instruments

Perception of Success Questionnaire (POSQ). The Spanish version [34] adapted to PE [35] of the original POSQ [36] was used. The scale consists of 12 items that measure the students' dispositional goal orientations in PE classes: task orientation (six items) and ego orientation (six items). Responses are collected on a five-point Likert scale ranging from 1 (strongly disagree) to 5 (strongly agree). A higher score indicates higher dispositional goal orientation (i.e., task orientation, ego orientation). In the present study, the CFA model presented the following goodness-of-fit indices: $\chi^2/df = 3.45$, $p < 0.001$; CFI = 0.97; TLI = 0.97; RMSEA = 0.059 (90%CI = 0.052, 0.066), SRMR = 0.038. The reliability obtained was: task orientation (McDonald's Omega, ω) = 0.88; ego orientation, $\omega = 0.91$.

Importance of Physical Education (IPE). The version by Moreno et al. [21] was used, which evaluates the importance and usefulness students give to PE through three items. Responses are collected on a four-point Likert scale ranging from 1 (strongly disagree) to 4 (strongly agree). A higher score indicates higher importance of PE. In the present study, the CFA model presented the following goodness-of-fit indices: $\chi^2/df = 2.45$, $p = 0.126$; CFI = 0.98; TLI = 0.96; RMSEA = 0.045 (90%CI = 0.038, 0.051), SRMR = 0.042. The reliability obtained was: $\omega = 0.78$.

Satisfaction with Physical Education (SSI-PE). The satisfaction/fun subscale from the Spanish version of the SSI-PE [26], part of the original Sport Satisfaction Instrument [25,37,38], was used. It consists of five items that measure satisfaction/fun with PE classes. Responses are collected on a five-point Likert scale ranging from 1 (strongly disagree) to 5 (strongly agree). A higher score indicates higher satisfaction with PE. In the present study, the CFA model (satisfaction/fun) presented the following goodness-of-fit indices: $\chi^2/df = 1.01$, $p = 0.400$; CFI = 0.99; TLI = 0.99; RMSEA = 0.013 (90%CI = 0.002, 0.042), SRMR = 0.004. The reliability obtained was: $\omega = 0.94$.

Intention to partake in leisure time physical activity (Intention-PLTPA). The Spanish version [39], based on Chatzisarantis et al. [36], Ajzen and Madden [16], and Ajzen and Fishbein [40], was used. This one-dimensional scale, consisting of three items, measures the intention of secondary school students to be physically active in their leisure time (outside of school). Responses are collected on a 7-point Likert scale ranging from 1 (very unlikely) to 7 (very likely). A higher score indicates higher intention to partake in leisure time physical activity. In the present study, the CFA model presented the following goodness-of-fit indices: $\chi^2/\text{df} = 1.78$, $p = 0.171$; CFI = 0.99; TLI = 0.99; RMSEA = 0.003 (90%CI = 0.001, 0.039), SRMR = 0.003. The reliability obtained was: $\omega = 0.93$.

2.4. Data Analysis

Descriptive statistics and correlations between the analysed variables were estimated with the SPSS v.28 programme. In addition, the McDonald's omega coefficient was calculated for each of the variables, considering values >0.70 as being indicative of good reliability [41]. The SEM was controlled for gender, age, and sports competition outside of school during the week a two-step structural equation model (SEM) was carried out with AMOS v.26 [42] to analyse the predictive relationships between the students' dispositional goal orientation and the intention to perform physical sports practice in their leisure time, analysing the mediating role of importance and satisfaction with Physical Education. In the first step, referred to as the measurement model, the robustness of the bidirectional relationships between the variables that make up the model was analysed. In the second step, the predictive effects between the variables were determined. In the event of the multivariate normality assumption being violated (Mardia's coefficient = 217.23; $p < 0.001$), the analysis was performed using the maximum likelihood method and the 5000-iteration bootstrapping procedure [42]. To evaluate the model's goodness of fit, chi-square and degrees of freedom (χ^2/df) values <5.0 were considered acceptable, as were CFI (Comparative Fit Index) and TLI (Tucker–Lewis Index) values >0.90, in conjunction with values up to 0.80 for the SRMR (Standardized Root Mean Square Residual) and RMSEA (Root Mean Square Error of Approximation) [43,44]. To analyse the direct and indirect effects, the proposal of Shrout and Bolger [45] was followed; thus, the indirect effects (i.e., mediated) and their 95% CI were estimated with the bootstrapping technique, and the significant indirect effect ($p < 0.05$) was considered if its 95% CI did not include the zero value.

3. Results

3.1. Preliminary Results

The descriptive statistics and the correlations between the latent study variables are presented in Table 1. First, it is notable that the mean values for the dispositional goal orientation are higher in the task orientation than in the ego orientation. The rest of the variables present moderately high values, considering the measurement range. Regarding the correlations, all are statistically significant, and the task orientation presents closer relationships with the ego orientation and satisfaction with PE. The correlations between the ego orientation and the rest of the variables are low, while the close relationship between satisfaction with PE and the intention to PLTPA is striking, as is the relationship between the importance of PE and satisfaction with PE. Regarding the reliability of the scales, all presented values are higher than 0.70.

Table 1. Descriptive statistics and correlation between variables.

Variable	Range	M	SD	Q1	Q2	2	3	4	6
1. Task-orientation	1–5	4.18	0.68	−1.20	1.27	0.34 **	0.18 **	0.32 **	0.28 **
2. Ego-orientation	1–5	3.25	1.13	−0.22	−0.59		0.09 **	0.15 **	0.15 **
3. Importance of PE	1–4	3.05	0.77	−0.79	0.36			0.45 **	0.3 **
4. Satisfaction with PE	1–5	4.19	0.93	−1.08	1.03				0.56 **
5. Intention-PLTPA	1–7	4.97	1.85	−0.64	−0.74				

Note. ** The correlation is significant at the 0.01 level. M = mean; SD = standard deviation; Q1 = skewness; Q2 = Kurtosis; PE = Physical Education; Intention-PLTPA = intention to partake in leisure time physical activity.

3.2. Main Results

In step 1, the model presented acceptable goodness-of-fit values: $\chi^2/\text{df} = 3.024$, $p < 0.001$; CFI = 0.97; TLI = 0.97; RMSEA = 0.040 (90%CI = 0.037; 0.043), SRMR = 0.056. In step 2, the predictive SEM model also presented an acceptable fit: $\chi^2/\text{df} = 4.580$, $p < 0.001$;

CFI = 0.95; TLI = 0.94; RMSEA = 0.053 (90%CI = 0.050; 0.056), SRMR = 0.069. The SEM achieved an explained variance of 37% for intention-PLTPA, 20% for satisfaction with PE, and 9% for importance of PE. After controlling for gender, age, and sports competition outside school lessons during the week, in the SEM (Figure 2), it was observed that ego orientation has no effect on the intention to practice physical activity in one's leisure time or outside the PE classes, since it does not present statistically significant relationships with any of the variables (i.e., importance of PE, satisfaction with PE, intention-PLTPA). In contrast, the direct relationships were significant between task orientation and the importance of PE, satisfaction with PE, and intention-PLTPA, as were the direct effects of the importance of PE and satisfaction with PE on intention-PLTPA. It is worth noting the mediating effect of satisfaction with PE between task orientation and intention-PLTPA (0.19), as it increases the effect between the latter two variables. The mediating effect of the importance of PE between task orientation and intention-PLTPA is less relevant (0.05) than that of satisfaction with PE. In addition, the overall effects of task orientation on intention-PLTPA were higher with the mediating effect of satisfaction with PE (see Table 2).

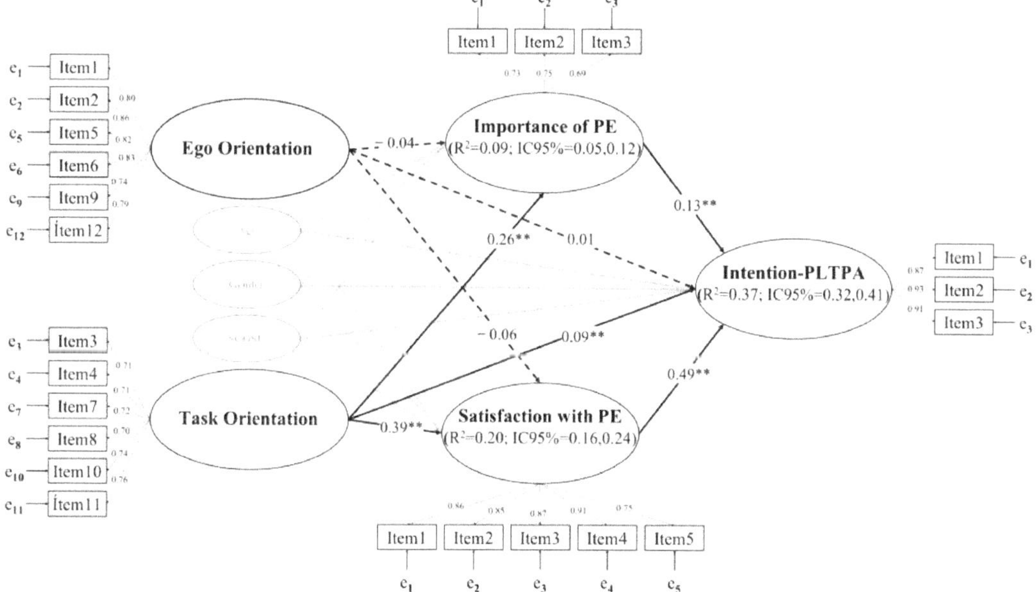

Figure 2. Predictive relationships of dispositional goal orientations on the intention to partake in leisure time physical activity through the mediation of the importance and usefulness of Physical Education and satisfaction/fun with Physical Education. Note: ** $p < 0.01$; SCOSL = sports competition outside of school lessons; PE = Physical Education; Intention-PLTPA = intention to partake in leisure time physical activity; R^2 = explained variance; CI = confidence interval. Dashed lines represent non-significant relationships. The SEM was controlled for gender, age, and sports competition outside school lessons during the week.

Table 2. Estimation of significant standardized parameters and statistics of the mediation model.

	Independent Variable	Dependent Variable	Mediator	β	SE	95%CI	
						Inf	Sup
Direct effects							
	Task orientation	Importance PE		0.26 **	0.04	0.19	0.33
	Task orientation	Intention-PLTPA		0.09 **	0.03	0.04	0.154
	Task orientation	Satisfaction PE		0.39 **	0.04	0.33	0.45
	Importance PE	Intention-PLTPA		0.13 **	0.04	0.08	0.20
	Satisfaction PE	Intention-PLTPA		0.49 **	0.04	0.42	0.54
Indirect effects							
	Task orientation	Intention-PLTPA	Importance PE	0.05 *	0.05	0.02	0.08
	Task orientation	Intention-PLTPA	Satisfaction PE	0.19 **	0.06	0.12	0.25
Total Effects							
	Task orientation	Intention-PLTPA	Importance PE	0.14 *	0.04	0.05	0.15
	Task orientation	Intention-PLTPA	Satisfaction PE	0.28 *	0.04	0.23	0.36

Note. β = estimation of standardized parameters; SE = standard error; 95%CI = 95% confidence interval; Inf = lower limit of 95%CI; Sup = upper limit of 95%CI; PE = Physical Education; Intention-PLTPA = intention to partake in leisure time physical activity; ** $p < 0.01$; * $p < 0.05$.

4. Discussion

The objective of the present research was to analyse the mediation of the importance and usefulness of PE and satisfaction with PE between dispositional goal orientations and the intention of secondary school students to practise physical activity in their leisure time. The main results show that task orientation has a direct and positive effect on the intention to practice physical activity in one's leisure time. The total effects on this variable are increased, above all, with the mediation of the satisfaction with PE that the student experiences. Ego orientation is not significantly related to any of the variables studied.

Few studies have analysed the prediction of goal orientations on the intention of adolescents to practice physical activity in their leisure time. The present work shows that task orientation directly and positively predicts the intention to practice physical activity during their leisure time in the future, while no significant results were obtained in those students with an ego orientation. Similar results were obtained by Franco et al. [28]—in their study, only task orientation predicted the intention to practice physical activity in the future. Several studies have highlighted how important it is for PE teachers to design their classes to encourage task-oriented learning climates—this can be decisive, both for achieving positive results in PE classes and for increasing active lifestyles outside of school [20,46]. These results could be due to the fact that when students have a task-oriented disposition to achieve their goals, they believe the success of the final result depends on their own effort, interest, and self-learning [47–49], with these behaviours being related to a lower probability of them abandoning sports practice in their leisure time [31].

In addition, the results of the present study show that the total effects of dispositional goal orientation on the intention to practice physical activity in one's leisure time are increased with the mediation of adolescents finding PE important and useful but, above all, of them enjoying the subject. We are not aware of any research that has analysed the mediating effect of these variables. Focusing on the mediating role of satisfaction with PE, several studies have found that when a student has fun and feels satisfied with PE classes, their intention to be active in their leisure time will increase [10,20,28], with task orientation being a strong predictor of satisfaction with PE [12]. This relationship could be due to the fact that when adolescents are task-oriented, they tend to make an effort in their personal development, focusing their satisfaction on self-improvement [47,50,51], thus increasing their satisfaction with PE [12,48]. Various authors have highlighted the importance of strengthening student satisfaction and motivation in order to continue physical activity outside the school classroom [3,32,52–54]. In this way, students will be enriched by positive experiences in the PE classroom, a fundamental aspect for their commitment to healthy and active lifestyle habits [55,56]. Conversely, the accumulation of negative physical activity experiences can lead to habitual physical inactivity [3,53–55,57].

In the case of the importance and usefulness of PE for students, we are likewise unaware of any research that has analysed the mediating effect of these variables. In this regard, we can say that our results agree with those obtained by other studies, which related this variable with the intention to be physically active [20–23] and with task orientation [58,59]. This could be due to the fact that when students are task oriented, they use internal sources to judge their motor ability, and when they fail to achieve the results, they respond with increased effort and greater persistence to perform the activity until they are successful [60]. In this way, they will give greater importance to PE, finding it more useful for solving everyday situations [61].

Finally, a series of limitations and strengths should be mentioned. In terms of limitations, mention that the present study only measured the intention, not the actual behaviours, that is, the intentions of adolescents to be physically active do not always translate into active behaviours. It is worth noting the cross-sectional research design, since when it is carried out in a determined time, it is not possible to analyse the behaviour of the subjects during a determined time. Another limitation of the study was that the measurements and the results obtained were carried out through questionnaires, this could condition the responses of the participants due to the search for social desirability, causing a bias in the research. Finally, it is worth mentioning that the sample was not representative, so the results cannot be generalized to Spanish adolescents. However, regarding the study's strengths, the size of the sample and the type of statistical analysis can be highlighted, since the SEM has been carried out with latent variables. As future research, we suggest that longitudinal or experimental studies should be carried out, in which it is analysed over time if the intention to be physically active becomes behaviour, and to analyse which are the most influential variables in this transformation. Finally, the research topic addressed in this study is of great interest, namely, proposing a possible solution to physical inactivity approached through the subject of PE.

Practical Implications

The results of the present study underline the importance of task orientation among adolescent students and, in turn, that they feel satisfied with PE and that they understand its importance and usefulness, thus encouraging the intention to continue with physical activity in their leisure time. For this reason, it is recommended that secondary school PE teachers highlight the importance of PE in their classes and the usefulness that students can get from it in their day-to-day lives—for example, by creating an educational blog about PE [62], encouraging walking or cycling to school [63], or creating fun classes and new content [64,65], among others. Furthermore, it is recommended that they design their classes to support student autonomy, with sessions that are motivating and that focus on self-improvement, since this can help increase enjoyment and satisfaction with PE, and thus the intention of leading active lifestyles [66–70].

5. Conclusions

In summary, it can be stated that a task-oriented disposition in students increases their intention to practise physical activity in their leisure time. Furthermore, this intention to lead an active lifestyle increases especially when students have fun and are satisfied with PE, although it is also necessary that students consider the subject important and useful. Finally, secondary school PE teachers should incorporate methodological strategies in their classes that can be extrapolated to their adolescent students practising physical sports activity in their leisure time, promoting the use of the surrounding environment (whether natural or urban), as well as encouraging student autonomy in deciding what activities to carry out, while motivating them to practice sports for the enjoyment and satisfaction of experiencing positive emotions and feelings, rather than to achieve an award, demonstrate ability, or seek social approval from their peers.

Author Contributions: Conceptualization, F.J.P.-Q. and A.G.-G.; methodology, A.G.-G. and A.B.-E.; formal analysis, A.G.-G. and A.B.-E.; investigation, F.J.P.-Q., A.G.-G., A.B.-E. and R.B.; data curation, F.J.P.-Q.; writing—original draft preparation, F.J.P.-Q. and R.B.; writing—review and editing, A.G.-G. and A.B.-E. All authors have read and agreed to the published version of the manuscript.

Funding: This research received no external funding.

Institutional Review Board Statement: The study was conducted in accordance with the Declaration of Helsinki and approved by the University Ethics Committee (protocol code Ref: 19002018; 19 February 2018).

Informed Consent Statement: Informed consent was obtained from all subjects involved in the study.

Data Availability Statement: The data presented in this study are available on request from the corresponding author. The data are not publicly available due to privacy.

Conflicts of Interest: The authors declare no conflict of interest.

References

1. Ley Orgánica 3/2020, de 29 de diciembre, por la que se modifica la Ley Orgánica 2/2006, de 3 de mayo, de Educación. Available online: https://www.boe.es/buscar/doc.php?id=BOE-A-2020-17264 (accessed on 1 September 2022).
2. Ley Orgánica 2/2006, de 3 de mayo, de Educación. Available online: https://www.boe.es/buscar/act.php?id=BOE-A-2006-7899 (accessed on 1 September 2022).
3. Aznar-Ballesta, A.; Vernetta, M. Influence of the satisfaction and importance of physical education on sports dropout in secondary school. *Espiral* **2023**, *16*, 18–28. [CrossRef]
4. Gómez-López, M.; Granero-Gallegos, A.; Baena-Extremera, A. Perceived barriers by university students in the practice of physical activities. *J. Sports. Sci. Med.* **2010**, *9*, 374–381. [PubMed]
5. Raudsepp, L.; Vink, K. Longitudinal associations between sedentary behavior and depressive symptoms in adolescent girls followed 6 years. *J. Phys. Act. Health* **2019**, *16*, 191–196. [CrossRef] [PubMed]
6. Tambalis, K.D.; Panagiotakos, D.B.; Psarra, G.; Sidossis, L.S. Concomitant Associations between Lifestyle Characteristics and Physical Activity Status in Children and Adolescents. *J. Res. Health Sci.* **2019**, *19*, 1–7. [CrossRef]
7. UNICEF (2021). The State of the World's Children. 2021. Available online: https://www.unicef.org/reports/state-worlds-children-2021 (accessed on 27 December 2022).
8. de Bont, J.; Bennett, M.; León-Muñoz, L.M.; Duarte-Salles, T. Prevalencia e incidencia de sobrepeso y obesidad en 2, 5 millones de niños y adolescentes en España. *Rev. Esp. Cardiol.* **2022**, *75*, 300–307. [CrossRef]
9. Instituto Nacional de Estadística. Ejercicio Físico Regular y Sedentarismo en el Tiempo Libre. 2020. Available online: https://bit.ly/3CruIuV (accessed on 3 November 2022).
10. Baños, R. Intention of practice, satisfaction with physical education and life by gender in Mexican and Spanish students. *Retos* **2020**, *37*, 412–418. [CrossRef]
11. Aznar-Ballesta, A.; Vernetta, M. Satisfaction and importance of physical education in Secondary schools. *Rev. Iberoam. Cienc. Act. Fis. Dep.* **2022**, *11*, 44–57. [CrossRef]
12. Baena-Extremera, A.; Granero-Gallegos, A.; Ponce de León Elizondo, A.; Sanz-Arazuri, E.; Valdemoros San Emeterio, M.Á.; Martínez-Molina, M. Psychological factors related to physical education classes as predictors of students' intention to partake in leisure-time physical activity. *Cienc. Saude Coletiva* **2016**, *21*, 105–112. [CrossRef]
13. Granero-Gallegos, A.; Baena-Extremera, A.; Sánchez-Fuentes, J.A.; Martínez-Molina, M. Motivational profiles of autonomy support, self-determination, satisfaction, importance of physical education and intention to partake in leisure time physical activity. *Cuad. Psicol. Dep.* **2014**, *14*, 59–70. [CrossRef]
14. Muñoz-González, V.; Gómez-López, M.; Granero-Gallegos, A. Relación entre la satisfacción con las clases de Educación Física, su importancia y utilidad y la intención de práctica del alumnado de Educación Secundaria Obligatoria. *Rev. Complut. Educ.* **2019**, *30*, 149–161. [CrossRef]
15. Ajzen, I. The theory of planned behavior. *Organ. Behav. Hum. Decis. Process.* **1991**, *50*, 179–211. [CrossRef]
16. Ajzen, I.; Madden, T. Prediction of goal-directed behaviour: Attitudes, intentions and perceived behavioural control. *J. Exp. Soc. Psychol.* **1986**, *22*, 453–474. [CrossRef]
17. Conner, M.; Norman, P. Understanding the intention-behavior gap: The role of intention strength. *Front. Psychol.* **2022**, *42*, 49. [CrossRef] [PubMed]
18. Fung, X.C.; Pakpour, A.H.; Wu, Y.K.; Fan, C.W.; Lin, C.Y.; Tsang, H.W. Psychosocial variables related to weight-related self-stigma in physical activity among young adults across weight status. *Int. J. Environ. Res. Public Health* **2020**, *17*, 64. [CrossRef] [PubMed]
19. Rhodes, R.E.; de Bruijn, G.J. How big is the physical activity intention–behaviour gap? A meta-analysis using the action control framework. *Br. J. Health Psychol.* **2013**, *18*, 296–309. [CrossRef] [PubMed]
20. Moreno-Murcia, J.A.; Hernández, E.H. Effect of a teaching intervention on motivation, enjoyment, and importance given to Physical Education. *Motricidade* **2019**, *15*, 21–31.

21. Moreno, J.A.; González-Cutre, D.; Ruiz, L.M. Self-determined motivation and physical education importance. *Hum. Mov.* **2009**, *10*, 5–11. [CrossRef]
22. Green, K. Mission impossible? Reflecting upon the relationship between physical education, youth sport and lifelong participation. *Sport. Educ. Soc.* **2014**, *19*, 357–375. [CrossRef]
23. Baños, R.; Marentes, M.; Zamarripa, J.; Baena-Extremera, A.; Ortiz-Camacho, M.; Duarte-Félix, H. Influence of satisfaction, boredom and importance of physical education with the intention of performing extracurricular exercise amongst Mexican teenagers. *Cuad. Psicol. Dep.* **2019**, *19*, 205–215. [CrossRef]
24. Diener, E.; Emmons, R.A. The independence of positive and negative affect. *J. Pers. Assess.* **1985**, *99*, 91–95. [CrossRef]
25. Balaguer, I.; Atienza, F.L.; Castillo, I.; Moreno, Y.; Duda, J.L. Factorial structure of measures of satisfaction/interest in sport and classroom in the case of Spanish adolescents. In *Abstracts of 4th European Conference of Psychological Assessment*; University of Lisbon: Lisbon, Portugal, 1997; p. 76.
26. Baena-Extremera, A.; Granero-Gallegos, A.; Bracho-Amador, C.; Pérez-Quero, F.J. Spanish Version of the Sport Satisfaction Instrument (SSI) Adapted to Physical Education. *Rev. Psicodidact.* **2012**, *17*, 377–396. [CrossRef]
27. Jang, H.; Kim, E.; Reeve, J. Why students become more engaged or more disengaged during a semester: A self-determination theory dual-process model. *Lear Instr.* **2016**, *43*, 27–38. [CrossRef]
28. Franco, E.; Coterón, J.; Huéscar, E.; Moreno-Murcia, J.A. A person-centered approach in physical education to better understand low-motivation students. *J. Teach. Phys. Educ.* **2020**, *39*, 91–101. [CrossRef]
29. Nicholls, J.G. *The Competitive Ethos and Democratic Education*; Harvard University Press: Cambridge, MA, USA, 1989.
30. Fernández-Río, F.; Méndez-Giménez, A.; Cecchini, J.A.; González, C. Achievement Goals and Social Goals' Influence on Physical Education Students' Fair Play. *Rev. Psicodidact.* **2012**, *17*, 73–91.
31. Ntoumanis, N. A prospective study of participation in optional school physical education using a self-determination theory framework. *J. Educ. Psychol.* **2005**, *97*, 444. [CrossRef]
32. Granero-Gallegos, A.; Baena-Extremera, A. Prediction of self-determined motivation as goal orientations and motivational climate in Physical Education. *Retos* **2014**, *25*, 23–27. [CrossRef]
33. Soper, D.S. A-priori Sample Size Calculator for Structural Equation Models. 2023. Available online: https://bit.ly/3oxyIQS (accessed on 21 December 2022).
34. Cervelló, E.M.; Escartí, A.; Balagué, G. Relaciones entre la orientación de metas disposicional y la satisfacción con los resultados deportivos, las creencias sobre las cusas de éxito en deporte y la diversión con la práctica deportiva. *Rev. Psicol. Dep.* **1999**, *8*, 7–19.
35. Granero-Gallegos, A.; Baena-Extremera, A.; Gómez-López, M.; Abraldes, A. Psychometric Study and the Prediction of the Importance of Physical Education from Goal Guidance ("Perception of Success Questionnaire—POSQ"). *Psicol.-Reflex. Crit.* **2014**, *27*, 443–451. [CrossRef]
36. Martínez, C.; Alonso, N.; Moreno, J.A. Análisis factorial confirmatorio del "Cuestionario de Percepción de Éxito (POSQ)" en alumnos adolescentes de Educacion Fisica. In *IV Congreso de la Asociación Española de Ciencias del Deporte*, En, M.A., González, J.A., Sánchez, Y.A., Areces, Eds.; Xunta de Galicia: A Coruña, Spain, 2006; pp. 757–761.
37. Roberts, G.C.; Treasure, D.C.; Balagué, G. Achievement goals in sport: The development and validation of the Perception of Success Questionnaire. *J. Sport Sci.* **1998**, *16*, 337–347. [CrossRef]
38. Duda, J.L.; Nicholls, J.G. Dimensions of achievement motivation in schoolwork and sport. *J. Educ. Psychol.* **1992**, *84*, 290–299. [CrossRef]
39. Granero-Gallegos, A.; Baena-Extremera, A.; Pérez-Quero, F.J.; Ortiz-Camacho, M.M.; Bracho-Amador, C. Spanish validation of the scale «intention to leisure-time in partake physical activity». *Retos* **2014**, *26*, 40–45. [CrossRef]
40. Chatzisarantis, N.L.D.; Biddle, S.J.H.; Meek, G.A. A self-determination theory approach to the study of intentions and the intention-behaviour relationship in children's physical activity. *Br. J. Health Psychol.* **1997**, *2*, 343–360. [CrossRef]
41. Ajzen, I.; Fishbein, M. *Understanding Attitudes and Predicting Social Behaviour*; Prentice-Hall: Englewood Cliffs, NJ, USA, 1980.
42. Viladrich, C.; Angulo-Brunet, A.; Doval, E. A journey around alpha and omega to estimate internal consistency reliability. *An. Psicolo.* **2017**, *33*, 755. [CrossRef]
43. Kline, R.B. *Principles and Practice of Structural Equation Modeling*, 4th ed.; The Guilford Press: New York, NY, USA, 2016.
44. Hu, L.T.; Bentler, P.M. Cutoff criteria for fit indexes in covariance structure analysis: Conventional criteria versus new alternatives. *Struct. Equ. Model. A Multidiscip. J.* **1999**, *6*, 1–55. [CrossRef]
45. Marsh, H.W.; Hau, K.T.; Wen, Z. In search of golden rules: Comment on hypothesis-testing approaches to setting cutoff values for fit indexes and dangers in overgeneralizing Hu and Bentler's (1999) findings. *Struct. Equ. Model.* **2004**, *11*, 320–341. [CrossRef]
46. Shrout, P.E.; Bolger, N. Mediation in experimental and nonexperimental studies: New procedures and recommendations. *Psych. Methods* **2002**, *7*, 422. [CrossRef]
47. Leo, F.M.; Behzadnia, B.; López-Gajardo, M.A.; Batista, M.; Pulido, J.J. What kind of interpersonal need-supportive or need-thwarting teaching style is more associated with positive consequences in physical education? *J. Teach. Phys. Educ.* **2022**, *ahead of print*. [CrossRef]
48. González-Suárez, R.; Trillo López, A.; Díaz-Pita, L.; Gómez-Pulido, A.; Guisande Couñago, A. Motivacional profiles of Spanish students PISA 2018: Emotional characterization and differences in academic performance. *Psychol. Soc. Educ.* **2022**, *14*, 57–67. [CrossRef]

49. Nicolosi, S.; Ortega Ruiz, R.; Benítez Sillero, J.D. Achievement goal orientations and perceived physical competence profiles in adolescent physical activity. *Psychol. Soc. Educ.* **2021**, *13*, 27–47. [CrossRef]

50. Sánchez-Alcaraz, B.J.; Murcia, S.; Alfonso, M.; Hellín, M. Physical activity level in students regarding age, gender, sport and motivational climate. *Espiral* **2020**, *13*, 160–169. [CrossRef]

51. Ruiz-Juan, F.; Piéron, M.; Zamarripa, J. Versión española del «Task and Ego Orientation in Sport Questionnaire (TEOSQ) adaptado a Educación Física. *Est. Psicol.* **2011**, *32*, 179–193. [CrossRef]

52. Baños, R.; Arrayales-Millán, E.M. Prediction of boredom in physical education from the motivational climate. *Retos* **2020**, *38*, 83–88. [CrossRef]

53. Baños, R.; Ortiz-Camacho, M.M.; Baena-Extremera, A.; Tristán-Rodríguez, J.L. Satisfaction, motivation and academic performance in students of secondary and high school: Background, design, methodology and proposal of analysis for a research paper. *Espiral* **2017**, *10*, 40–50. [CrossRef]

54. Huéscar Hernández, E.; Moreno-Murcia, J.A.; Ruiz González, L.; León González, J. Motivational profiles of high school physical education students: The role of controlling teacher behavior. *Int. J. Environ. Res. Public Health* **2019**, *16*, 1714. [CrossRef]

55. Polet, J.; Hassandra, M.; Lintunen, T.; Laukkanen, A.; Hankonen, N.; Hirvensalo, M.; Tammelin, T.; Hagger, M.S. Using physical education to promote out-of school physical activity in lower secondary school students–a randomized controlled trial protocol. *BMC Public Health* **2019**, *19*, 1–15. [CrossRef]

56. Peiró-Velert, C.; Pérez-Gimeno, E.; Valencia-Peris, A. Facilitación de la autonomía en el alumnado dentro de un modelo pedagógico de educación física y salud. *Tándem* **2012**, *40*, 28–44.

57. Beltran-Carrillo, V.J.; Devis-Devis, J.; Peiro-Velert, C.; Brown, D.H. When physical activity participation promotes inactivity: Negative experiences of Spanish adolescents in physical education and sport. *Youth Soc.* **2012**, *44*, 3–27. [CrossRef]

58. Erturan-İlker, G.; Yu, C.; Alemdaroglu, U.; Köklü, Y. Basic psychological needs and self-determined motivation in physical education to predict health-related fitness level. *J. Sport Health Res.* **2018**, *10*, 91–100.

59. Granero-Gallegos, A.; Baena-Extremera, A.; Pérez-Quero, F.J.; Ortiz-Camacho, M.M.; Bracho-Amador, C. Analysis of motivational profiles of satisfaction and importance of physical education in high school adolescents. *J Sports Sci. Med.* **2012**, *11*, 614.

60. Moreno-Murcia, J.A.; Huéscar-Hernández, E.; Ruíz-González, L. Capacidad predictiva del apoyo a la autonomía en clases de educación física sobre el ejercicio físico. *Rev. Lat. Psicol.* **2019**, *51*, 30–37. [CrossRef]

61. Van Yperen, N.; Duda, J. Goal orientations, beliefs about success, and performance improvement among young elite Dutch soccer players. *Scandinvian J. Med. Sci. Sports* **1999**, *9*, 358–364. [CrossRef] [PubMed]

62. López-Postigo, L.; Burgueño, R.; González-Fernández, F.T.; Morente-Oria, H. Edublog in Physical Education as an instrument for coeducation, commitment and intention to be physically active. *Espiral* **2021**, *14*, 1–8. [CrossRef]

63. Cerro-Herrero, D.; Prieto-Prieto, J.; Tapia, M.A.; Vaquero-Solís, M.; Sánchez-Miguel, P.A. Relationship between the intention to be physically active and active commuting to school: Proposals for intervention to increase active commuting. *Espiral* **2022**, *15*, 51–58. [CrossRef]

64. Baños, R.; Toval-Sánchez, Á.; Morales-Delgado, N.; Ferrán, J.L. Analysis of movement during climbing as a strategy for learning the anatomy of the locomotor system in Sport Sciences. *Espiral* **2021**, *14*, 102–112. [CrossRef]

65. Baños, R.; Morán-Navarro, R.; Toval, Á.; del Lidón López-Iborra, M. Learning and evaluation of human anatomy content in Sports Sciences through Surf videos. *Espiral* **2022**, *15*, 1–10. [CrossRef]

66. Cheon, S.H.; Reeve, J.A. classroom-based intervention to help teachers decrease students' amotivation. *Contemp. Educ. Psychol.* **2015**, *40*, 99–111. [CrossRef]

67. Fernández-Espínola, C.; Almagro, B.J.; Tamayo Fajardo, J.A. Prediction of physical education students' intention to be physically active: A model mediated by the need for novelty. *Retos* **2020**, *37*, 442–448. [CrossRef]

68. Lochbaum, M.; Jean-Noel, J. Percepción de la formación de apoyo a la autonomía y resultados en estudiantes en educación física y tiempo libre: Una revisión meta-analítica de correlaciones. *RICYDE* **2015**, *12*, 29–47. [CrossRef]

69. Lonsdale, C.; Lester, A.; Owen, K.B.; White, R.L.; Peralta, L.; Kirwan, M.; Diallo, T.M.O.; Maeder, A.J.; Bennie, A.; MacMillan, F.; et al. An internet-supported school physical activity intervention in low socioeconomic status communities: Results from the Activity and Motivation in Physical Education (AMPED) cluster randomised controlled trial. *Br. J. Sports Med.* **2019**, *53*, 341–347. [CrossRef] [PubMed]

70. Moreno-Murcia, J.A.; Huéscar Hernández, E.; Ruiz, L. Perceptions of controlling teaching behaviors and the effects on the motivation and behavior of high school physical education students. *Int. J. Environ. Res. Public Health* **2018**, *15*, 2288. [CrossRef] [PubMed]

 *healthcare*

Article

Comparison of Knee Muscular Strength Balance among Pre- and Post-Puberty Adolescent Swimmers: A Cross-Sectional Pilot Study

Bruno Lombardi Amado [1], Claudio Andre Barbosa De Lira [2], Rodrigo Luiz Vancini [3], Pedro Forte [4,5,6], Taline Costa [7], Katja Weiss [8], Beat Knechtle [8,*] and Marilia Santos Andrade [7]

[1] Sports Medicine Residency Program, Department of Orthopedics and Traumatology, Federal University of São Paulo, São Paulo 04021-001, Brazil
[2] Human and Exercise Physiology Division, Faculty of Physical Education and Dance, Federal University of Goiás, Goiânia 74690-900, Brazil
[3] Center for Physical Education and Sports, Federal University of Espírito Santo, Vitória 29075-910, Brazil
[4] Department of Sport, Higher Institute of Educational Sciences of the Douro, 4560-547 Porto, Portugal
[5] Department of Sports, Instituto Politécnico de Bragança, 5300-253 Bragança, Portugal
[6] Research Center in Sports, Health and Human Development, 7000-671 Covilhã, Portugal
[7] Department of Physiology, Federal University of São Paulo, São Paulo 04021-001, Brazil
[8] Institute of Primary Care, University of Zurich, 8091 Zürich, Switzerland
* Correspondence: beat.knechtle@hispeed.ch

Abstract: Muscular weakness and strength imbalance between the thigh muscles are considered risk factors for knee injuries. Hormonal changes, characteristic of puberty, strongly affect muscle strength; however, it is unknown whether they affect muscular strength balance. The present study aimed to compare knee flexor strength, knee extensor strength, and strength balance ratio, called the conventional ratio (CR), between prepubertal and postpubertal swimmers of both sexes. A total of 56 boys and 22 girls aged between 10 and 20 years participated in the study. Peak torque, CR, and body composition were measured using an isokinetic dynamometer and dual-energy X-ray absorptiometry, respectively. The postpubertal boys group presented significantly higher fat-free mass ($p < 0.001$) and lower fat mass ($p = 0.001$) than the prepubertal group. There were no significant differences among the female swimmers. Peak torque for both flexor and extensor muscles was significantly greater in postpubertal male ($p < 0.001$, both) and female swimmers ($p < 0.001$ and $p = 0.001$, respectively) than in prepubertal swimmers. The CR did not differ between the pre- and postpubertal groups. However, the mean CR values were lower than the literature recommendations, which brings attention to a higher risk of knee injuries.

Keywords: puberty; isokinetic; peak torque; conventional ratio; muscle balance

Citation: Amado, B.L.; De Lira, C.A.B.; Vancini, R.L.; Forte, P.; Costa, T.; Weiss, K.; Knechtle, B.; Andrade, M.S. Comparison of Knee Muscular Strength Balance among Pre- and Post-Puberty Adolescent Swimmers: A Cross-Sectional Pilot Study. *Healthcare* **2023**, *11*, 744. https://doi.org/10.3390/healthcare11050744

Academic Editor: Abdel-Latif Mohamed

Received: 20 December 2022
Revised: 28 February 2023
Accepted: 1 March 2023
Published: 3 March 2023

1. Introduction

Since the beginning of the 20th century, strength training for swimmers has been discussed among coaches and researchers, but there is evidence of the importance of propulsion in water and swimming performance [1]. Several previous studies have shown a positive correlation between muscular strength and swimming performance in short-distance (shorter than 400 m) [2] and long-distance events (longer than 400 m) [3,4].

Despite the undoubted importance of upper limb strength for performance during the free-swimming phase, lower limb strength is of fundamental importance at the start and the turns. Indeed, a significant correlation between lower-limb strength and turn times [5–7] or time to reach 5 or 10 m after the start [7,8] has already been demonstrated.

In addition, there are strength training recommendations for swimmers to improve bone mineral density (BMD), especially among children and adolescent athletes [9,10], because swimmers have been demonstrated to have lower BMD than other athletes who perform weight-bearing exercises, such as rhythmic gymnasts [11] or footballers [12].

Strength training has also been recommended as a part of an injury prevention program [13]. Tooth et al. [14] conducted a review of risk factors for sports injuries and demonstrated that muscular weakness is a very important risk factor for injuries. Muscular strength balance plays an important role in joint stability [1,15]. Thus, the isokinetic strength profile and strength balance ratios between antagonistic muscles have been described for several sports and joints [16–19]. Therefore, lower limb strength plays a significant role, as the knees are one of the most affected joints in swimmers, in addition to the shoulder and lumbar spine [13].

For the knee joint, the strength balance ratio between the flexor and extensor muscles has been measured to assess joint stability [15,20–22]. Traditionally, the strength balance has been calculated from the isokinetic peak torque value of knee flexor muscles divided by the isokinetic peak torque value of knee extensor muscles, both in concentric action (flexor-to-extensor ratio), and this ratio has been called the conventional balance ratio (CR) [23]. Values for CR higher than 60% have been generally accepted as preventive measures for injuries [23]. However, athletes from different sports commonly present some degree of strength imbalance because of their unique and repetitive muscular demands or lateral preferences [22].

Another factor that could affect strength balance among adolescent athletes is the pubertal phase. From puberty, a marked gain in muscle strength is evident [24,25]. However, it is unknown whether the abrupt muscular strengthening that occurs during puberty is a balanced gain in all the muscles of the lower limbs or whether the specific and repetitive muscular demand (numerous turns during swimming training) added to the abrupt strength gain characteristic of puberty can generate an imbalance between knee flexor and extensor muscles.

Therefore, the aim of the present study was to compare knee muscle strength, the ratio between the knee flexor and extensor muscles, and body composition between prepubertal and postpubertal swimmers of both sexes. It was hypothesized that the prepubertal participants would present less imbalance in comparison to the postpubertal group.

2. Materials and Methods

2.1. Design and Procedure

This was a cross-sectional study. Participants attended the laboratory once and were asked to refrain from strenuous workouts on the day before the visit. Sexual maturity was evaluated using the Tanner scale [24], and an isokinetic strength test for thigh muscles and body composition tests were performed. All the tests were performed in the morning between 9 a.m. and 11 a.m. Athletes and guardians were informed of the potential risks and benefits of the study and signed an informed consent form to participate in this study. All experimental procedures were approved by the University Human Research Ethics Committee (no.0503/2017) and conformed to the principles of the Declaration of Helsinki.

2.2. Participants

Seventy-eight swimmers (56 boys and 22 girls) participated in this study. All the athletes from the Olympic Center of Training and Research swimming team were recruited to participate in the study. The volunteers were divided into two groups according to sexual maturity and sex (pre- and postpuberty). Group 1 consisted of athletes presenting Tanner scale G1, G2, or G3 for boys and M1, M2, or M3 for girls. Group 2 consisted of athletes with Tanner scale G4 or G5 for boys and M4 or M5 for girls. The cut-off for separating the groups was supported by the fact that Tanner scale 4 is the point of puberty development that causes a significant increase in strength [25]. The inclusion criteria were as follows: participation in swimming training for at least 1 year, five times a week, and at least 1 h per day. The exclusion criteria were as follows: taking any medication, presenting pain in the lower limbs in the last 6 months, having previous lower limb surgery, or presenting any disease that could impact muscular strength. No athletes were excluded.

2.3. Evaluation of Sexual Maturity

The sexual maturity of the athletes was evaluated by a pediatrician based on the criteria defined by Tanner [24]. These criteria include pubic hair development (P1, P2, P3, P4, and P5) in both sexes and the quantity and distribution of secondary sexual characteristics according to the developmental stage of the genitals in boys (G1, G2, G3, G4, and G5) and breast development in girls (M1, M2, M3, M4, and M5). Stage 1 (P1, G1, and M1) corresponds to the prepubertal stage, and stage 5 (P5, G5, and M5) corresponds to the complete development stage.

2.4. Body Composition

The body composition of the participants was evaluated using dual-energy X-ray absorptiometry (with the software 12.3, Lunar DPX, Madison, WI, USA), which has already been shown to be a reliable method [26] for obtaining data about fat-free mass (FFM) mass and fat mass (FM). These variables were expressed in kilograms (kg) and percentages (%). The values were obtained with the participants in a supine position, centrally aligned with 10 cm between the feet and 5 cm between the hands and trunk. Participants were instructed to wear comfortable clothes without metal objects such as snaps, belts, jewelry, zippers, etc. and to eat and drink as they normally would. They were all evaluated after bladder voiding. The same experienced examiner performed all tests.

2.5. Isokinetic Strength Test

Before the isokinetic strength test, all participants performed a 5-min warm-up on a cycle ergometer (Cybex Inc., Ronkonkoma, NY, USA) at a resistance of 25 W, followed by low-intensity dynamic stretching exercises to avoid muscular strength decrease [27].

After the warm-up, an isokinetic strength test for the knee flexor and extensor muscles of the dominant lower limb was performed using an isokinetic dynamometer (Biodex Medical Systems Inc., Shirley, NY, USA) Lower limb dominance was determined by asking participants which limb they preferred to use when kicking a ball. The test position was determined according to the manufacturer's instructions. The participants remained in the sitting position with their hips flexed at approximately $85°$, and standard stabilization straps were placed around the trunk, waist, and distal femur of the dominant limb to minimize additional movement, secure the test, and perform all the tests under the same conditions. The axis of the dynamometer was visually aligned with the lateral femoral condyle while the knees were flexed at $90°$. The resistance pad was placed as distally as possible on the lower limb, and the knee was tested from $5°$ to $95°$ of knee flexion (full knee extension defined as $0°$). Gravity correction was performed according to the manufacturer's specifications.

The participants performed five sequenced maximal repetitions of knee flexion and extension at $60°/s$ to measure the peak torque (PT) (Nm). The conventional ratio (CR) ratio was calculated from the values of the peak torque of knee flexor muscles/peak torque of knee extensor muscles, both in concentric action at 60 deg/sec. Data were stored for analysis. An angular speed of $60°/s$ was selected as the lowest speed to avoid high joint pressure while producing the highest torque. The same verbal encouragement was provided to each participant throughout the test [28].

2.6. Statistics

The analyses were performed using Jamovi (Jamovi project, version 2.2.5 [Computer Software]). All variables presented normal distribution and homogeneous variability according to the Shapiro–Wilk and Levene tests, respectively. Data are presented as the mean and standard deviation. To compare anthropometric and isokinetic strength data between the two groups, Student's *t*-test for independent samples was used. The measures of Cohen's effect size (d) for differences between the groups were determined by calculating the mean difference between the two sexes and then dividing the result by the pooled standard deviation. Calculating effect sizes, the magnitude of any difference was judged according to the following criteria: d = 0.2–0.4 was considered a "small" effect size; 0.5 to 0.7 represented a "medium" effect size; and higher than 0.8 a "large" effect size [29]. The G*Power version 3.1.9.2 (a program written, conceptualized, and designed by Franz, Universitat Kiel, Germany; freely available windows application software) was used for the power analysis calculation [30]. For the power analysis, the *t*-test family was selected, and the mean and standard deviation values for all measured variables for both groups, as well as the effect size values (Cohen d), were included for power analysis. Statistical power ranged from 0 to 1, with higher values indicating a higher chance of correctly rejecting the null hypothesis when a specific alternative hypothesis is true. The basis desired power is usually set to 0.80 as a convention [29]. The level of significance was set at $p < 0.05$.

3. Results

As expected, the age (years), body mass (kg), and height (cm) of the prepubertal boys group were significantly lower than the values for the postpubertal group (Table 1).

Table 1. Anthropometric data of the boys in the study.

	Pre Puberty (n = 22)	Post Puberty (n = 34)	*p*-Value	Power	Effect Size (d)	CI for Effect Size
Age (years)	11.3 ± 1.2	15.3 ± 2.8	<0.001	>0.99	1.676	0.9 to 2.4
Body mass (kg)	39.7 ± 8.8	61.5 ± 11.4	<0.001	>0.99	2.086	1.2 to 3.0
Height (cm)	149.9 ± 9.91	169.7 ± 9.5	<0.001	>0.99	2.066	1.2 to 3.0

Data are expressed as mean ± standard deviation. CI: confidence interval.

The girls also presented values for age, body mass, and height significantly higher for postpuberal than for prepubertal groups (Table 2).

Table 2. Anthropometric data of the girls in the study.

	Pre Puberty (n = 11)	Post Puberty (n = 11)	*p*-Value	Power	Effect Size (d)	CI for Effect Size
Age (years)	11.3 ± 1.2	15.0 ± 1.8	<0.001	>0.99	2.401	1.3 to 3.5
Body mass (kg)	42.4 ± 7.4	54.8 ± 7.4	<0.001	>0.99	1.673	0.9 to 2.4
Height (cm)	153.6 ± 7.4	161.1 ± 4.9	0.012	0.97	1.182	0.4 to 1.8

Data are expressed as mean ± standard deviation. CI: confidence interval.

The FFM (kg) was significantly higher, and FM (%) was significantly lower in the boys postpubertal group than in the prepubertal group (Table 3).

Table 3. Body composition and strength values for the boys' groups.

	Pre Puberty (n = 22)	Post Puberty (n = 34)	*p*-Value	Power	Effect Size (d)	CI for Effect Size
Fat-free mass (kg)	29.8 ± 4.8	50.6 ± 9.8	<0.001	>0.99	2.540	1.4 to 3.6
Fat mass (%)	20.4 ± 8.7	14.3 ± 4.6	0.001	0.58	0.941	0.5 to 1.3
PT Ext 60°/s (N·m)	87.2 ± 21.4	155.8 ± 40.9	<0.001	>0.99	1.977	1.1 to 2.9
PT Flex 60°/s (N·m)	45.3 ± 12.3	84.2 ± 24.0	<0.001	>0.99	1.920	1.1 to 2.8
CR (Flex/Ext) (%)	51.9 ± 7.3	53.9 ± 6.7	0.292	0.70	0.291	−0.2 to 0.7

Data are expressed as mean ± standard deviation. CI: confidence interval. PT: peak torque; Ext: extensors; Flex: flexors; CR: conventional ratio.

Conversely, puberty presented a different effect in the girls group. The FFM (kg) and FM (%) were not different between the prepubertal and postpubertal groups (Table 4).

Table 4. Body composition and strength values for the girls' groups.

	Pre Puberty (n = 22)	Post Puberty (n = 34)	*p*-Value	Power	Effect Size (d)	CI for Effect Size
Fat-free mass (kg)	29.9 ± 4.1	33.6 ± 10.0	0.278	0.87	0.475	−0.3 to 1.2
Fat mass (%)	29.8 ± 17.4	26.1 ± 6.8	0.516	0.86	0.282	−0.5 to 1.0
PT Ext 60°/s (N·m)	87.9 ± 14.5	120.0 ± 24.7	0.001	>0.99	1.585	0.9 to 2.3
PT Flex 60°/S (N·m)	40.6 ± 7.2	57.8 ± 8.9	<0.001	>0.99	2.112	1.2 to 3.0
CR (Flex/Ext) (%)	46.4 ± 4.8	49.5 ± 9.4	0.332	0.87	0.424	−0.3 to 1.2

Data are expressed as mean ± standard deviation. CI: confidence interval. PT: peak torque; Ext: extensors; Flex: flexors; CR: conventional ratio.

In the boys groups, the PT values for the knee extensor muscles were significantly higher in the postpubertal group than in the prepubertal group. In the same way, the PT for the knee flexor muscles also was higher in the postpubertal group than in the prepubertal group. In addition, the effect size of the PT differences was similar for the knee extensor and flexor muscles. Considering the CR between the flexor and extensor PT values, there was no significant difference between the boys groups (Table 3).

In the girls groups, the PT for the knee extensor muscles was also higher in the postpubertal group than in the prepubertal group. A similar pattern was observed in the knee flexor muscles, where the postpubertal group presented significantly higher values than the prepubertal group. However, it could be seen that the effect size of the knee flexor muscles difference was higher. Consequently, the CR between the flexor and extensor PT values tended to be higher in the postpubertal group than in the prepubertal group, although the difference did not reach the significance threshold (Table 4).

4. Discussion

The aim of this study was to evaluate and compare knee muscle strength and CR between thigh muscles between prepubertal and postpubertal swimmers. The main results of the present study were as follows: (i) postpubertal male swimmers presented higher FFM (kg) and lower FM (%) than prepubertal swimmers; (ii) FFM (kg) and FM (%) were not different between prepubertal and postpubertal girls groups; (iii) postpubertal swimmers from both sexes showed greater knee flexor and extensor PT values than prepubertal swimmers; (iv) prepubertal and postpubertal swimmers, from both sexes, were not different in CR. Therefore, the initial hypothesis of the present study was not confirmed as the CR was not affected by puberty, neither in male nor female athletes.

FFM is a variable that is affected by puberty, especially in boys. Male puberty is characterized by increased levels of circulating testosterone, which is the hormone responsible for promoting sexual function development and male somatic characteristics, including higher muscle mass [31,32]. In contrast to boys, puberty in girls is affected by the sex hormone estrogen, which triggers the release of gonadotropin-releasing hormone and the subsequent activation of the hypothalamic-pituitary-gonadal axis [31]. A higher estrogen level is associated with an increase in FM but has no effect on FFM. This justifies the fact that the boys presented an expressive increase in FFM after puberty, whereas the girls did not, as previously demonstrated by Fukunaga et al. [33] and Costa et al. [25].

Considering FM (%) in the boys groups, a significant decrease was observed, which can be mathematically explained by the significant increase in absolute values of FFM (kg); therefore, the percentage of FM decreased. In addition, the literature supports these data because of the intense activity of the hypothalamic-pituitary-gonadal axis during this period of life [31]. Schneider et al. also reported this change in body composition, with the

FM percentage varying from 21.0% to 13.6% in male swimmers after puberty, which is in congruence with the present data (14.3% FM in the postpubertal boys group) [34].

FM (%) did not differ between the girls groups. For female nonathletes, an increase in FM (%) after puberty is expected because of the higher estrogen levels, which directly influence fat accumulation [35]. However, the participants in the present study were athletes, and despite the higher estrogen levels, the participants were used to training at least 5 days a week, 1 h per day, which certainly helps to maintain FM (%) at lower levels, despite the present results (26.1% for FM in the postpubertal girls group) being higher than the values reported by Schneider et al., which were 20.7–18.7% in the postpubertal female swimmers [34].

According to the strength values, the present results showed significantly higher values for knee flexor and extensor muscle PT in the postpubertal groups than in the prepubertal boys group. There were significant differences between the groups in PT, with the postpubertal groups having higher values in both extension and flexion at 60°/s. These data are consistent with the acknowledgment of the strength increase that occurs during puberty, partly due to the elevation in the serum levels of anabolic hormones, such as growth hormone and testosterone [36,37]. Of course, these hormones play an important role in the development of neuromuscular maturation, but the improvement of the central nervous system during the child's development (especially in those who have more physical abilities) is another variable to be considered [25]. These findings corroborate previous studies that compared PT between prepubertal and postpubertal adolescents [33,34]. It is important to note that the effect size of puberty on the difference in the strength of the extensor and flexor muscles is quite similar, indicating that both muscle groups became stronger in a similar way. Consequently, no difference was observed in the CR between the flexor and extensor muscles. Nevertheless, the mean CR values presented by these swimmers deserve further attention. Although the optimum CR between the flexor and extensor muscles has not been reported, values approaching 60% (measured at 60°/sec during concentric action) are recommended for injury prevention during dynamic movements [23,38]. Values below 60% were previously associated with a higher risk of knee injuries (such as muscular injuries, especially in the hamstrings, anterior cruciate ligament injury, and tendon injuries) [39–41]. Thus, the mean CR values for the prepubertal (51.9 ± 7.3%) and postpubertal (53.9 ± 6.7%) groups can be described as relatively low.

Similar to the boys groups, the CR between the flexor and extensor muscles was not different between the girls groups. However, the situation requires much more attention because prepubertal, and postpubertal mean values for CR were even lower than 60% (46.4 ± 4.8% and 49.5 ± 9.4%, respectively). Similar results were presented by Secchi et al. (2011) for elite swimmers and Dalamitros et al. (2015) for adolescent swimmers, with CR values ranging from 46.5% to 53.4% [42].

This study had some limitations. First, the reduced sample size may have contributed to the lack of significant difference between CR for the male groups since the power of statistical analysis was lower than 0.80 (power = 0.70). Conversely, for the female groups, the power for all the analyses, including for CR, were higher than 0.80, which ensures sufficient security to draw conclusions resulting from the results of statistical analyzes. Therefore, the authors recommend that other studies should be done with a larger sample size, mainly for the male group. Second, it was a cross-sectional study, and for a better understanding of how puberty affects muscular strength and balance ratios, a longitudinal study should be performed. Third, sexual maturity was measured using Tanner levels, and more accurate and precise methods should be used to precisely measure this variable.

In conclusion, the flexor and extensor muscles were significantly stronger in postpubertal female and male swimmers than in prepubertal swimmers of the same sex, and the strengthening was balanced between agonist and antagonist muscles once the CR did not change between the groups of the same sex. However, the CR for the female swimmers was lower than the literature recommendation; therefore, this situation deserves attention. The authors of the present study recommend that coaches focus on strengthening the hamstring muscles of female adolescent swimmers to maintain a CR higher than 60%, which is recommended to minimize the risk of knee injury.

Author Contributions: All authors made a significant contribution to the work reported as follows: Conceptualization: M.S.A. and B.L.A.; methodology: M.S.A., T.C. and B.L.A.; validation: M.S.A. and C.A.B.D.L.; formal analysis: M.S.A. and B.L.A.; investigation: M.S.A. and T.C.; data curation: C.A.B.D.L. and R.L.V.; writing—original draft preparation: M.S.A., K.W., B.L.A. and P.F.; writing—review and editing: B.L.A., B.K. and P.F.; visualization: M.S.A. and B.L.A.; supervision: M.S.A. All authors have read and agreed to the published version of the manuscript.

Funding: This research received no external funding.

Institutional Review Board Statement: The study was conducted in accordance with the Declaration of Helsinki and approved by the Research Ethics Committee of the Federal University of Sâo Paulo (protocol 0503/2017).

Informed Consent Statement: Informed consent was obtained from all subjects before their inclusion in the study.

Data Availability Statement: The data presented in this study are available on request from the corresponding author. The data are not publicly available due to privacy restrictions.

Conflicts of Interest: The authors declare no conflict of interest.

References

1. Wirth, K.; Keiner, M.; Fuhrmann, S.; Nimmerichter, A.; Haff, G.G. Strength Training in Swimming. *Int. J. Environ. Res. Public Health* **2022**, *19*, 5369. [CrossRef] [PubMed]
2. Aspenes, S.; Kjendlie, P.L.; Hoff, J.; Helgerud, J. Combined Strength and Endurance Training in Competitive Swimmers. *J. Sport. Sci. Med.* **2009**, *8*, 357–365.
3. Born, D.-P.; Kuger, J.; Polach, M.; Romann, M. Start and Turn Performances of Elite Male Swimmers: Benchmarks and Underlying Mechanisms. *Sport. Biomech.* **2021**, *4*, 1–19. [CrossRef] [PubMed]
4. Morais, J.E.; Barbosa, T.M.; Neiva, H.P.; Marinho, D.A. Stability of Pace and Turn Parameters of Elite Long-Distance Swimmers. *Hum. Mov. Sci.* **2019**, *63*, 108–119. [CrossRef] [PubMed]
5. Cronin, J.; Jones, J.; Frost, D. The Relationship between Dry-Land Power Measures and Tumble Turn Velocity in Elite Swimmers. *J. Swim. Res.* **2007**, *17*, 17–23.
6. Jones, J.V.; Pyne, D.B.; Greg Haff, G.; Newton, R.U. Comparison Between Elite and Subelite Swimmers on Dry Land and Tumble Turn Leg Extensor Force-Time Characteristics. *J. Strength Cond. Res.* **2018**, *32*, 1762–1769. [CrossRef]
7. Keiner, M.; Wirth, K.; Fuhrmann, S.; Kunz, M.; Hartmann, H.; Haff, G.G. The Influence of Upper- and Lower-Body Maximum Strength on Swim Block Start, Turn, and Overall Swim Performance in Sprint Swimming. *J. Strength Cond. Res.* **2021**, *35*, 2839–2845. [CrossRef]
8. West, D.J.; Owen, N.J.; Cunningham, D.J.; Cook, C.J.; Kilduff, L.P. Strength and Power Predictors of Swimming Starts in International Sprint Swimmers. *J. Strength Cond. Res.* **2011**, *25*, 950–955. [CrossRef]
9. Fagundes, U.; Vancini, R.L.; Seffrin, A.; de Almeida, A.A.; Nikolaidis, P.T.; Rosemann, T.; Knechtle, B.; Andrade, M.S.; de Lira, C.A.B. Adolescent female handball players present greater bone mass content than soccer players: A cross-sectional study. *Bone* **2022**, *154*, 116217. [CrossRef]
10. Fagundes, U.; Vancini, R.L.; de Almeida, A.A.; Nikolaidis, P.T.; Weiss, K.; Knechtle, B.; Andrade, M.S.; de Lira, C.A.B. Reference values for bone mass in young athletes: A cross-sectional study in São Paulo, Brazil. *Sci. Rep.* **2023**, *13*, 286. [CrossRef]
11. Maïmoun, L.; Coste, O.; Mura, T.; Philibert, P.; Galtier, F.; Mariano-Goulart, D.; Paris, F.; Sultan, C. Specific Bone Mass Acquisition in Elite Female Athletes. *J. Clin. Endocrinol. Metab.* **2013**, *98*, 2844–2853. [CrossRef] [PubMed]
12. Vlachopoulos, D.; Barker, A.R.; Williams, C.A.; Arngrímsson, S.A.; Knapp, K.M.; Metcalf, B.S.; Fatouros, I.G.; Moreno, L.A.; Gracia-Marco, L. The Impact of Sport Participation on Bone Mass and Geometry in Male Adolescents. *Med. Sci. Sport. Exerc.* **2017**, *49*, 317–326. [CrossRef] [PubMed]
13. Wanivenhaus, F.; Fox, A.J.S.; Chaudhury, S.; Rodeo, S.A. Epidemiology of Injuries and Prevention Strategies in Competitive Swimmers. *Sport. Health* **2012**, *4*, 246–251. [CrossRef] [PubMed]
14. Tooth, C.; Gofflot, A.; Schwartz, C.; Croisier, J.L.; Beaudart, C.; Bruyère, O.; Forthomme, B. Risk Factors of Overuse Shoulder Injuries in Overhead Athletes: A Systematic Review. *Sport. Health* **2020**, *12*, 478–487. [CrossRef] [PubMed]
15. Motta, C.; de Lira, C.A.B.; Vargas, V.Z.; Vancini, R.L.; Andrade, M.S. Profiling the Isokinetic Muscle Strength of Athletes Involved in Sports Characterized by Constantly Varied Functional Movements Performed at High Intensity: A Cross-Sectional Study. *PM & R* **2019**, *11*, 354–362. [CrossRef]

16. De Lira, C.; Vargas, V.; Vancini, R.; Andrade, M. Profiling Isokinetic Strength of Shoulder Rotator Muscles in Adolescent Asymptomatic Male Volleyball Players. *Sports* **2019**, *7*, 49. [CrossRef] [PubMed]

17. Andrade, M.D.S.; Fleury, A.M.; de Lira, C.A.B.; Dubas, J.P.; da Silva, A.C. Profile of Isokinetic Eccentric-to-Concentric Strength Ratios of Shoulder Rotator Muscles in Elite Female Team Handball Players. *J. Sport. Sci.* **2010**, *28*, 743–749. [CrossRef] [PubMed]

18. Andrade, M.S.; Vancini, R.L.; de Lira, C.A.B.; Mascarin, N.C.; Fachina, R.J.F.G.; da Silva, A.C. Shoulder Isokinetic Profile of Male Handball Players of the Brazilian National Team. *Braz. J. Phys.* **2013**, *17*, 572–578. [CrossRef]

19. Andrade, M.S.; Silva, W.A.; Lira, C.A.B.L.; Mascarin, N.C.; Vancini, R.L.; Nikolaidis, P.N.; Beat, K. Isokinetic Muscular Strength and Aerobic Physical Fitness in Recreational Long-Distance Runners: A Cross-Sectional Study. *J. Strength Cond. Res.* **2022**, *36*, e73–e80. [CrossRef]

20. Lee, J.W.Y.; Mok, K.M.; Chan, H.C.K.; Yung, P.S.H.; Chan, K.M. Eccentric Hamstring Strength Deficit and Poor Hamstring-to-Quadriceps Ratio Are Risk Factors for Hamstring Strain Injury in Football: A Prospective Study of 146 Professional Players. *J. Sci. Med. Sport* **2018**, *21*, 789–793. [CrossRef]

21. Vieira, A.; Alex, S.; Martorelli, A.; Brown, L.E.; Moreira, R.; Bottaro, M. Lower-Extremity Isokinetic Strength Ratios of Elite Springboard and Platform Diving Athletes. *Phys. Sportsmed.* **2017**, *45*, 87–91. [CrossRef]

22. De Lira, C.A.B.; Mascarin, N.C.; Vargas, V.Z.; Vancini, R.L.; Andrade, M.S. Isokinetic Knee Muscle Strength Profile in Brazilian Male Soccer players: A Cross-sectional Study. *Int. J. Sport. Phys. Ther.* **2017**, *12*, 1103–1110. [CrossRef] [PubMed]

23. Coombs, R.; Garbutt, G. Developments in the Use of the Hamstring/Quadriceps Ratio for the Assessment of Muscle Balance. *J. Sport. Sci. Med.* **2002**, *1*, 56–62.

24. Tanner, J.M. Growth and Maturation during Adolescence. *Nutr. Rev.* **1981**, *39*, 43–55. [CrossRef]

25. Costa, T.; Murara, P.; Vancini, R.L.; de Lira, C.A.B.; Andrade, M.S. Influence of Biological Maturity on the Muscular Strength of Young Male and Female Swimmers. *J. Hum. Kinet.* **2021**, *78*, 67–77. [CrossRef] [PubMed]

26. Colyer, S.L.; Roberts, S.P.; Robinson, J.B.; Thompson, D.; Stokes, K.A.; Bilzon, J.L.J.; Salo, A.I.T. Detecting Meaningful Body Composition Changes in Athletes Using Dual-Energy X-ray Absorptiometry. *Physiol. Meas.* **2016**, *37*, 596–609. [CrossRef] [PubMed]

27. Mascarin, N.C.; Vancini, R.L.; Lira, C.A.B.; Andrade, M.S. Stretch-Induced Reductions in Throwing Performance Are Attenuated by Warm-up Before Exercise. *J. Strength Cond. Res.* **2015**, *29*, 1393–1398. [CrossRef]

28. Vargas, V.Z.; Motta, C.; Peres, B.; Vancini, R.L.; Andre Barbosa De Lira, C.; Andrade, M.S. Knee Isokinetic Muscle Strength and Balance Ratio in Female Soccer Players of Different Age Groups: A Cross-Sectional Study. *Physician Sportsmed.* **2020**, *48*, 105–109. [CrossRef]

29. Cohen, J. *Statistical Power Analysis for the Behavioral Sciences Statistical Power Analysis for the Behavioral Sciences*, 2nd ed.; Lawrence Erlbaum Associates: Hillsdale, NJ, USA, 1988.

30. Kang, H. Sample Size Determination and Power Analysis Using the G*Power Software. *J. Educ. Eval. Health Prof.* **2021**, *18*, 17. [CrossRef]

31. Chulani, V.L.; Gordon, L.P. Adolescent Growth and Development. *Prim. Care* **2014**, *41*, 465–487. [CrossRef]

32. Lourenço, B.; Queiroz, L.B. Crescimento e Desenvolvimento Puberal Na Adolescência. *Rev. Med. (Rio J.)* **2010**, *89*, 70–75. [CrossRef]

33. Fukunaga, Y.; Takai, Y.; Yoshimoto, T.; Fujita, E.; Yamamoto, M.; Kanehisa, H. Effect of Maturation on Muscle Quality of the Lower Limb Muscles in Adolescent Boys. *J. Physiol. Anthr.* **2014**, *33*, 30. [CrossRef] [PubMed]

34. Schneider, P.; Meyer, F. Avaliação Antropométrica e Da Força Muscular Em Nadadores Pré-Púberes e Púberes. *Rev. Bras. Med. Esporte* **2005**, *11*, 209–213. [CrossRef]

35. Malina, R. *Growth, Maturation and Physical Activity*, 2nd ed.; Human Kinetics: Champaign, IL, USA, 2004.

36. Ramos, E.; Frontera, W.R.; Llopart, A.; Feliciano, D. Muscle Strength and Hormonal Levels in Adolescents: Gender Related Differences. *Int. J. Sport. Med.* **1998**, *19*, 526–531. [CrossRef]

37. Round, J.M.; Jones, D.A.; Honour, J.W.; Nevill, A.M. Hormonal Factors in the Development of Differences in Strength between Boys and Girls during Adolescence: A Longitudinal Study. *Ann. Hum. Biol.* **1999**, *26*, 49–62. [CrossRef] [PubMed]

38. Dalamitros, A.A.; Manou, V.; Christoulas, K.; Kellis, S. Knee Muscles Isokinetic Evaluation after a Six-Month Regular Combined Swim and Dry-Land Strength Training Period in Adolescent Competitive Swimmers. *J. Hum. Kinet.* **2015**, *49*, 195–200. [CrossRef] [PubMed]

39. Croisier, J.L.; Ganteaume, S.; Binet, J.; Genty, M.; Ferret, J.M. Strength Imbalances and Prevention of Hamstring Injury in Professional Soccer Players: A Prospective Study. *Am. J. Sport. Med.* **2008**, *36*, 1469–1475. [CrossRef]

40. Delextrat, A.; Gregory, J.; Cohen, D. The Use of the Functional H:Q Ratio to Assess Fatigue in Soccer. *Int. J. Sport. Med.* **2010**, *31*, 192–197. [CrossRef]

41. Heiser, T.M.; Weber, J.; Sullivan, G.; Clare, P.; Jacobs, R.R. Prophylaxis and Management of Hamstring Muscle Injuries in Intercollegiate Football Players. *Am. J. Sport. Med.* **1984**, *12*, 368–370. [CrossRef]
42. Secchi, L.L.B.; Muratt, M.D.; Ciolac, E.G.; Greve, J.M.D. Knee Muscles Isokinetic Evaluation in Short Distance Elite Swimmers: A Comparison between Symmetric and Asymmetric Swimming Styles. *Isokinet. Exerc. Sci.* **2011**, *19*, 261–264. [CrossRef]

Article

Physical Activity, Mental Health, and Quality of Life among School Students in the Jazan Region of Saudi Arabia: A Cross-Sectional Survey When Returning to School after the COVID-19 Pandemic

Mohamed Salih Mahfouz [1,*], Ahmad Y. Alqassim [1], Nasser Hadi Sobaikhi [2], Abdulaziz Salman Jathmi [2], Fahad Omar Alsadi [2], Abdullah Mohammed Alqahtani [2], Mohammed Mohalhil Shajri [2], Ibrahim Darwish Sabi [2], Ahmed M. Wafi [3] and Jonathan Sinclair [4]

[1]   Family and Community Medicine Department, Faculty of Medicine, Jazan University, Jazan 45142, Saudi Arabia
[2]   Faculty of Medicine, Jazan University, Jazan 45142, Saudi Arabia
[3]   Physiology Department, Faculty of Medicine, Jazan University, Jazan 45142, Saudi Arabia
[4]   Research Centre for Applied Sport, Physical Activity and Performance, Faculty of Allied Health and Wellbeing, School of Sport & Health Sciences, University of Central Lancashire, Preston PR1 2HE, UK
*   Correspondence: mm.mahfouz@gmail.com

Abstract: Increasing evidence suggests that physical activity (PA) can reduce depression and anxiety in adolescents. At the same time, quality of life (QoL) is sensitive to both mental health and PA, but little is known about the mechanism between these three variables among adolescents. This study aimed to assess the physical activity, mental health, and quality of life of school students when they returned to school after two years of distance learning in the Jazan region. This current investigation represented an observational cross-sectional survey conducted in January 2022 among a random sample of 601 students from intermediate and high schools in the Jazan region, Saudi Arabia. Three standardized questionnaires were used for data collection; the Arabic version of the Pediatric Quality of Life Inventory (PedsQL), Depression Anxiety Stress Scales (DASS21), and the Fels PAQ for children. The analysis revealed a moderate level of physical activity, decreased HRQoL, and symptoms of mental health problems (anxiety, depression, and stress) among the schools' students when they returned to school following COVID-19 lockdown. The overall Pediatric Quality of Life mean score was (81.4 ± 16.4), which differed significantly according to gender, age groups, and grade levels ($p < 0.05$ for all). There was a negative correlation between the overall quality of life and mental health domains. Sport was negatively correlated with mental illness symptoms and positively correlated ($p < 0.05$) with Pediatric Quality of Life. The regression models revealed that stress was a significant predictor for the quality of life of male and female adolescents ([$\beta = -0.30$, (95% CI (-0.59) to (-0.02), $p < 0.05$)] and [$\beta = -0.40$, (95% CI (-0.70) to (-0.01), $p < 0.05$)], respectively). The analysis revealed a moderate level of physical activity among the schools' students when they returned to school following COVID-19 lockdown. Children's involvement in physical activity was associated with improved quality of life and mental health. The results call for the need to develop appropriate intervention programs to increase school students' physical activity levels.

Keywords: depression; pediatric quality of Life; environmental health; sport and children

Citation: Mahfouz, M.S.; Alqassim, A.Y.; Sobaikhi, N.H.; Jathmi, A.S.; Alsadi, F.O.; Alqahtani, A.M.; Shajri, M.M.; Sabi, I.D.; Wafi, A.M.; Sinclair, J. Physical Activity, Mental Health, and Quality of Life among School Students in the Jazan Region of Saudi Arabia: A Cross-Sectional Survey When Returning to School after the COVID-19 Pandemic. *Healthcare* **2023**, *11*, 974. https://doi.org/10.3390/healthcare11070974

Academic Editor: Rafael Oliveira

Received: 10 February 2023
Revised: 21 March 2023
Accepted: 22 March 2023
Published: 29 March 2023

1. Introduction

The COVID-19 pandemic has significantly altered people's daily lives, compelling countries to implement school closures and replacements with virtual classes, lockdowns, isolation, and social distancing to contain the virus's spread and mitigate its adverse health consequences [1,2]. Additionally, COVID-19 has greatly impacted daily life, influencing how individuals live, work, learn, interact, engage in physical activity, and manage their mental health

Physical activity refers to any body movement generated by skeletal muscles that require energy expenditure [3]. How to encourage children and adolescents to perform an appropriate level of PA has always been a major public health concern that was worsened by the pandemic [4]. According to the recommendation given by the WHO, to achieve health benefits, children and adolescents should do 60 min or more of moderate-to-vigorous-intensity PA each day [5,6].

Recent research [7,8] suggests that the COVID-19 pandemic may exacerbate physical inactivity and prolong sedentary time in children, which may be attributed to the lack of equal opportunities to participate in PA and the unavailability of sports facilities caused by school closures. It has been shown that students have lower levels of PA and more sedentary time during the weekends than in school [9].

The COVID-19 pandemic also poses a significant mental health threat among children, adolescents, and all students generally [10]. It has caused tremendous stress levels, primarily due to schools being closed and home quarantining. As a result, children have become physically less active, have much-prolonged screen times, have irregular sleep schedules, and have less healthy diets, resulting in excess weight and a lack of cardiorespiratory performance [10]. Students exposed to these incidents can trigger the development of anxiety, panic attacks, depression, and mood disorders, and it may also lead to a deterioration in mental health [11].

A prior study of parents in Italy and Spain discovered that more than 85% of their children decreased their physical activity, increased their screen times, and increased maladaptive emotional and behavioral indicators [11]. According to the Centers for Disease Control and Prevention, 1.9 million children between the ages of 3 and 17 years have been diagnosed with depression, and 4.4 million have been diagnosed with anxiety as a result of home quarantine in the USA [12]. Moreover, mental illness and physical inactivity among students are not just limited to COVID-19 lockdown but extend to the post-lockdown period as well. Recent research from China suggests that students and adolescents reported significantly more mental health problems after the COVID-19 school lockdowns [13]

Many studies have documented that PA is positively associated with mental health, while an inverse association has been reported between psychiatric distress (depression, stress, and anxiety) and quality of life among adolescents [14,15]. Additionally, physical inactivity has been found to be inversely associated with quality of life. Despite the inter-variable associations between these three important variables, the mechanisms underlying these associations have not been investigated after the COVID-19 pandemic lockdown. Therefore, the objectives of this study are: (i) to assess the physical activity, mental health, and quality of life of school students when they returned to school after two years of distance learning in the Jazan region, southwest Saudi Arabia; (ii) to investigate the association between physical activity, mental health, and quality of life among the students; and (iii) to determine whether any differences exist between students in terms of physical activities, mental health, and quality of life according to gender. The study results will shed light on adolescents' quality of life, physical activity, and mental health post COVID-19 lockdown and offer policymakers and school administrations guidance regarding these essential indicators.

2. Materials and Methods

2.1. Study Design, Setting, and Population

A cross-sectional survey was conducted among students of intermediate and high schools in the Jazan region. School-aged children and adolescents (intermediate and secondary schools) are an important segment of the population, and additional information was needed to determine their mental health, QoL, and PA statuses post COVID-19 period. The Jazan region is one of the KSA's thirteen regions. It is located on the tropical Red Sea Coast in southwestern KSA. The region covers an area of 11,671 km^2, including some

5000 villages and towns, with a population of 1.5 million. The intermediate educational stage in Saudi Arabia lasts three years, while secondary education lasts for the same period. During the COVID-19 pandemic, around 22,000 schools were closed from 9 March 2020. Classrooms opened again for students in intermediate and high schools on 29 April 2021. Inclusion criteria involved school-age children (intermediate and secondary schools) who registered for the academic year 2021/2022, lived in the Jazan region, and were aged between 12 to 18 years old. This research was conducted in January 2022.

2.2. Sampling Procedures

A sample size of 640 students was determined for conducting this survey. The sample size estimation was based on Cochran's formula [16] for cross-sectional surveys: initial sample size n = $[(z^2 * p * q)]/d^2$, where p is the (estimated) proportion of the population, which has the attribute in question; q = $(1 - p)$; d is the desired level of precision; and Z is the critical value of the normal distribution at the required confidence level; usually, Z = 1.96 for 95% C.I. This research utilized the following indicators: p = prevalence of the required, studied phenomenon, which was set at 50% (as no information related to the main research outcome was available); Z = 1.96 confidence interval; and d = error not more than 4%, with a 10% non-response rate. Due to the pandemic situations of COVID-19, the research combined both the random and non-random sampling designs to recruit the students for this study. In the first stage, eight schools (two intermediate and two high schools for both genders) were selected randomly from each educational sector. The Jazan region, administratively, is divided into two educational sectors. Second, a convenience sample was utilized to select the pupils from each selected school to participate in this study. The survey link was sent to the child's parent/guardian, and the parents/guardians were asked to assent to their child's participation in the survey before the child could participate.

2.3. Data Collection and Instrumentation

The study was conducted using a self-administered questionnaire prepared in Arabic and distributed via an anonymous online survey instrument. The first part of the questionnaire covered the socio-demographic information of the study participants, such as age, gender, nationality, residence, and parent's education level. The second part of the questionnaire involved the validated Arabic version of the Fels PAQ for children (7–19 years). The instrument was a self-administered questionnaire assessing habitual physical activity in an eight-item questionnaire containing three "open" questions for which some activities required listing by the study participants. The frequency of participation for each activity was also obtained [17]. The third part of the questionnaire was the Arabic version of the Depression Anxiety Stress Scales (DASS21) [18]. DASS is a 21-item instrument measuring current ("over the past week") depression, anxiety, and stress symptoms. Participants were asked to use a 4-point combined severity/frequency scale to rate the extent to which they experienced each item over the past week. The scale ranged from 0 (did not apply to me at all) to 3 (applied to me very much or most of the time). Scores for depression, anxiety, and stress were calculated by summing the relevant item scores of depression, anxiety, and stress. The fourth part of the questionnaire was the Arabic version of the Pediatric Quality of Life Inventory (PedsQL) [19], which is a 23-item generic health status instrument with child forms that assesses five domains of health (physical functioning, emotional functioning, psychosocial functioning, social functioning, and school functioning) in children and adolescents aged 13 to 18. A pilot study was conducted among 30 participants from one school to test the questionnaire's applicability and understanding before starting the actual research. The assessment of the internal consistency for the three measures showed Cronbach's α = 0. 932 for the PedsQL; DASS21 showed a Cronbach's α of 0.954; and finally, the FELS PAQ for children recorded a Cronbach's α = 0.867.

2.4. Study Ethics

The research was conducted according to the ethics guidelines of Saudi Arabia. Ethical clearance was obtained from the Standing Committee for Scientific Research Ethics-Jazan University (HAPO-10-Z-001) (REF# REC-43/05/088). All study participants read and signed the consent form following a short introduction about the study objectives. Participants were told that they had the right to withdraw at any time and that there would not be any harm or loss of benefits if they continued or withdrew from participating. Finally, we strictly ensured that each participant's personal information was preserved and confidentiality was maintained.

2.5. Statistical Analysis

The IBM SPSS Statistics for Windows, Version 24.0. Armonk, NY: IBM Corp program was used for data analysis. The analysis involved descriptive as well as inferential statistics according to the essential purpose of each relationship. Categorical variables were described as frequencies and percentages. The normality of the continuous variables was assessed using the Kolmogorov–Smirnov test. To assess the differences in the demographic characteristics in the FELS PAQ and PedsQL, we compared the means of continuous variables using the Student's t-test and one-way ANOVA, respectively. The effect size, based on Cohen's d and omega squared, was calculated. Cohen's d was interpreted as follows: <20 as very small; 0.20–0.49 as small; 0.51–0.80 as intermediate; and >0.8 as a large effect. An omega-squared effect size of 0.01–0.06 was considered small, 0.06–0.14 was considered medium, and >0.14 was considered large. A multiple linear regression model was used to assess predictors of the overall score of quality of life among the pupils. The final model was assessed for multicollinearity and assumptions of the ordinary least squares (OLS) technique. The Pearson correlation coefficient was computed to assess the correlation between the overall quality of life, sport index, and the DASS21 domains among the adolescents. The correlation coefficient was defined as weak (0–0.29), moderate (0.30–0.50), and strong (>0.50). A p-value less than 0.05 was considered significant.

3. Results

Table 1 summarizes the study participants' characteristics and the levels of the HRQoL and FELS PAQ scores. Among the pupils, males constituted 337 (56.1%). Most of them were in the age group of 17–18 years, and 460 (76.5%) lived in rural areas. Regarding their parents' education levels, 273 (53.8%) of the fathers were of secondary and university education, compared to 218 (42.9%) of the mothers. Almost 377 (74.2%) of the mothers were housewives, while more than one-third of the fathers, 196 (38.6%), were working in governmental positions. The table further showed that the PedsQoL scores differed significantly according to gender, age groups, and grade levels, whereas the FELS PAQ total score was significantly different according to grades only ($p < 0.05$ for all).

Table 1. Background characteristic, health-related quality of life, and physical activity scores among the study participants (n = 601).

Characteristics		N	%	PedsQoL Scores		FELS PAQ Total Score	
				Mean ± SD	p-Value	Mean ± SD	p-Value
Gender	Male	337	(56.1)	84.1 ± 16.0	<0.001	9.5 ± 1.8	0.203
	Female	264	(43.9)	77.8 ± 17.5		9.4 ± 1.6	
Age groups	12–14 year	86	(14.3)	76.7 ± 17.8	0.010	9.6 ± 1.6	0.304
	15–16 years	189	(31.4)	80.9 ± 16.2		9.6 ± 1.7	
	17–18 years	326	(54.2)	82.9 ± 17.0		9.4 ± 1.8	
Residence	Village	460	(76.5)	80.8 ± 17.1	0.148	9.4 ± 1.7	0.279
	Town	141	(23.5)	83.2 ± 16.3		9.6 ± 1.9	

Table 1. *Cont.*

Characteristics		N	%	PedsQoL Scores		FELS PAQ Total Score	
				Mean ± SD	*p*-Value	Mean ± SD	*p*-Value
Grades	1st Intermediate	43	(7.2)	72.0 ± 16.6	<0.001	9.9 ± 1.6	0.300
	2nd Intermediate	43	(7.2)	79.7 ± 21.5		9.5 ± 1.5	
	3rd Intermediate	52	(8.7)	77.3 ± 16.2		9.6 ± 1.9	
	1st High School	164	(27.3)	82.5 ± 15.2		9.6 ± 1.8	
	2nd High School	158	(26.3)	85.0 ± 15.0		9.1 ± 1.6	
	3rd High School	141	(23.5)	80.9 ± 18.4		9.5 ± 1.9	
Father's Level of Education	Primary	118	(23.2)	80.0 ± 16.3	0.508	9.4 ± 1.7	0.079
	Intermediate	68	(13.4)	80.2 ± 17.4		10.1 ± 1.7	
	Secondary	140	(27.6)	80.9 ± 17.9		9.4 ± 1.7	
	University	133	(26.2)	82.5 ± 15.2		9.5 ± 1.6	
	Postgraduate	49	(9.6)	77.8 ± 17.7		9.5 ± 1.9	
Mother's Level of Education	Primary	184	(36.2)	80.7 ± 16.5	0.922	9.5 ± 1.7	0.850
	Intermediate	72	(14.2)	80.3 ± 16.9		9.5 ± 1.7	
	Secondary	112	(22.0)	80.7 ± 18.5		9.5 ± 1.8	
	University	106	(20.9)	81.7 ± 14.2		9.6 ± 1.6	
	Postgraduate	34	(6.7)	78.7 ± 19.1		9.2 ± 1.9	
Father's Job	Own Business	45	(8.9)	81.7 ± 16.4	0.173	9.7 ± 1.8	0.751
	Government Sector	196	(38.6)	80.5 ± 15.8		9.5 ± 1.6	
	Private Sector	26	(5.1)	73.7 ± 14.6		9.9 ± 1.6	
	Retired	167	(32.9)	80.7 ± 18.9		9.5 ± 1.8	
	Not Working	74	(14.6)	83.1 ± 14.4		9.5 ± 1.6	
Mother's Job	Own Business	21	(4.1)	77.4 ± 20.5	0.773	9.0 ± 1.8	0.391
	Government Sector	81	(15.9)	82.0 ± 14.2		9.3 ± 1.6	
	Private Sector	10	(2.0)	76.9 ± 18.0		9.9 ± 1.6	
	Retired	19	(3.7)	80.6 ± 21.1		9.7 ± 1.6	
	HW	377	(74.2)	80.8 ± 16.8		9.6 ± 1.7	

Abbreviations: SD = standard deviation, HRQol = health-related quality of life based on the PedsQL questionnaire. *p*-value is based on the independent sample t-test or the one-way ANOVA test.

Table 2 presents descriptive statistics for mental health, physical activity, and quality of life for boys and girls. According to the table, girls scored significantly higher than boys on all three DASS subscales. Boys had a significantly higher sports index than girls ($p < 0.001$). According to the Pediatric Quality of Life (PedsQL), the table showed that boys had significantly higher scores in all PedsQL domains ($p < 0.05$), with a small effect size ranging between 0.169–0.450.

Regarding mental health among the students, 341 (56.7%) were assessed to be free of depression symptoms; 67 (11.1%) suffered from mild depression; 99 (16.5%) had moderate depression; and only 45 (7.5%) and 49 (8.2%) were considered to have severe and extremely severe depression symptoms, respectively. Regarding the anxiety subscale, 41 (51.6%) were normal; 41 (6.8%) suffered from mild anxiety; 105 (17.5%) had moderate anxiety; and 49 (8.2%) and 96 (16.0%) were considered to suffer from severe and extremely severe anxiety, respectively. Finally, for the self-reported stress majority, 415 (69.1%) were normal; 63 (10.5%) were considered to have mild stress; 49 (8.2%) were classified as having moderate stress symptoms; and only 43 (7.2%) and 31 (5.2%) were considered to suffer from severe and extremely severe stress, respectively. The PedsQL scores were found to be significantly different across all DASS21 self-reported subscales ($p < 0.05$) [Table 3].

Table 2. Descriptive statistics of mental health, physical activity, and quality of life for boys (n = 337) and girls (n = 264).

Scale	Component	Total		Boys		Girls		*p*-Value *	Cohen's d
		Mean	SD	Mean	SD	Mean	SD		
DASS21 scores	Anxiety scores	4.8	5.0	3.9	4.5	5.9	5.5	<0.001	−0.414
	Depression scores	5.0	5.3	4.2	4.6	6.1	5.8	<0.001	−0.360
	Stress scores	5.5	5.5	4.7	4.9	6.4	6.1	<0.001	−0.309
FELS PAQ Total score	Leisure Index	3.1	0.8	3.1	0.9	3.1	0.7	0.922	−0.008
	Sports index	3.2	0.8	3.4	0.7	3.0	0.7	<0.001	0.553
	Work Index	3.2	1.1	3.1	1.1	3.3	1.0	0.032	−0.178
	Fles Total score	9.5	1.7	9.5	1.8	9.4	1.6	0.203	0.109
Pediatric Quality of Life (PedsQL)	Physical Functioning	83.2	17.9	86.3	17.2	79.3	18.0	<0.001	0.402
	Emotional Functioning	76.1	24.3	80.8	21.5	70.1	26.3	<0.001	0.450
	Social Functioning	85.9	19.0	87.7	17.4	83.6	20.7	0.008	0.217
	School Functioning	80.2	20.4	81.8	20.5	78.3	20.2	0.040	0.169
	Overall Qol Scores	81.4	16.9	84.1	16.0	77.8	17.5	<0.001	0.379

* *p*-value is based on the independent sample *t*-test.

Table 3. The Depression, Anxiety, and Stress Scale—21 items (DASS21) (n = 601).

DASS Symptoms		N	%	HRQoL Scores		*p*-Value *	Omega Squared
				Mean	SD		
Depression	Normal	341	56.7	89.8	12.1	<0.001	0.400
	Mild	67	11.1	77.9	13.9		
	Moderate	99	16.5	72.0	14.2		
	Severe	45	7.5	70.7	9.1		
	Extremely Severe	49	8.2	56.3	18.9		
Anxiety	Normal	310	51.6	90.1	12.4	<0.001	0.348
	Mild	41	6.8	81.2	12.8		
	Moderate	105	17.5	76.5	13.4		
	Severe	49	8.2	70.9	13.5		
	Extremely Severe	96	16.0	63.9	17.9		
Stress	Normal	415	69.1	87.4	13.2	<0.001	0.322
	Mild	63	10.5	73.6	14.3		
	Moderate	49	8.2	72.5	14.4		
	Severe	43	7.2	62.8	15.9		
	Extremely Severe	31	5.2	56.7	18.7		

* *p*-value is based on one-way ANOVA.

Table 4 shows the correlations between adolescents' quality of life, sports, and mental health (DASS21 domains). According to the table, there was a significant, strong negative correlation between the overall quality of life and all mental health domains ($p < 0.05$). Sports showed weak negative correlations with all mental health domains ($p > 0.05$) and a weak positive correlation with the overall quality of life ($p > 0.05$).

Table 5 shows the results of the multiple linear regression models, which demonstrated that stress was a significant predictor for the low quality of life of male and female adolescents ([β = −0.30, (95% CI (−0.59) to (−0.02), $p < 0.05$)] and [β = −0.40, (95% CI (−0.70) to (−0.01), $p < 0.05$), respectively]).

Table 4. Correlations between the overall quality of life, sport index, and the DASS21 domains among adolescents.

	Depression Score	Anxiety Score	Stress Score	Sport Index	Overall Quality of Life
Depression Score		0.826 **	0.849 **	−0.061	−0.643 **
Anxiety Score	0.826 **		0.839 **	−0.002	−0.603 **
Stress Score	0.849 **	0.839 **		−0.010	−0.628 **
Sport index	−0.061	−0.002	−0.010		0.062
Overall Quality of Life	−0.643 **	−0.603 **	−0.628 **	0.062	

** Correlation is significant at the 0.01 level (2-tailed).

Table 5. Multiple linear regression models for the predictors of students' quality of life.

	Models	Estimate Beta	Estimate SE	*p*-Value	95.0% Confidence Interval for B Lower Bound	95.0% Confidence Interval for B Upper Bound	Model Fitness
Male	(Constant)	97.61	4.40	<0.001	88.95	106.27	
	Depression Score	−0.53	0.15	<0.001	−0.83	−0.24	
	Anxiety Score	−0.28	0.15	0.065	−0.57	0.02	
	Stress Score	−0.30	0.14	0.036	−0.59	−0.02	$R^2 = 0.424$
	Leisure Index	−0.84	0.95	0.379	−2.70	1.03	
	Sport index	1.11	0.98	0.260	−0.83	3.04	
	Work index	−1.71	0.71	0.017	−3.10	−0.31	
Female	(Constant)	101.98	5.30	<0.001	91.54	112.41	
	Depression Score	−0.63	0.15	<0.001	−0.93	−0.34	
	Anxiety Score	0.02	0.16	0.906	−0.29	0.33	
	Stress Score	−0.40	0.15	0.010	−0.70	−0.10	$R^2 = 0.462$
	Leisure Index	−1.81	1.23	0.141	−4.23	0.61	
	Sport index	−0.03	1.14	0.979	−2.27	2.21	
	Work index	−1.77	0.91	0.052	−3.56	0.01	
All	(Constant)	99.06	3.28	<0.001	92.61	105.50	
	Depression Score	−0.59	0.11	<0.001	−0.80	−0.38	
	Anxiety Score	−0.15	0.11	0.168	−0.35	0.06	
	Stress Score	−0.33	0.10	0.001	−0.54	−0.13	$R^2 = 0.457$
	Leisure Index	−1.25	0.74	0.092	−2.70	0.21	
	Sport index	0.88	0.71	0.216	−0.52	2.28	
	Work index	−1.81	0.55	0.001	−2.89	−0.72	

Abbreviation: Beta = regression coefficient; SE = standard error of the coefficient.

Figure 1 shows the sports played in school by the adolescents. It was found that the percentage of those who never practiced sports was 78.7% for tennis, 74.4% for basketball, 69.2% for volleyball, and 51.3% for football. The percentage who practiced football regularly was 25.0%, compared to 13.3% for volleyball, 14.3% for basketball, and 12.5% for tennis.

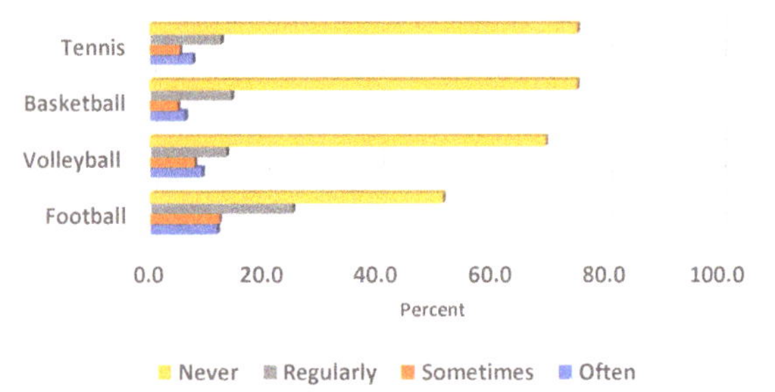

Figure 1. Sports played in school by adolescents.

4. Discussion

The present study aimed to assess the mental health, physical activity, and quality of life of school students when they returned to school after two years of distance learning in the Jazan region, southwest Saudi Arabia. Children and adolescents were at higher risk of mental problems due to restrictions during the pandemic. For that, families were greatly challenged to adapt to these restrictions and to create a healthy environment, particularly during online classes.

During the pandemic, the government established quarantine measures to control the spread of the disease; this led to restrictions on freedom and significantly increased stress among the population. These restrictions also led to violence, in some instances, among families and may have had an impact on the mental health of children and adolescents [20,21], eventually leading to long-term adverse outcomes. Our findings suggested that the COVID-19-related lockdown had a negative effect on children's and adolescents' mental health and physical activity. Additionally, the results concluded that this period had a negative impact on the pupils' quality of life.

Regular physical activity and exercise are related to an improved sense of well-being, physical health [22], life satisfaction, and cognitive performance. Our results revealed a high proportion of pupils who never practiced any type of sport. This pattern of low level of PA was not surprising, as the COVID-19 period affected the involvements of the students in all types of sports. This pattern of PA behavior was entirely consistent with expectations and the adolescents still affected by the COVID-19 period, which is characterized by a low level of physical activity [23]. In contrast to a previous study [23], we found significant gender differences in physical activity, with girls being less active than boys, but the reasons for this remain unclear. Previously, these gender differences were explained by non-modifiable elements, such as a girl's biology, as well as by certain modifiable variables, such as a girl's psychology, social support, and cultural and physical environmental factors. As is the case in most Arab countries, girls receive less social support than boys, as many families limit their girls from exercising outdoors, due to traditional cultural and religious norms [6]. As a result, authorities, schools, health, exercise providers, and families must be aware of the severity of the problem and quickly establish appropriate physical activity interventions to mitigate the detrimental effects of the COVID-19 outbreak on children's and adolescents' health [24].

A significant finding of our study was that depression, stress, and anxiety levels all had negative effects on the overall quality of life and physical activity; these findings were consistent with a study in Italy and Spain that reported that 85.7% observed a change in the behaviors and emotions of their children during the pandemic, based on parents' reports [11]. Another cross-sectional study was conducted in China, which targeted adolescents during the pandemic, and they assessed depression using the Patient Health Questionnaire (PHQ-9) and anxiety using Generalized Anxiety Disorder (GAD-7), both of which were not used in our study. The prevalence of depression and anxiety were 34.7% and 37.4%, respectively [25].

The results suggested that 43.3% of participants were depressed, and overall quality of life correlated with depression. These outcomes were also found in a study conducted in China among adolescents, which reported that 39.5% of the participants experienced depression, which was linked with less physical activity as a factor of depression during the COVID-19 pandemic [26].

In this study, quality of life involved the physiological integrity of individuals, as well as mental health and social comfort. The analysis revealed that parameters of quality of life were affected negatively by psychological issues. Additionally, males and females with higher physical activity levels had considerably higher quality of life scores. Regular physical activity was established in literature to considerably improve the quality of life of healthy individuals and chronic disease patients. In line with this approach, the results of our study's multiple linear regression analysis revealed that students' mental health had a negative effect on their quality of life, particularly depression, which is a major predictor of both male and female teenage quality of life. These findings highlighted the critical significance of including physical exercise and mental health programs in community-based rehabilitation programs to safeguard and maintain patients' quality of life during the ongoing pandemic [23]. This study highlighted the adverse impact of COVID-19 on physical activity and the mental health of adolescents. The results can help inform decision-makers about the relationship between the PA, HRQol, and mental health and pave the way for the formulation of intervention programs that can improve adolescents' quality of life.

The present study had some limitations. This was a cross-sectional-based study. Therefore, the temporal correlation between explanatory variables and outcomes could not be assessed. As this was an online survey-based study, response and selection biases are possible. Finally, the results of our study cannot be generalized to the whole population in the Jazan Region.

5. Conclusions

The analysis revealed a moderate level of physical activity among the schools' students when they returned to school following COVID-19 lockdown. Children's physical activity involvement is linked to the improvement in quality of life and mental health. The results call for the need to develop appropriate intervention programs to increase school students' physical activity levels. Future research focusing on the inter-variable associations between PA, mental health, and quality of life, using higher study designs, such as a cohort study, is highly needed.

Author Contributions: Conceptualization, M.S.M., A.Y.A., N.H.S., A.S.J., F.O.A., A.M.A., M.M.S. and I.D.S.; methodology, M.S.M., A.Y.A., N.H.S. and A.S.J.; software, M.S.M.; validation, N.H.S., A.S.J., F.O.A., A.M.A., M.M.S. and I.D.S.; formal analysis, M.S.M. and J.S.; investigation, N.H.S., A.S.J., F.O.A., A.M.A., M.M.S. and I.D.S.; resources, M.S.M. and A.Y.A.; data curation, M.S.M.; writing—original draft preparation, M.S.M., A.M.W. and J.S.; writing—review and editing M.S.M., A.M.W. and J.S.; visualization, M.S.M., A.Y.A., N.H.S. and A.S.J.; supervision, M.S.M., A.Y.A. and A.M.W.; project administration, N.H.S., A.S.J., F.O.A., A.M.A., M.M.S. and I.D.S. All authors have read and agreed to the published version of the manuscript.

Funding: This research received no external funding.

Institutional Review Board Statement: Ethical issues were addressed in accordance with the Helsinki Declaration and Saudi Bioethics standards' guidelines. Approval was obtained from the Standing Committee for Scientific Research Ethics-Jazan University (HAPO-10-Z-001), reference (REF# REC-43/05/088).

Informed Consent Statement: Informed consent was obtained from all subjects involved in the study.

Data Availability Statement: This study has no additional supporting data to share.

Acknowledgments: Our sincerest thanks go out to Jaber Ahmed Majrashi, Office Manager, Ministry of Education, Al Darb Governorate, Jizan Region, Saudi Arabia for his help and support regarding data collection. We would also like to thank all the students who participated in this study for their collaboration. We thank the Mapi Research Trust, France for using the Pediatric Quality of Life Inventory™ Version 4.0-Arabic (Saudi Arabia).

Conflicts of Interest: The authors declare no conflict of interest.

References

1. Aloqbi, H.S.; Shannaq, G.Y.; Douf, T.A.; Maashi, M.M.; Aljadani, F.M.; Alsultan, A.A.; Asiri, K.J.; Alotaibi, N.A. Prevalence of Clinical and Radiological Manifestations among COVID-19 Children and Adolescents: Systematic Review and Meta-Analysis. Available online: https://www.amhsr.org/articles/prevalence-of-clinical-and-radiological-manifestations-among-covid19-children-and-adolescents-systematic-review-and-metaanalysis-7397.html (accessed on 16 March 2023).
2. Pfefferbaum, B.; North, C.S. Mental health and the COVID-19 pandemic. *N. Engl. J. Med.* **2020**, *383*, 510–512. [CrossRef] [PubMed]
3. Coronavirus Disease (COVID-19). Available online: https://www.who.int/emergencies/diseases/novel-coronavirus-2019 ?adgroupsurvey=%7Badgroupsurvey%7D&gclid=CjwKCAjw_MqgBhAGEiwAnYOAelxdyf2owuyrXzoJECFRiMTcBMxM0r1 TEqixww67Q_ueTcjFEIiXQRoC5YsQAvD_BwE (accessed on 16 March 2023).
4. Neville, R.D.; Lakes, K.D.; Hopkins, W.G.; Tarantino, G.; Draper, C.E.; Beck, R.; Madigan, S. Global changes in child and adolescent physical activity during the COVID-19 pandemic. *JAMA Pediatr.* **2022**, *176*, 886–894. [CrossRef] [PubMed]

5.   WHO Guidelines on Physical Activity and Sedentary Behaviour. Available online: https://www.who.int/publications/i/item/9789240015128 (accessed on 16 March 2023).

6.   Abdelghaffar, E.-A.; Hicham, E.K.; Siham, B.; Samira, E.F.; Youness, E.A. Perspectives of adolescents, parents, and teachers on barriers and facilitators of Physical Activity Among School-Age Adolescents: A qualitative analysis. *Environ. Health Prev. Med.* **2019**, *24*, 21. [CrossRef] [PubMed]

7.   Mittal, V.A.; Firth, J.; Kimhy, D. Combating the dangers of sedentary activity on child and adolescent mental health during the time of COVID-19. *J. Am. Acad. Child Adolesc. Psychiatry* **2020**, *59*, 1197–1198. [CrossRef]

8.   Pfefferbaum, B.; Van Horn, R.L. Physical activity and sedentary behavior in children during the COVID-19 pandemic: Implications for mental health. *Curr. Psychiatry Rep.* **2022**, *24*, 493–501. [CrossRef]

9.   Kallio, J.; Hakonen, H.; Syväoja, H.; Kulmala, J.; Kankaanpää, A.; Ekelund, U.; Tammelin, T. Changes in physical activity and sedentary time during adolescence: Gender differences during weekdays and Weekend Days. *Scand. J. Med. Sci. Sport.* **2020**, *30*, 1265–1275. [CrossRef]

10.   Liang, L.; Ren, H.; Cao, R.; Hu, Y.; Qin, Z.; Li, C.; Mei, S. The effect of COVID-19 on Youth Mental Health. *Psychiatr. Q.* **2020**, *91*, 841–852. [CrossRef]

11.   Orgilés, M.; Morales, A.; Delvecchio, E.; Mazzeschi, C.; Espada, J.P. Immediate psychological effects of the COVID-19 quarantine in youth from Italy and Spain. *Front. Psychol.* **2020**, *11*, 579038. [CrossRef]

12.   Yeasmin, S.; Banik, R.; Hossain, S.; Hossain, M.N.; Mahumud, R.; Salma, N.; Hossain, M.M. Impact of COVID-19 pandemic on the mental health of children in Bangladesh: A cross-sectional study. *Child. Youth Serv. Rev.* **2020**, *117*, 105277. [CrossRef]

13.   Zhang, L.; Zhang, D.; Fang, J.; Wan, Y.; Tao, F.; Sun, Y. Assessment of Mental Health of Chinese primary school students before and after school closing and opening during the COVID-19 pandemic. *JAMA Netw. Open* **2020**, *3*, e2021482. [CrossRef]

14.   Marconcin, P.; Werneck, A.O.; Peralta, M.; Ihle, A.; Gouveia, É.R.; Ferrari, G.; Sarmento, H.; Marques, A. The association between Physical Activity and Mental Health during the first year of the COVID-19 pandemic: A systematic review. *BMC Public Health* **2022**, *22*, 209. [CrossRef] [PubMed]

15.   Freire, T.; Ferreira, G. Health-related quality of life of adolescents: Relations with positive and negative psychological dimensions. *Int. J. Adolesc. Youth* **2018**, *23*, 11–24. [CrossRef]

16.   Cochran, W.G. *Sampling Techniques*, 3rd ed.; Wiley: New York, NY, USA, 1977.

17.   Treuth, M.S.; Hou, N.; Young, D.R.; Maynard, L.M. Validity and reliability of the Fels Physical Activity Questionnaire for Children. *Med. Sci. Sport. Exerc.* **2005**, *37*, 488–495. [CrossRef] [PubMed]

18.   Moussa, M.T.; Lovibond, P.; Laube, R.; Megahead, H.A. Psychometric Properties of an Arabic version of the Depression Anxiety Stress Scales (DASS). *Res. Soc. Work Pract.* **2016**, *27*, 375–386. [CrossRef]

19.   Varni, J.W.; Seid, M.; Rode, C.A. The pedsql™: Measurement Model for the pediatric quality of Life Inventory. *Med. Care* **1999**, *37*, 126–139. [CrossRef]

20.   Holt, S.; Buckley, H.; Whelan, S. The impact of exposure to domestic violence on children and young people. A review of the literature. *Child Abus. Negl.* **2008**, *32*, 797–810. [CrossRef]

21.   MacMillan, H.L.; Wathen, C.N.; Varcoe, C.M. Intimate partner violence in the family: Considerations for Children's safety. *Child Abus. Negl.* **2013**, *37*, 1186–1191. [CrossRef]

22.   Ozdemir, F.; Cansel, N.; Kizilay, F.; Guldogan, E.; Ucuz, I.; Sinanoglu, B.; Colak, C.; Cumurcu, H.B. The role of physical activity on mental health and quality of life during COVID-19 outbreak: A cross-sectional study. *Eur. J. Integr. Med.* **2020**, *40*, 101248. [CrossRef]

23.   Castañeda-Babarro, A.; Arbillaga-Etxarri, A.; Gutiérrez-Santamaría, B.; Coca, A. Physical activity change during COVID-19 confinement. *Int. J. Environ. Res. Public Health* **2020**, *17*, 6878. [CrossRef] [PubMed]

24.   Xiang, M.; Zhang, Z.; Kuwahara, K. Impact of COVID-19 pandemic on children and adolescents' lifestyle behavior larger than expected. *Prog. Cardiovasc. Dis.* **2020**, *63*, 531–532. [CrossRef]

25.   Zhou, S.-J.; Zhang, L.-G.; Wang, L.-L.; Guo, Z.-C.; Wang, J.-Q.; Chen, J.-C.; Liu, M.; Chen, X.; Chen, J.-X. Prevalence and socio-demographic correlates of psychological health problems in Chinese adolescents during the outbreak of COVID-19. *Eur. Child Adolesc. Psychiatry* **2020**, *29*, 749–758. [CrossRef] [PubMed]

26.   Zhou, J.; Yuan, X.; Qi, H.; Liu, R.; Li, Y.; Huang, H.; Chen, X.; Wang, G. Prevalence of depression and its correlative factors among female adolescents in China during the coronavirus disease 2019 outbreak. *Glob. Health* **2020**, *16*, 69. [CrossRef] [PubMed]

 *healthcare*

Article

Effects of 10-Week Online Moderate- to High-Intensity Interval Training on Body Composition, and Aerobic and Anaerobic Performance during the COVID-19 Lockdown

Lorena Rodríguez-García [1,2], Halil Ibrahim Ceylan [3], Rui Miguel Silva [4,5,*], Ana Filipa Silva [4,5,*], Amelia Guadalupe-Grau [6,7] and Antonio Liñán-González [8]

[1] Department of Physical Activity and Sport Sciences, Pontifical University of Comillas, 07013 Palma, Spain; lrodriguez@cesag.org
[2] SER Research Group, Pontifical University of Comillas, 07013 Palma, Spain
[3] Physical Education and Sports Teaching Department, Kazim Karabekir Faculty of Education, Ataturk University, 25240 Erzurum, Turkey; halil.ibrahimceylan60@gmail.com
[4] Escola Superior Desporto e Lazer, Instituto Politécnico de Viana do Castelo, Rua Escola Industrial e Comercial de Nun'Álvares, 4900-347 Viana do Castelo, Portugal
[5] Delegação da Covilhã, Instituto de Telecomunicações, 1049-001 Lisboa, Portugal
[6] GENUD Toledo Research Group, Universidad Castilla-La Mancha, 45002 Toledo, Spain; amelia.guadalupe@uclm.es
[7] CIBER of Frailty and Healthy Aging (CIBERFES), Instituto de Salud Carlos III (ISCIII), 28029 Madrid, Spain
[8] Department of Nursing, Faculty of Health Sciences, Melilla Campus, University of Granada, 52005 Melilla, Spain; antoniolg@ugr.es
* Correspondence: rui.s@ipvc.pt (R.M.S.); anafilsilva@gmail.com (A.F.S.)

Citation: Rodríguez-García, L.; Ceylan, H.I.; Silva, R.M.; Silva, A.F.; Guadalupe-Grau, A.; Liñán-González, A. Effects of 10-Week Online Moderate- to High-Intensity Interval Training on Body Composition, and Aerobic and Anaerobic Performance during the COVID-19 Lockdown. *Healthcare* **2024**, *12*, 37. https://doi.org/10.3390/healthcare12010037

Academic Editor: Ines Aguinaga-Ontoso

Received: 23 November 2023
Revised: 19 December 2023
Accepted: 21 December 2023
Published: 23 December 2023

Abstract: The present study aimed to investigate the effects of a 10-week online high-intensity interval training (HIIT) program on body composition and aerobic and aerobic performance in physically sedentary women. A parallel, two-group, longitudinal (pre, post) design was used with physical tests performed before (preintervention) and after (postintervention) the 10-week intervention period. A total of forty-eight healthy and physically sedentary women (defined as an individual who lacks regular exercise or a structured fitness routine) were recruited to participate in this study. The participants were distributed in two groups: the experimental group (EG) with 24 women (mean ± SD: age 21.21 ± 2.15 years; weight: 61.16 ± 8.94 kg; height: 163.96 ± 4.87 cm; body mass index (BMI): 22.69 ± 2.49 kg/m^2) and the control group (CG) with another 24 women (mean ± SD: age 20.50 ± 1.29 years; weight: 62.0 ± 6.65 kg; height: 163.92 ± 4.89 cm; body mass index: 23.04 ± 1.74 kg/m^2). The EG performed an online HIIT program for 10 weeks, while the CG continued with their daily life routines. The repeated measures ANCOVA indicated a significant effect in the within-group analysis for weight ($p = 0.001$; $d = -0.96$) and for BMI ($p = 0.001$; $d = 0.24$), with a significant decrease in the experimental group (EG). The control group (CG) did not show any significant decrease in either body weight or BMI. Regarding the maximal oxygen uptake (VO2 max) values, the EG exhibited a significant improvement ($p = 0.001$; $d = -1.07$), whereas the CG did not demonstrate a significant improvement ($p = 0.08$; $d = -0.37$). The EG's power output (W) ($p = 0.001$; $d = -0.50$) and power output standardized by body weight (W/kg) ($p = 0.001$; $d = -0.96$) were significantly improved. The CG did not show a significant improvement in either power output (W/kg) or power output. Lastly, the within-group analysis with load revealed that the EG significantly improved ($p = 0.001$; $d = -0.50$), while CG did not show a significant improvement in load ($p = 0.10.$; $d = -0.10$). The present study showed that 10 weeks of HIIT in an online environment during the COVID-19 lockdown significantly improved maximum oxygen consumption and caused weight loss and a significant decrease in body mass index in physically sedentary women. These results suggest that HIIT may be used as a time-efficient strategy to improve body composition and cardio-respiratory fitness in sedentary women.

Keywords: performance; sedentary; women; anthropometric; interval training

1. Introduction

The COVID-19 pandemic has ushered in unprecedented challenges, significantly disrupting daily routines and prompting profound changes in lifestyle patterns worldwide [1]. Government-imposed lockdowns and restrictions on mobility have led to a surge in sedentary behaviors among individuals, exacerbating concerns about health and well-being [2]. Sedentary lifestyles are associated with various health issues, including increased risks of cardiovascular diseases, obesity, and mental health disorders [3]. The pandemic's impact on physical activity underscores the critical need for innovative and accessible exercise solutions that can be implemented under restrictive conditions.

High-intensity interval training (HIIT) is estimated to be one of the most popular trends in fitness according to the American College of Sports Medicine (ACSM)'s Annual Fitness Trend Forecast [4]. HIIT is characterized by repeated short-term explosive-intensity anaerobic activities ($\geq$85–90% maximal oxygen uptake (VO2 max) for health subjects or $\geq$80% VO2 max for clinical populations), interspersed by periods of passive or low-intensity exercise recovery, and it typically takes less than 30 min per training session [5,6]. HIIT has gained recognition as a popular alternative exercise approach compared to traditional moderate-intensity continuous exercise. This method aims to reduce the body fat percentage, lower the resting heart rate, and improve aerobic fitness in a shorter timeframe, particularly among sedentary individuals [7–9].

Similarly, HIIT provides significant enhancements in VO2 max. This contributes to improvements in both aerobic and anaerobic fitness within a brief period, surpassing the benefits offered by alternative training methods [10]. Therefore, previous studies suggested that HIIT could be a solution to improve health and reduce morbidity in the adult population [11,12]. This type of training is tolerable and acceptable to physically sedentary people, although at first, it generated a lot of controversy in people with little training due to the high intensity that this mode of training entails. However, in recent years, it has been prescribed to the elderly, young people, and adolescents [13]. Therefore, it must be taken into account that these HIIT programs can cause a greater feeling of fatigue due to a higher ratio of perceived exertion (RPE) in comparison with the classic continuous training of moderate intensity [14].

The context of the COVID-19 pandemic, which disrupted regular exercise routines, underscores the relevance of understanding the impact of online HIIT during periods of restricted mobility and social distancing. While previous literature has predominantly delivered HIIT in person, rather than conducting it online [7], the present study explored the novel approach of online delivery and considered the unique challenges and opportunities it presents. Therefore, the present study may contribute valuable insights into promoting accessible and effective exercise strategies, particularly during times of restricted mobility and social isolation.

Understanding how an online HIIT approach can impact individuals during a period of restricted mobility and social distancing is highly relevant. For the above reasons, the purpose of this study was to investigate the effects of a 10-week online HIIT intervention on body composition and aerobic performance variables in healthy, physically sedentary women.

2. Materials and Methods

2.1. Experimental Approach to the Problem

A parallel, two-group, longitudinal (pre, post) design was used with physical tests performed before (preintervention) and after (postintervention) the 10-week intervention period. The participants were assigned and matched into two groups, an experimental group (EG) and a control group (CG). The participants from the CG were asked to maintain their daily life routines, while those from the EG were introduced to a one-hour HIIT familiarization session before the start of the intervention. Regarding the day of training, it was preceded by 48 h of absence of high effort activities. The study was conducted between January and March 2021. (See Table 1, for more information).

Table 1. Timeline of this study.

		2021											
Months	January		February						March				
Week		1	2	3	4	5	6	7	8	9	10		
EG	Pre	High-intensity interval training											Post
CG	Pre	Maintain routines											Post

2.2. Participants

A total of forty-eight physically sedentary women were recruited to participate in this study. There were two groups: the EG with twenty-four women (n = 24; age: 21.21 ± 2.15 years; weight: 61.16 ± 8.94 kg; height: 163.96 ± 4.87 cm; body mass index: 22.69 ± 2.49 kg/m^2) and the CG with another twenty-four women (n = 24; age: 20.50 ± 1.29 years; weight: 62.0 ± 6.65 kg; height: 163.92 ± 4.89 cm; body mass index: 23.04 ± 1.74 kg/m^2). The randomization sequence was generated electronically (https://www.randomizer.org, accessed on 5 January 2021) and was concealed until the interventions were assigned.

Moreover, a priori sample size calculation was performed using a free online tool, G*Power (www.gpower.hhu.de, accessed on 3 January 2021), with a power level of 95% and an α level of 0.05 and based on previous and similar studies [15]; it revealed that a sample size of >36 would be sufficient for conducting a randomized controlled trial. The sample group for this study was selected using the criterion sampling method, which falls under the category of purposive sampling techniques [16]. The inclusion criteria for the participants in this study were (i) being physically sedentary (not engaging in 150 min per week of moderate-intensity exercise or 75 min per week of vigorous-intensity exercise or a combination equivalent to these two different intensities); (ii) not presenting any injuries; (iii) giving their consent; and (iv) participated in at least 90% of the training sessions during the intervention.

Finally, the participants obtained information on the main objectives of the research and signed the informed consent form. All the participants in this research were treated according to the guidelines of the American Psychological Association (APA), and therefore the anonymity of the participants' responses was guaranteed. The study was carried out in accordance with the ethical principles of the Declaration of Helsinki for research in humans and was approved by the Research Ethics Committee of the Pontifical University of Comillas (2021/85).

2.3. Procedures

The forty-eight physically sedentary women enrolled in this study visited the laboratory twice, both for pre-test and post-test assessments, between 9:30 a.m. and 5:00 p.m. During the pre- and post-assessments, the participants underwent a sub-maximal incremental fitness test on a stationary ergometer. Adhering to the guidelines outlined by the American College of Sports Medicine (2018), measures were taken to ensure the safety of the physically sedentary women. Body mass, without shoes, was measured utilizing a bioelectrical impedance analysis (BIA) device (Tanita BC-730) accurate to the nearest 0.1 kg. Height was measured using a stadiometer (Type SECA 225, Hamburg, Germany) accurate to the nearest 0.1 cm. Body mass index was computed as mass (in kilograms) divided by the square of height (in meters).

All participants performed a sub-maximal incremental fitness test on a cycle ergometer (Viasprint 150 P cycle-ergometer) connected to a Jaeger Master Screen gas analyzer. Determination of the ventilatory anaerobic threshold (VAT) was based on the respiratory gas exchange method (RER) (RER = CO_2 production/O_2 consumption), which detected the VAT at the point at which RER exceeds the cut-off value of 1.0 [17], as well as the obtained maximal oxygen uptake. First, an RS800CX Polar monitor (Polar Electro, Helsinki,

Finland) was used to monitor and record the maximal heart rate (HRmax) during the pre- and post-assessments. In this sense, the protocol consisted of a submaximal incremental test with a fixed cadence of 60 revolutions per minute (rpm). The warm-up started at 0 W and the workload was increased by 10 W every min until min 5. The participant began the exercise phase pedaling at 50 W, and the workload was increased to 25 W every two min. After each increase, the workload remained stable for the next 2 min. The submaximal test ended once the VAT was reached. The highest power output (W) reached during the cycle ergometer test was recorded. Finally, the highest power output reached was normalized by the participant's body weight (W/kg).

2.4. Training Program

The experimental group (EG) engaged in a 10-week online program, involving three sessions per week. The sessions comprised high-intensity interval training (HIIT) following the Tabata method, involving 4 min intervals with 8 intensive training blocks, each followed by 1 min of recovery [18]. The exercises involved body weights and functional movements, aligning with a protocol that demonstrated favorable outcomes in female university students [8]. Notably, the last session of each week focused on continuous running, as outlined in Table 2.

In the first month, they performed 20 s intervals of intense exercise with 10 s of recovery, repeating this for 4 min. In the second month, they increased the technical difficulty of the exercises and increased the volume of the sessions until they achieved 20 min of Tabata in the last week of the intervention.

The HIIT sessions were carried out twice a week and their intensity was monitored daily using the Borg's Scale (6–20) of the rate of perceived exertion (RPE) [19]. Each participant was asked, "How did you perceive exertion during the exercise execution?". The first four weeks had an RPE of 12–13, indicating a somewhat hard level. The following four weeks were at an RPE of 15–16, signifying a hard level. In the last two weeks, the RPE was 18–19, representing an extremely hard level. However, the HR was not monitored during the intervention.

The participants were instructed to perform the exercise at their maximum exertion while maintaining the correct technique. The structure of the session consisted of a 10–15 min warm-up, a 35 min main part, and a 10 min cooldown. On the third day of each training week, the participants completed a moderate-intensity session outdoors, combining continuous running with walking.

Table 2. Example of HIIT and running session and work set progression during the HIIT program.

Example of Weekly Session		
1st Session	**2nd Session**	**3rd Session**
HIIT. Example 1	**HIIT. Example 2**	**Running**
Push-ups	Jumping jacks	5 min of running at 65–75% HRmax
Mountain Climbing	Box squat	5 min of walking at 40–45% HRmax
Forearm plank	Bench step-ups	5 min of running at 65–75% HRmax
Dips using a bench or chair	Deadlift w/elastic band	5 min of walking

Exercises		
1st		
2nd		
3rd		
4th		

Training progression during the HIIT program		
Week	**Set/session**	**Recovery between sets (s)**
1st and 2nd	2	60
3rd to 5th	3	60
6th to 8th	4	60
9th and 10th	4	45

Technical progressions in exercise complexity were systematically implemented throughout the experimental period. The progression for push-ups involved starting with hands against the wall, followed by positioning knees on the floor, and advancing to performing the exercise without resting the knees. Planks transitioned from forearms on the floor to arms fully extended. Similarly, jumping jacks initially began as static and progressed to dynamic, while squats advanced from using a box to incorporating a jump. This systematic approach extended to all exercises. During the first two weeks of the intervention, the participants followed an 8 min Tabata protocol, adhering to 20 s of intense exercise with 10 s of recovery, repeated for 4 min. Subsequently, we systematically increased the Tabata duration in the following weeks. More precisely, we implemented a progressive overload strategy by incorporating an additional minute every two weeks. By the conclusion of the 10-week intervention, the Tabata duration had reached a total of 20 min.

Due to the COVID-19 confinement regulations mandated by the Spanish government, the training program had to be conducted inside participants' homes, where the absence of equipment led to the selection of calisthenics exercises.

2.5. Statistical Procedures

The data were analyzed using Statistica software (version 10.0; Statsoft, Inc., Tulsa, OK, USA). Finally, the significance level was set at $p < 0.05$. Normal distribution and homogeneity tests (Kolmogorov–Smirnov and Levene's, respectively) were conducted on all metrics. A paired sample *t*-test was used for determining differences as a repeated measures analysis (pre and post). Cohen's d was used as the effect size indicator. To interpret the magnitude of the effect size, we adopted the following criteria: d $\leq$ 0.20, small; d $\leq$ 0.50, medium; and d $\leq$ 0.80, large. To elucidate the between-group differences, an ANCOVA test was performed using the pre-test as a covariate and the pre and post times as factors. To interpret the magnitude of the effect size of ANCOVA, we adopted the following criteria: $\eta p^2 = 0.02$, small; $\eta p^2 = 0.06$, medium; and $\eta p^2 = 0.14$, large.

3. Results

Descriptive statistics were calculated for each variable (Table 3). No significant baseline between-group differences were recorded for all measurements ($p > 0.05$, d = 0.04–0.31). Moreover, the adherence rate to the online HIIT program in the experimental group was 96.3% $\pm$ 2.2%, and no withdrawals were reported.

Table 3. Performance variables before (pre-test) and after (post-test) the intervention period (mean $\pm$ SD).

	Participants ($n = 48$)								
	Control Group ($n = 24$)				Experimental Group ($n = 24$)				Differences Between Groups (ANCOVA Test)
	Pre-Test	Post-Test	%	RM *t*-Test (p)	Pre-Test	Post-Test	%	RM *t*-Test (p)	
	Anthropometric measures								
Weight (kg)	62.00 ± 6.65	62.53 ± 7.07	0.75 ± 2.89	$p = 0.15$ $d = -0.08$	61.16 ± 8.94	59.26 ± 8.73	−3.22 ± 1.98	$p = 0.001$ ** $d = 0.41$	$F(1,47) = 1244.36$; $p = 0.000$; $\eta p^2 = 0.97$
Body Mass Index (%)	23.04 ± 1.74	23.24 ± 1.97	0.75 ± 2.89	$p = 0.15$ $d = -0.11$	22.69 ± 2.49	22.10 ± 2.47	−2.72 ± 1.98	$p = 0.001$ ** $d = 0.24$	$F(1,47) = 685.01$; $p = 0.000$; $\eta p^2 = 0.35$
	Incremental test								
VO2 max (mL/kg/min)	20.17 ± 2.03	21.03 ± 2.59	3.37 ± 10.37	$p = 0.08$ $d = -0.37$	20.58 ± 3.65	24.67 ± 3.98	15.94 ± 12.08	$p = 0.001$ ** $d = -1.07$	$F(1,47) = 32.38$; $p = 0.000$; $\eta p^2 = 0.42$
Power output (W/kg)	1.52 ± 0.16	1.54 ± 0.12	1.18 ± 3.84	$p = 0.20$ $d = -0.11$	1.48 ± 0.26	1.73 ± 0.26	13.50 ± 11.72	$p = 0.001$ ** $d = -0.96$	$F(1,47) = 43.21$; $p = 0.000$; $\eta p^2 = 0.49$
Power output (W)	94.71 ± 17.32	96.48 ± 15.04	1.90 ± 5.21	$p = 0.10$ $d = -0.10$	91.26 ± 23.11	102.92 ± 23.86	10.67 ± 12.57	$p = 0.001$ ** $d = -0.50$	$F(1,47) = 148.58$; $p = 0.000$; $\eta p^2 = 0.77$

Note: %: percent change; RM: repeated measures ** denotes significance at $p < 0.01$.

Significant group*time interactions were observed through repeated measures AN-COVA for weight (p = 0.001; ηp^2 = 0.97), body mass index (p = 0.001; ηp^2 = 0.35), VO2 max (p = 0.001; ηp^2 = 0.42), power output (W/kg) (p = 0.001; ηp^2 = 0.49), and power output (W) (p = 0.001; ηp^2 = 0.87). In the EG, weight significantly decreased (p = 0.001; d = -0.96), while the CG did not show a significant change (p = 0.15; d = -0.11) in body mass index. In terms of VO2 max, the EG demonstrated significant improvement (p = 0.001; d = -1.07), while the CG did not exhibit a significant change (p = 0.08; d = -0.37). A separate analysis of power output (W) revealed a significant improvement in the EG (p = 0.01; d = -0.96), while the CG did not show a significant change in power output (W/kg) (p = 0.20; d = -0.11). Lastly, the analysis of the power output (W) showed there was significant improvement in the EG (p = 0.001; d = -0.50), whereas the CG did not show a significant change (p = 0.10; d = -0.10). Refer to Figures 1 and 2 for a graphical representation.

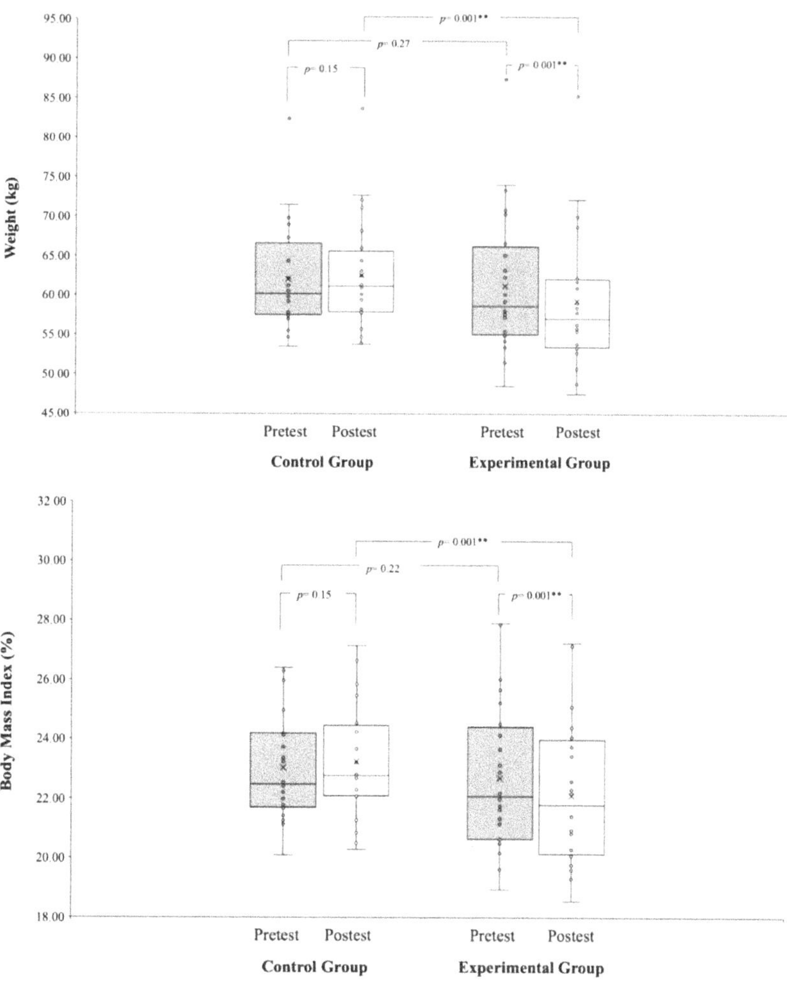

Figure 1. Pre- and post-tests anthropometrical measures (weight and body mass index) of CG and EG. ** denotes significance at $p < 0.01$.

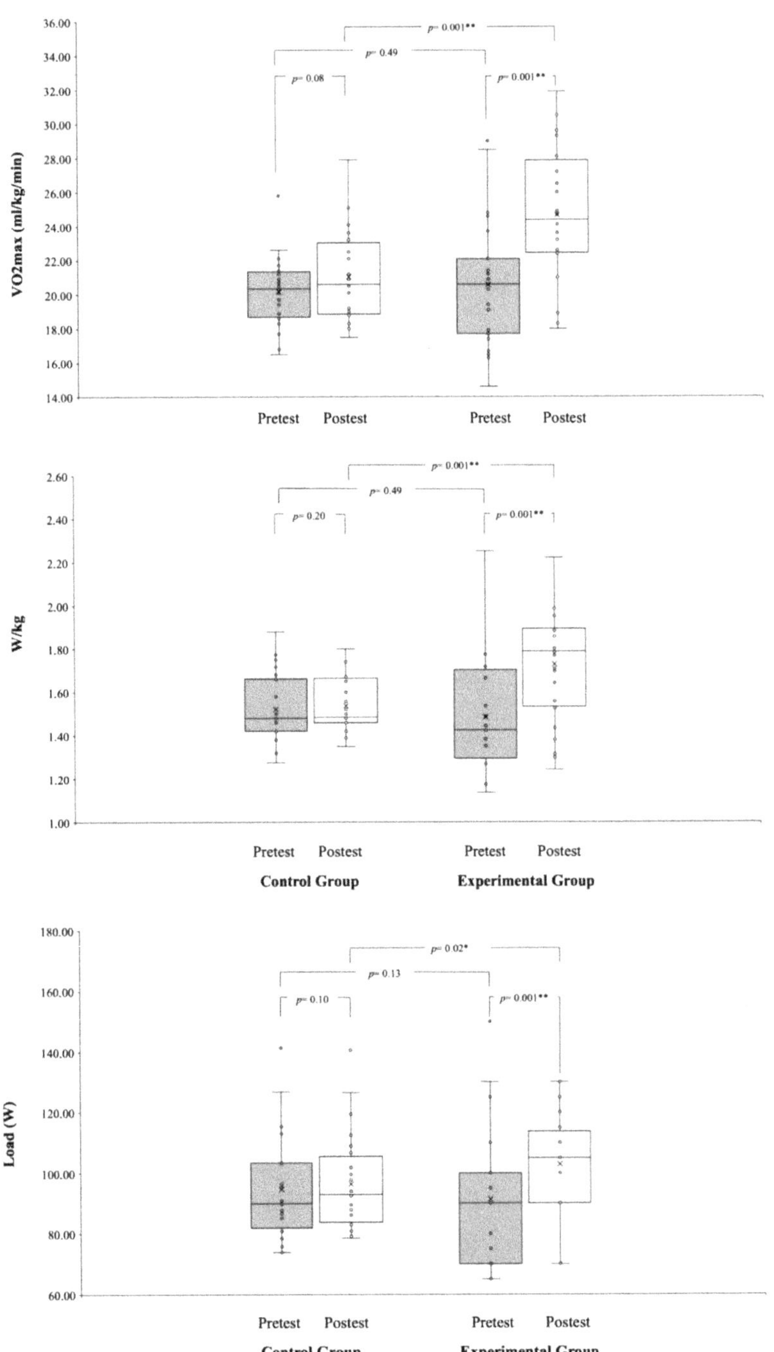

Figure 2. Pre- and post-tests performance variables (VO2 max, power output (W), and power output (W/kg)) of CG and EG. * denotes significance at $p < 0.05$. ** denotes significance at $p < 0.01$.

4. Discussion

The purpose of this study was to investigate the impact of a 10-week online high-intensity interval training (HIIT) program on body composition and aerobic performance during the COVID-19 lockdown. The main evidence of the present study revealed that significant reductions in body weight were observed after the 10-week HIIT program compared to the CG. Moreover, there were significant improvements in aerobic fitness after the HIIT program compared to the CG.

The results of the present study showed that online HIIT can be an effective stimulus for weight loss and decreasing BMI. Significant weight loss (effect size: 0.41, moderate, −3.22%) and a decrease in body mass index (effect size: 0.24, moderate, −2.72%) compared to the control group. This aligns with recent studies reporting body composition improvements within 8 weeks [8] and 15 weeks [20] of HIIT in sedentary young females. The effectiveness of online HIIT in promoting weight loss and reducing BMI in our study may be attributed to increased post-exercise energy expenditure. HIIT is known to induce excess post-exercise oxygen consumption (EPOC), leading to elevated oxygen consumption and calorie expenditure, contributing to a negative energy balance and, consequently, weight loss [21]. Contrary to our findings, previous studies [22–25] reported no significant changes in body weight, BMI, and body fat percentage after HIIT in sedentary young females. Differences in participant characteristics and exercise program content (duration, frequency, intensity, volume, and progression) could contribute to these varying results [25,26].

A 10-week online HIIT intervention led to significant improvements in VO2 max, power output (W/kg), and power output (W) when compared to the control group. This is in line with prior research, such as Lu et al. [8], that reported a substantial increase in VO2 max (+18.8%) following an 8-week HIIT in sedentary young females. Another study demonstrated noteworthy enhancements in VO2 max (approximately 22%) with cycling HIIT performed for 18–30 min per week over 12 weeks [22]. Moreover, evidence suggests that both low-volume (1 × 4 min treadmill running at 85–95% HRmax) and high-volume HIIT (4 × 4 min treadmill running at 85–95% HRmax), executed thrice weekly over 6 weeks, exhibit equal effectiveness in enhancing aerobic capacity (VO2 max) among sedentary young women [15]. A systematic review and meta analysis on adults indicated that short-term HIIT (<12 weeks) and long-term HIIT (≥12 weeks) performed at least 3 times a week for 12 weeks can result in an increased VO2 max. Additionally, the increase in aerobic capacity was reported to be greater at longer training times [7]. Another meta-analysis demonstrated that HIIT ranging from 2 to 8 weeks lead to a substantial increase in VO2 max by 4.2–13.4% in healthy, sedentary/recreationally active adults, emphasizing its positive impact on aerobic capacity [27]. A recent systematic review underscores that HIIT, with a minimum of 4 weeks of exercise training (3 times per week, 18–30 min per session, 85–95% HRR), is effective for augmenting maximal aerobic capacity in women with a sedentary lifestyle [28].

Considering the above studies, it can be said that longer durations of HIIT are required for greater improvements in VO2 max. However, it was observed that HIIT exercises for 2–12 weeks improved cardiorespiratory fitness or aerobic performance. In addition, in our study, a 15.94% VO2 max increase was detected after 10 weeks of the HIIT protocol. Our results are in agreement with previous studies, which indicated that improvements of 9% to 13% were found in VO2 max following 8-week [29] and 12-week [30] HIIT protocols in sedentary individuals. Moreover, the observed improvement was attributed to the training intensity, emphasizing that aerobic exercise intensity plays a pivotal role in enhancing VO2 max. These findings underscore the significance of HIIT conducted at higher training intensities, specifically within the range of 90–95% HRmax, for achieving optimal training-induced enhancements in VO2 max, especially concerning age-related considerations [22,29]. The considerable enhancement in VO2 max observed can be ascribed to the exercise intensity implemented in the HIIT program. Additionally, the existing literature, which aligns with our study, highlights that individuals with a more sedentary lifestyle exhibit the most substantial VO2 max response [22,28].

Furthermore, previous studies have suggested several physiological mechanisms underlying the improvement in VO2 max after HIIT. For instance, in one study, it was suggested that the improvement in VO2 max or cardiorespiratory fitness after HIIT could be associated with several central adaptations, such as increases in systolic volume and cardiac output due to the increased cardiac contractility and oxygen availability. Also, it was asserted that peripheral changes such as increased muscle oxidative potential, increased skeletal muscle diffusion capacity, increased number and size of mitochondria, increased mitochondrial enzyme activity, arterial vasodilation, increased nitric oxide bioavailability, and reduced oxidative stress might be responsible for the improvement in VO2 max following the HIIT intervention [22,31]. Moreover, in another study, it was stated that HIIT-induced increases in VO2 max and improved maximal aerobic performance could partly be explained by specific oxidative adaptations in type II fibers [27].

This study has some limitations that should be considered. The first limitation of the study is related to the use of the cycle ergometer evaluation, which may not precisely reflect the intensities prescribed for running and jumping exercises in the online high-intensity interval training program. Secondly, it is important to acknowledge that relying solely on participants' RPE without concurrent heart rate monitoring introduces a potential limitation. While RPE is a valuable subjective measure of effort, incorporating heart rate data would have provided an additional objective parameter to quantify the actual physiological intensity experienced by participants during each session. This absence of heart rate monitoring restricts the precision of intensity characterization and may limit the generalizability of our findings to a broader context where objective intensity measures are commonly employed. Thirdly, we did not control for other influencing factors that could influence our results. Furthermore, given that the sample consisted of sedentary individuals, the outcomes may have been impacted by the duration of exercise progressions. Therefore, future studies employing similar training protocols in sedentary populations might benefit from extending the intervention period.

5. Conclusions

The present study aimed to analyze the effects of a 10-week online HIIT program on body composition and aerobic performance during the COVID-19 lockdown. The main findings showed that the 10-week online HIIT intervention yielded significant improvements in maximum oxygen consumption and weight loss, and a noteworthy decrease in body mass index among sedentary young females. These findings underscore the efficacy of HIIT as a time-efficient strategy for enhancing both body composition and cardiorespiratory fitness in sedentary women, even under limited circumstances such as a confinement period.

Author Contributions: Conceptualization, L.R.-G.; methodology, A.F.S. and R.M.S.; formal analysis, A.F.S. and R.M.S.; writing—original draft preparation, A.F.S., R.M.S., H.I.C., A.G.-G. and A.L.-G.; writing—review and editing, A.G.-G. and H.I.C.; supervision, L.R.-G. All authors have read and agreed to the published version of the manuscript.

Funding: This research received no funding.

Institutional Review Board Statement: The study was carried out in accordance with the ethical principles of the Declaration of Helsinki for research in humans and was approved by the Research Ethics Committee of the Pontifical University of Comillas (2021/85).

Informed Consent Statement: Informed consent was obtained from all subjects involved in the study.

Data Availability Statement: The raw data used to support this article's conclusions will be made freely available.

Acknowledgments: We thank the participants for their collaboration with and participation in the study.

Conflicts of Interest: The authors declare no conflict of interest.

References

1. Caroppo, E.; Mazza, M.; Sannella, A.; Marano, G.; Avallone, C.; Claro, A.E.; Janiri, D.; Moccia, L.; Janiri, L.; Sani, G. Will Nothing Be the Same Again? Changes in Lifestyle during COVID-19 Pandemic and Consequences on Mental Health. *Int. J. Environ. Res. Public Health* **2021**, *18*, 8433. [CrossRef] [PubMed]
2. Musa, S.; Dergaa, I.; Bachiller, V.; Saad, B.H. Global Implications of COVID-19 Pandemic on Adults' Lifestyle Behavior: The Invisible Pandemic of Noncommunicable Disease. *Int. J. Prev. Med.* **2023**, *8*, 1–9.
3. Park, J.H.; Moon, J.H.; Kim, H.J.; Kong, M.H.; Oh, Y.H. Sedentary Lifestyle: Overview of Updated Evidence of Potential Health Risks. *Korean J. Fam. Med.* **2020**, *41*, 365–373. [CrossRef] [PubMed]
4. Thompson, W.R. Worldwide survey of fitness trends for 2019. *ACSM's Health Fit. J.* **2018**, *22*, 10–17. [CrossRef]
5. Gibala, M.J.; Jones, A.M. Physiological and performance adaptations to high-intensity interval training. *Nestle Nutr. Inst. Workshop Ser.* **2013**, *76*, 51–60. [CrossRef] [PubMed]
6. Norton, K.; Norton, L.; Sadgrove, D. Position statement on physical activity and exercise intensity terminology. *J. Sci. Med. Sport* **2010**, *13*, 496–502. [CrossRef] [PubMed]
7. Batacan, R.B., Jr.; Duncan, M.J.; Dalbo, V.J.; Tucker, P.S.; Fenning, A.S. Effects of high-intensity interval training on cardiometabolic health: A systematic review and meta-analysis of intervention studies. *Br. J. Sports Med.* **2017**, *51*, 494–503. [CrossRef]
8. Lu, M.; Li, M.; Yi, L.; Li, F.; Feng, L.; Ji, T.; Zang, Y.; Qiu, J. Effects of 8-week high-intensity interval training and moderate-intensity continuous training on bone metabolism in sedentary young females. *J. Exerc. Sci. Fit.* **2022**, *20*, 77–83. [CrossRef]
9. Ramos, J.S.; Dalleck, L.C.; Tjonna, A.E.; Beetham, K.S.; Coombes, J.S. The Impact of High-Intensity Interval Training Versus Moderate-Intensity Continuous Training on Vascular Function: A Systematic Review and Meta-Analysis. *Sports Med.* **2015**, *45*, 679–692. [CrossRef]
10. Bartlett, D.B.; Willis, L.H.; Slentz, C.A.; Hoselton, A.; Kelly, L.; Huebner, J.L.; Kraus, V.B.; Moss, J.; Muehlbauer, M.J.; Spielmann, G.; et al. Ten weeks of high-intensity interval walk training is associated with reduced disease activity and improved innate immune function in older adults with rheumatoid arthritis: A pilot study. *Arthritis Res.* **2018**, *20*, 127. [CrossRef]
11. Dorneles, G.P.; da Silva, I.; Boeira, M.C.; Valentini, D.; Fonseca, S.G.; Lago, P.D.; Peres, A.; Romão, P.R.T. Cardiorespiratory fitness modulates the proportions of monocytes and T helper subsets in lean and obese men. *Scand. J. Med. Sci. Sports* **2019**, *29*, 1755–1765. [CrossRef] [PubMed]
12. Reljic, D.; Lampe, D.; Wolf, F.; Zopf, Y.; Herrmann, H.J.; Fischer, J. Prevalence and predictors of dropout from high-intensity interval training in sedentary individuals: A meta-analysis. *Scand. J. Med. Sci. Sports* **2019**, *29*, 1288–1304. [CrossRef] [PubMed]
13. Osawa, Y.; Azuma, K.; Tabata, S.; Katsukawa, F.; Ishida, H.; Oguma, Y.; Kawai, T.; Itoh, H.; Okuda, S.; Matsumoto, H. Effects of 16-week high-intensity interval training using upper and lower body ergometers on aerobic fitness and morphological changes in healthy men: A preliminary study. *Open Access J. Sports Med.* **2014**, *5*, 257–265. [CrossRef] [PubMed]
14. Oliveira, B.R.R.; Slama, F.A.; Deslandes, A.C.; Furtado, E.S.; Santos, T.M. Continuous and High-Intensity Interval Training: Which Promotes Higher Pleasure? *PLoS ONE* **2013**, *8*, e79965. [CrossRef] [PubMed]
15. Bhati, P.; Bansal, V.; Moiz, J.A. Comparison of different volumes of high intensity interval training on cardiac autonomic function in sedentary young women. *Int. J. Adolesc. Med. Health* **2021**, *31*, 1–13. [CrossRef] [PubMed]
16. Palinkas, L.A.; Horwitz, S.M.; Green, C.A.; Wisdom, J.P.; Duan, N.; Hoagwood, K. Purposeful Sampling for Qualitative Data Collection and Analysis in Mixed Method Implementation Research. *Adm. Policy Ment. Health Ment. Health Serv. Res.* **2015**, *42*, 533–544. [CrossRef] [PubMed]
17. Yeh, M.P.; Gardner, R.M.; Adams, T.D.; Yanowitz, F.G.; Crapo, R.O.; Soller, B.R.; Yang, Y.; Lee, S.M.C.; Wilson, C.; Hagan, R.D.; et al. "Anaerobic threshold": Problems of determination and validation. *J. Appl. Physiol.* **1983**, *55*, 1178–1186. [CrossRef]
18. Alonso-Fernández, D.; Fernández-Rodríguez, R.; Taboada-Iglesias, Y.; Gutiérrez-Sánchez, Á. Impact of High-Intensity Interval Training on Body Composition and Depressive Symptoms in Adults under Home Confinement. *Int. J. Environ. Res. Public Health* **2022**, *19*, 6145. [CrossRef]
19. Borg, G. Psychophysical scaling with applications in physical work and the perception of exertion. *Scand. J. Work Environ. Health* **1990**, *16* (Suppl. S1), 55–58. [CrossRef]
20. Dupuit, M.; Maillard, F.; Pereira, B.; Marquezi, M.L.; Lancha, A.H., Jr.; Boisseau, N. Effect of high intensity interval training on body composition in women before and after menopause: A meta-analysis. *Exp. Physiol.* **2020**, *105*, 1470–1490. [CrossRef]
21. Panissa, V.L.G.; Fukuda, D.H.; Staibano, V.; Marques, M.; Franchini, E. Magnitude and duration of excess of post-exercise oxygen consumption between high-intensity interval and moderate-intensity continuous exercise: A systematic review. *Obes. Rev.* **2021**, *22*, e13099. [CrossRef] [PubMed]
22. Astorino, T.A.; Schubert, M.M.; Palumbo, E.; Stirling, D.; McMillan, D.W.; Cooper, C.; Godinez, J.; Martinez, D.; Gallant, R. Magnitude and time course of changes in maximal oxygen uptake in response to distinct regimens of chronic interval training in sedentary women. *Eur. J. Appl. Physiol.* **2013**, *113*, 2361–2369. [CrossRef] [PubMed]
23. Nybo, L.; Sundstrup, E.; Jakobsen, M.D.; Mohr, M.; Hornstrup, T.; Simonsen, L.; Bülow, J.; Randers, M.B.; Nielsen, J.J.; Aagaard, P.; et al. High-Intensity Training versus Traditional Exercise Interventions for Promoting Health. *Med. Sci. Sports Exerc.* **2010**, *42*, 1951–1958. [CrossRef] [PubMed]
24. Chidnok, W.; Wadthaisong, M.; Iamsongkham, P.; Mheonprayoon, W.; Wirajalarbha, W.; Thitiwuthikiat, P.; Siriwittayawan, D.; Vachirasrisirikul, S.; Nuamchit, T. Effects of high-intensity interval training on vascular function and maximum oxygen uptake in young sedentary females. *Int. J. Health Sci. (Qassim)* **2020**, *14*, 3–8. [PubMed]

25. Nunes, P.R.; Martins, F.M.; Souza, A.P.; Carneiro, M.A.; Orsatti, C.L.; Michelin, M.A.; Murta, E.F.; de Oliveira, E.P.; Orsatti, F.L. Effect of high-intensity interval training on body composition and inflammatory markers in obese postmenopausal women: A randomized controlled trial. *Menopause* **2019**, *26*, 256–264. [CrossRef]

26. Greer, B.K.; O'brien, J.; Hornbuckle, L.M.; Panton, L.B. EPOC Comparison Between Resistance Training and High-Intensity Interval Training in Aerobically Fit Women. *Int. J. Exerc. Sci.* **2021**, *14*, 1027–1035.

27. Sloth, M.; Sloth, D.; Overgaard, K.; Dalgas, U. Effects of sprint interval training on VO_{2max} and aerobic exercise performance: A systematic review and meta-analysis. *Scand. J. Med. Sci. Sports* **2013**, *23*, e341–e352. [CrossRef]

28. Syamsudin, F.; Wungu, C.D.K.; Qurnianingsih, E.; Herawati, L. High-intensity interval training for improving maximum aerobic capacity in women with sedentary lifestyle: A systematic review and meta-analysis. *J. Phys. Educ. Sport* **2021**, *21*, 1788–1797. [CrossRef]

29. Støren, Ø.; Helgerud, J.; Sæbø, M.; Støa, E.M.; Bratland-Sanda, S.; Unhjem, R.J.; Hoff, J.; Wang, E. The Effect of Age on the V-O2max Response to High-Intensity Interval Training. *Med. Sci. Sports Exerc.* **2017**, *49*, 78–85. [CrossRef]

30. Amaro-Gahete, F.J.; De-La-O, A.; Jurado-Fasoli, L.; Dote-Montero, M.; Gutiérrez, Á.; Ruiz, J.R.; Castillo, M.J. Changes in Physical Fitness After 12 Weeks of Structured Concurrent Exercise Training, High Intensity Interval Training, or Whole-Body Electromyostimulation Training in Sedentary Middle-Aged Adults: A Randomized Controlled Trial. *Front. Physiol.* **2019**, *10*, 451. [CrossRef]

31. Serna, V.H.A.; Vélez, E.F.A.; Arias, R.D.G.; Feito, Y. Effects of a high-intensity interval training program versus a moderate-intensity continuous training program on maximal oxygen uptake and blood pressure in healthy adults: Study protocol for a randomized controlled trial. *Trials* **2016**, *17*, 413. [CrossRef] [PubMed]

Article

Feasibility of a Supervised Postpartum Exercise Program and Effects on Maternal Health and Fitness Parameters—Pilot Study

Carla Brites-Lagos [1], Liliana Ramos [1,2], Anna Szumilewicz [3] and Rita Santos-Rocha [1,4,*]

[1] ESDRM, Department of Physical Activity and Health, Sport Sciences School of Rio Maior, Polytechnic Institute of Santarem, 2040-413 Rio Maior, Portugal; 200500015@esdrm.ipsantarem.pt (C.B.-L.); lilianaramos@esdrm.ipsantarem.pt (L.R.)

[2] CIEQV, Life Quality Research Center, Polytechnic Institute of Santarem, 2040-413 Rio Maior, Portugal

[3] Department of Fitness, Faculty of Physical Culture, Gdansk University of Physical Education and Sport, 80-336 Gdansk, Poland; anna.szumilewicz@awf.gda.pl

[4] CIPER, Interdisciplinary Centre for the Study of Human Performance, Faculty of Human Kinetics, University of Lisbon, 1499-002 Cruz Quebrada, Portugal

* Correspondence: ritasantosrocha@esdrm.ipsantarem.pt

Abstract: The postpartum period is marked by profound changes in women at physical, psychological, and physiological levels. Many of these changes persist after four to six weeks postpartum, and most women do not resume their levels of physical activity, which increases the risk of remaining inactive for many years. It is crucial to implement effective programs that promote exercise during the postpartum period. The objective of this study was to test the feasibility and analyze the effects of a structured and supervised postpartum exercise program on maternal health and fitness parameters. To analyze the potential effects of the intervention, the level of physical activity, quality of life, pelvic girdle and low back pain, fatigue, depression, and the level of functional and physical fitness were assessed at baseline, after 8 weeks, and after 16 weeks of intervention. Feedback on the exercise program was collected after the final assessment. The results showed that a structured and supervised postpartum exercise program was feasible and safe and produced positive effects on selected maternal health and fitness parameters. These results will encourage a study protocol with a larger sample in order to prove its effectiveness, improve the guidelines for postpartum exercise, and incorporate this program into a routine healthcare setting.

Keywords: exercise; physical activity; postpartum; exercise prescription

Citation: Brites-Lagos, C.; Ramos, L.; Szumilewicz, A.; Santos-Rocha, R. Feasibility of a Supervised Postpartum Exercise Program and Effects on Maternal Health and Fitness Parameters—Pilot Study. *Healthcare* **2023**, *11*, 2801. https://doi.org/10.3390/healthcare11202801

Academic Editor: Masafumi Koshiyama

Received: 5 September 2023
Revised: 27 September 2023
Accepted: 19 October 2023
Published: 23 October 2023

1. Introduction

After delivery and after completing the anatomical and functional changes of pregnancy, the reversal process begins. Many of the physiological and morphological changes of pregnancy persist for four to six weeks after delivery [1]. The puerperium or postpartum period can last up to a year and is marked by a great emotional vulnerability for the woman, as it is a transition period. It implies profound changes at the physical, psychological, and sociological levels. Some physical and psychological health conditions resulting from pregnancy persist after this period [1]. Among the most prevalent issues arising from pregnancy are postpartum weight retention, musculoskeletal complications such as pelvic pain, low back pain, abdominal diastasis, pelvic floor problems, and psychological complications such as postpartum depression [2]. The benefits of exercise for the general population are well known [3]; however, specifically for this period, the benefits include recovery from childbirth, promotion of return to pre-pregnancy weight, reduced risk of developing future chronic conditions of health, improvement of fitness parameters, interactions between mother and baby, and social interactions [4]. In addition to these benefits, recent systematic reviews have shown that exercise in the postpartum period is effective in weight loss [5] and in reducing symptoms of depression [5–9], musculoskeletal disorders [10,11], and fatigue [12].

The U.S. Department of Health and Human Services (USDHHS) [13], the World Health Organization (WHO) [14], and the American College of Sports Medicine (ACSM) [15] issued guidelines on physical activity and sedentary behavior for postpartum women, reinforcing the need for national policies to include and monitor this subpopulation. Specific guidelines for physical activity during the postpartum period are embedded in the documents supported by the American College of Obstetricians and Gynecologists (ACOG) [1], the Brazilian Society of Cardiology (SBC) [16], and Sports Medicine Australia (SMA) [17]. One consensus paper supported by the International Olympic Committee (IOC) addresses exercise and postpartum in recreational and elite athletes [2]. Since these international guidelines refer to the late postpartum and lack specific content regarding the implementation of postnatal exercise programs, an updated textbook chapter about "Exercise Prescription and Adaptations in Early Postpartum" [18] was previously published.

Even so, there is a lack of public health policies related to exercise in both early and late postpartum periods, as well as a scarcity of studies in this area. Moreover, it is important to understand and implement effective strategies that promote exercise during the postpartum period.

Considering the importance of exercise in the postpartum period, the characteristics of women in this period, and the recommendations of the main international organizations, a specific exercise program was developed and validated [19].

The aim of this pilot study was to test the feasibility and analyze the effects of the "Active Mums" postpartum exercise program on maternal health and fitness parameters.

2. Materials and Methods

2.1. Participants

The sample consisted of Portuguese postpartum women.

Participants were recruited through social networks (Facebook and Instagram), as well as for convenience. The inclusion criteria were postpartum women between 18 and 45 years old, with no medical contraindications for the practice of physical exercise, and understood the Portuguese language. Exclusion criteria were any medical contraindications for physical exercise. During the recruitment stage, it was explained that the exercise program could be performed in person or online, according to their residence.

2.2. Equipment and Materials

To implement this exercise-based intervention aimed at improving a comprehensive set of maternal health and fitness parameters, the following equipment and materials were used to support the promotion and planning of the program, to implement the program, and to collect and process data. These tools were chosen because they may comprehensively evaluate outpatient postpartum recovery.

Planning of the program:

- Informed consent.
- Participant's form with sociodemographic and clinical information.
- Checklist and individual registration sheets.
- Computer to insert and analyze data.

Collecting and analyzing health, physical activity, and quality of life parameters:

- Physical Activity Readiness Questionnaire for Everyone (PAR-Q+) [20], adapted to Portuguese [21].
- International Physical Activity Questionnaire (IPAQ) [22], adapted to Portuguese [23].
- World Health Organization Quality of Life Questionnaire (WHOQOL-Bref) [24], adapted to Portuguese [25].
- Pelvic Girdle Questionnaire (PGQ) [26], adapted to Portuguese [27].
- Roland-Morris Disability Questionnaire (RMDQ) [28], adapted to Portuguese [29].
- Fatigue Assessment Scale (FAS) [30], adapted to Portuguese [31].

- Edinburgh Postpartum Depression Scale (EPDS) [32], adapted to Portuguese [33].
- Blood pressure and resting heart rate, using monitors [34].
- Body Mass Index (BMI), using a scale and a stadiometer [34].
- Waist-hip ratio, using a tape measure [34].
- Body composition, using a TANITA RD 545 Bioimpedance scale [34].

Collecting and analyzing functional and fitness parameters:

- Postural assessment [34], using a camera.
- Battery of functional tests [35–37], using a mattress, chair, measuring tape, and a stick.
- Cardiorespiratory fitness, using a treadmill and a heart rate monitor [38,39].
- Muscular resistance, using a mattress and a chair [40].
- Flexibility, using a mattress and a measuring tape [41].

Implementation of the exercise program:

- Sports equipment used in exercise sessions: mattresses, Swiss balls™, elastic bands, free weights, kettlebells, TRX™, softballs, ergometers, pulleys.
- Computer for online provision of exercise sessions.

Promotion of the exercise program

- Support exercise manual [42]
- Free download guides "Active Pregnancy Guide-Physical activity, nutrition, and sleep" [43] (also available in English [44]) and "Promotion of physical activity and exercise during pregnancy and postpartum. Health professionals guide" [45] (also available in English [46]), and free access videos posted on the YouTube channel "Gravidez Ativa–Active Pregnancy": https://www.youtube.com/channel/UC0Vyo okwc0mcQ5T70imtoNA/playlists (accessed on 21 September 2023).

2.3. Intervention Program

The postpartum exercise program consisted of a supervised, personalized training program carried out in-person or online, individually or in small groups of up to four participants. Each participant was invited to attend 3 of the available exercise sessions per week for 16 weeks (48 sessions in total). The exercise program was periodized into 3 weekly sessions of 60 min, referring to 3 mesocycles: adaptation (first two weeks), improvement (six to eight weeks), and maintenance. A typical 60 min session includes low-impact cardiorespiratory training, postural and functional/resistance training (core, lower and upper limbs, back), neuromotor training (balance and coordination), pelvic floor muscle training, and stretching, breathing, and relaxation exercises. However, the exercises are personalized, considering the needs and specificities of each woman, especially taking into account the type of birth, the early postpartum stage (0–6 weeks), and the late postpartum stage.

The intervention program is described in the study "Development and Validation of the Physical Exercise Program "Active Mums" for Postpartum Recovery. Qualitative study with application of the CReDECI-2 Guidelines" [19], and in the exercise manual "Prescription of Physical Exercise in the Postpartum Period" [42].

2.4. Tasks, Procedures and Protocols

For the promotion and implementation of the exercise program, the following procedures were followed.

Before the intervention:

- Preparation and promotion of the educational materials [42–46].
- Training qualified exercise professionals [47].
- Preparation of the physical practice space: authorization request to the person responsible for the "Lateral Performance" training space to carry out the program, as well as the use of the respective equipment.
- Promotion of the exercise program on social networks (Facebook and Instagram).
- Recruitment of participants by completing an online form.

- Explanation to the participants about the objective and pertinence of the study and the importance of their collaboration and availability in this research.
- Informed consent: statement informing the study objectives, information about the assessment sessions, confidentiality, participation and abandonment, damages related to the investigation, exclusion criteria, and disclaimer.
- Medical clearance for the postpartum exercise program.

Initial assessments at baseline:

- Initial individual interview to fill out the participant form to obtain sociodemographic and clinical information, as well as potential barriers and preferences regarding the intervention.
- Measurement of parameters of physical activity, quality of life, health, physical fitness, and functionality (between 6 and 8 weeks postpartum).
- Measurement of blood pressure and resting heart rate, as well as calculation of reserve and maximum heart rate.
- Assessment of BMI and body composition.
- Measurement of waist and hip perimeters, calculation of the waist-hip ratio.
- Postural assessment, with static observation of the anatomical references in the frontal and sagittal planes, verifying the symmetry in relation to the imaginary midline and photographing the same.
- Application of some tests from the Battery of Functional Aptitude Tests Dynamic Neuromuscular Stabilization (DNS), namely the seated diaphragm test, intra-abdominal pressure test, quadruped rockforward test, and qualitative recording of the results.
- Application of some tests from the Battery of Functional Aptitude Tests Functional Movement Screen™ (FMS™), such as shoulder mobility, active straight leg raise, deep squat, rotary stability, and qualitative recording of results and scores.
- Assessment of cardiorespiratory fitness using the Rockport One-Mile Fitness Walking Test and recording time, heart rate, and calculation of maximum oxygen consumption (VO_{2max}).
- Assessment of muscular endurance, counting the maximum number of arm extensions (push-ups) and applying the "Chair Stand Test," registering the maximum number of repetitions in 30 s.
- Assessment of flexibility, with the "V-sit and reach test."
- Intermediate (after 8 weeks) and final assessment (after another 8 weeks):
- Assessment of parameters of physical activity, quality of life, health, physical fitness, and functionality.

After the final assessment:

- A form was sent to participants to collect feedback on the level of satisfaction with the exercise program, containing the following questions inspired by Haakstad et al. [48]:

  1. Level of satisfaction with the program.
  2. Level of satisfaction with the instructor.
  3. Do you consider that exercise in a group environment was/would be more motivating than if it were individual?
  4. In what parameter (s) did you feel improvements in terms of your physical fitness?
  5. Have you changed your physical activity levels?
  6. Do you feel more energy for daily activities and less stress?
  7. Would you recommend this program to a friend?
  8. Would you participate in the program again after another pregnancy?
  9. Would you like to leave other comments?

2.5. Data Processing

Data were recorded in Excel, where descriptive statistics were performed.

2.6. Ethical Considerations

A group of healthy postpartum women was invited to participate in the pilot intervention free of charge. Participants were informed about the purpose and nature of the study, the potential benefits for future programs, participation requirements, and their right to withdraw from the study. They were free to provide feedback or not, without any consequences, and their feedback was anonymous. All women were informed and agreed to participate in the program, in physical and functional assessments, and in assessments by questionnaires. An informed consent was signed prior to participation. The educational materials produced by the research team were made available to the participants free of charge.

All exercise sessions and assessments were conducted by a qualified exercise physiologist. All clinical appointments were conducted by a gynecologist. The study was conducted in accordance with the Helsinki Declaration. This study is part of the study protocol that has been approved by the Ethics Committee of the Polytechnic Institute of Santarém, Portugal (approval number 9-2021-ESDRM).

3. Results

3.1. Participants Characterization

The sample consisted of 11 Portuguese postpartum women. At the beginning of the program, 8 of them were between 6 to 8 weeks postpartum, and 3 of them were between 9 and 12 weeks postpartum, aged between 24 and 37 years old (mean age of 31 years). Two participants had a cesarean delivery by medical decision, while the remaining nine had vaginal deliveries, with two using suction cups. Eight women were primiparous, while the other three were multiparous. Nine women were residents in Leiria, Portugal, and carried out the program in person at the Lateral Performance training studio in Leiria, while the rest were residents in Alcochete and in Pombal, attending the online program but making themselves available to do in-person assessments. They were all married or in a relationship. Regarding the education of the participants, nine had higher education, and two had completed high school.

Except for one participant, all women in the sample practiced regular exercise before pregnancy (e.g., bodybuilding, group classes, handball, personalized training). During pregnancy, only seven women exercised (e.g., walking, personalized training for pregnancy, group classes, Pilates), and before starting the program, only one declared that she practiced exercises at home. As for the pre-gestational Body Mass Index (BMI), the values varied between 20.8 and 34.4 kg/m^2 (mean of 25.2 kg/m^2), with three women being overweight and one obese. Self-reported maximal gestational weight gain ranged from 8 to 25 kg (mean of 20.2 kg).

3.2. Level of Physical Activity, Quality of Life, and Other Health Parameters

Table 1 contains the results obtained at the three collection times.

3.2.1. Level of Physical Activity

Before starting the program, the mean of habitual physical activity of the group was 559 min/week. After eight weeks, it increased to 810 min/week, and at the end of the program, it increased to 1127 min/week. This trend was expected since it includes the time spent on the exercise program. The section in which the most time was spent in a normal/usual week was "Housework, House Maintenance, and Caring for Family," with a weekly mean of 327 min before the start of the program, rising to 676 min in the second evaluation and ending with 504 min.

Table 1. Results obtained at the three collection times.

	1 (Baseline) Before the Intervention		2 (Intermediate) After 8 Weeks of Intervention		3 (Final) After 16 Weeks of Intervention	
	Mean	Standard Deviation	Mean	Standard Deviation	Mean	Standard Deviation
IPAQ (min/wk)	559	±520	810	±1171	1127	±669
WHOQOL-BREF (0–100)	79.54	±8.42	80.63	±9.59	82.17	±9.71
PGQ (0–75)	5.8	±7	1.5	±3.2	0.4	±0.7
RMDQ (0–24)	1.2	±1.7	1.4	±1.7	1	±1.4
FAS (10–50)	22.4	±5.4	21.6	±5.4	21.6	±5.4
EPDS (0–30)	6.3	±3.7	5	±3	5	±3
Weight (Kg)	68.95	±13.83	68.4	±14.88	67.1	±15.18
BMI (Kg/m^2)	26.8	±4.67	26.5	±5.02	26.1	±5.17
Waist-hip Ratio	0.8	±0.05	0.8	±0.06	0.79	±0.06
Fat Mass (%)	34.7	±5.59	33.5	±6.3	28.3	±10.3
Muscular Mass (Kg)	42.18	±5.6	42.55	±5.84	43.08	±6.17
SBP (mmHg)	107.33	±9.90	100	±8.51	97.78	±9.56
DBP (mmHg)	71.33	±4.95	68.44	±4.33	68.11	±6.27
Rest HR (beats/min)	76	±16.16	74.89	±14.59	74.89	±11.34
VO$_{2max}$ (mL/kg/min)	29.39	±7.08	31.93	±4.53	37	±6.24
Push-ups (number)	16	±9.88	19	±8.25	25	±11.41
"Chair Stand Test" (number)	14	±2.38	16	±1.85	19	±3.38
"V-Sit and Reach" (cm)	43	±7.04	45	±7.09	47	±7.73

When analyzing the sections separately, it was observed that "Job-related Physical Activity" is the section to which the participants did not dedicate any time since they were all on maternity leave, a situation that remained similar until the end of the program.

In the "Transportation Physical Activity" section, it was observed that, initially, four participants did not use physical activity as a means of transport to move from one place to another, while the rest used walking as a means of transport. There was a similar situation in the remaining evaluative moments. Despite this, there was an increase in the weekly travel time from 127 to 193, ending at 180 min.

In the physical activities carried out in the "Housework, House Maintenance, and Caring for Family" section, it was found that only one participant did not perform such activities, with a minimum duration of ten continuous minutes, while the rest were dedicated to domestic tasks with moderate intensity, before starting the program. This situation was not maintained, with all participants starting to carry out these moderate activities and increasing their weekly means, as already mentioned.

Regarding the section that considers "Recreation, Sport, and Leisure-time Physical Activity," it is noteworthy that only three of the participants did not perform these activities weekly with a minimum duration of 10 continuous minutes before the exercise program, totaling a mean time weekly of 106 min. Walking was the physical exercise adopted during their free time, with only one participant who revealed that she performed moderate and vigorous activities. In the second evaluation, and considering that they participated in the exercise program, the same section started to present 289 min/week of physical activity. At the end of the program, the participants stated that they had performed a mean of 443 min of physical activity per week, thus quadrupling their initial value.

The "Time Spent Sitting" section is related to the time that the participants remain seated during the week and at the weekend. All participants in this study spent a lot of time in the sitting position during the week and on the weekend before starting the program, but during the weekend, the mean was 324 min/day, while during the week, the mean was 345 min/day. In the next evaluation, the mean time for each one decreased to 257 and 197 min/day, respectively. The downward trend continued, and after the program, the participants reported a mean of 165 min/day on the weekend and 156 min/day during the week. Thus, on weekends, the time spent sitting was greater than on weekdays.

3.2.2. Quality of Life

In this parameter, there was no change between domains through the WHOQOL-Bref analysis. Before starting the program, the highest score in terms of quality of life, from 0 to 100, was observed in the social relationships domain (81.98 ± 10.42), followed by the environment domain (81.3 ± 11.6), the physical domain (79.4 ± 6.45), and the psychological domain (75.4 ± 12.4). The general mean was 79.54, indicating that the participants have a good quality of life. In the following evaluation, the domain with the highest score was the psychological (82.4 ± 10.1), followed by the physical domain (81.2 ± 9.15), the environment domain (80.8 ± 12.7), and the social relationships domain (78.1 ± 9.99), inverting the disposition of the first evaluation and achieving an overall mean of 80.6. At the end of the program, it was again the psychological domain that obtained the highest score (86 ± 9.74), followed this time by the social relationships domain (82.66 ± 12.27), the environment domain (80.28 ± 11.97), and the physical domain (79.72 ± 9.74), indicating an improvement in almost all domains compared to the first assessment. Thus, the mean was 82.17 (±9.71).

In the initial assessment, when asked about their quality of life, most participants (91%) reported that it was "good." There was a similar situation in the following evaluative moments. As for satisfaction with their health, most participants (91%) reported being satisfied, a trend that continued until the end of the program.

3.2.3. Pelvic Pain and Low Back Pain

Regarding pelvic pain, with the application of the PGQ, a large difference was noted among women, ranging from zero (four participants) to 18 points (out of a possible 75), with a mean of 5.8 (±7). None of the participants reported a critical situation. There was a significant improvement in the intermediate evaluation, reaching a mean of 1.5 (±3.2), in which six participants no longer had pain. At the end of the intervention, pelvic pain became almost non-existent, having had a mean of 0.4 (±0.7), with only two participants reporting some type of pain.

As for low back pain, with the application of the RMDQ, no differences were reported during and after the exercise intervention. The initial mean score before starting the exercise program was only 1.2 (±1.7) out of 24, with five women not reporting any type of pain in the daily activities described, and the maximum score obtained was 5. Similar scores were reported in the following evaluations.

3.2.4. Fatigue

In the FAS questionnaire, each item is evaluated on a five-point Likert scale, where 1 corresponds to "never," and 5 corresponds to "always," ranging from 10 points (less fatigue) to 50 points (more fatigue). Before starting the program, the mean was 22.4 (±5.4). In the following evaluations, the mean slightly decreased to 21.6 (±5.4). These FAS scores indicated that women were not very fatigued in any of the evaluation moments.

3.2.5. Depression

Before starting the program, in accordance with the EDPS, only two participants reached a score equal to or greater than 12, which is considered a risk factor for the onset of postpartum depression, according to Cox et al. [32]. The other women had scores between two and eight, considered within the normal range, thus indicating a lower risk for the onset of postpartum depression. The baseline mean was 6.3 (±3.7). In the following assessments, the mean decreased to 5 (±3); therefore, none of the participants presented a risk of postpartum depression since the maximum score was eight.

3.3. Functional and Health-Related Components of Physical Fitness
3.3.1. Body Composition

There was no change in the mean values of postpartum body weight between the first, second, and last evaluations. Of the participants in this study, five were overweight, while in subsequent assessments, there were only four, according to the BMI classification. Thus, their mean BMI at baseline was 26.8 kg/m^2 ($\pm$4.7), decreasing to 26.5 kg/m^2 ($\pm$5.0) after 8 weeks and ending with 26.1 kg/m^2 ($\pm$5.2).

Regarding other anthropometric variables, such as waist and hip perimeters and their respective ratios, all participants showed differences in the different assessments, losing a mean of 4.1 cm in the waist from the initial assessment to the second and losing a mean of 1.6 cm from the second to the last assessment. In the hip perimeters, the differences between the initial and the following evaluations were smaller, with a mean loss of 3.3 cm, and the same trend continued for the following evaluation, with a mean loss of 3 cm. In this way, the waist-hip ratio did not suffer major changes, with only one person having a ratio greater than 0.85 and maintaining it until the end of the program.

Regarding the body composition variables assessed by bioimpedance, it is important to highlight the percentage of fat mass and muscle mass. The percentage of fat mass registered favorable differences in the different evaluations, showing mean values of 34.7% ($\pm$5.6) at the initial moment, changing to 33.5% ($\pm$6.3) in the intermediate evaluation and ending with 28.3% ($\pm$10.3). Muscle mass, on the other hand, registered an increasing trend in the different evaluations, initially having a mean of 42.2% ($\pm$5.6), 42.6% ($\pm$5.8) in the second evaluation, and ending with 43.1% ($\pm$6.2).

3.3.2. Cardiorespiratory Fitness

Regarding blood pressure, all participants showed normal values [15]. However, we noticed an apparent decrease in mean values throughout the program, both in Systolic Blood Pressure (SBP) and Diastolic Blood Pressure (DBP).

As for mean rest heart rate, there was a slight decrease from the first to the second assessment and maintenance between the second and the last assessment.

The Rockport 1-mile fitness walking test was used to assess aerobic fitness. The predicted VO$_{2max}$ values in the initial evaluation had a mean of 29.4 ($\pm$7.1) mL/kg/min, a value considered "average" for this age group, according to the ACSM [15]. In the second evaluation, the mean value of predicted VO$_{2max}$ increased to 31.9 ($\pm$4.5) mL/kg/min, considered "good," and in the last evaluation, the mean increased to 37 ($\pm$6.2) mL/kg/min, reaching the threshold of "excellent".

3.3.3. Muscular Endurance and Flexibility

The participants performed arm extensions until they were unable to perform more executions or until they lost body alignment. The mean of push-ups before starting the program was 16 ($\pm$9.9) repetitions. In the following evaluation, the mean was 19 ($\pm$8.3) repetitions, ending, in the final evaluation, with 25 ($\pm$11.4) repetitions. The "Chair Stand Test" was performed for 30 s, obtaining means of 14 ($\pm$2.4), 16 ($\pm$1.9) and 19 ($\pm$3.4) repetitions, respectively. In the "V-Sit and Reach" test, the participants performed a mean of 43 ($\pm$7.0) cm, 45 ($\pm$7.1) cm, and 47 ($\pm$7.7) cm in the first, second, and third evaluations, respectively.

These scores are considered "normal" for this age group.

3.3.4. Postural Assessment

A qualitative postural assessment was performed. In the postural analysis, in the anterior view, the puerperal women presented a good alignment of the anatomical references in the initial evaluation. Regarding the position of the visually evaluated body segments, the majority of the participants had the head aligned, knees and iliac crests aligned, and feet supinated. Most of them also had a smaller earlobe-shoulder distance on the side of their dominant hand. In the following evaluations, the postural pattern was maintained.

In the sagittal plane, in the initial evaluation, most of the participants had a forward head, cervical lordosis, protruding shoulders, hyperkyphosis, lumbar hyperlordosis, hip joint in anteversion, knees in hyperextension, and misaligned external malleoli. In this plane, several differences were noticed in the following evaluations, and most participants (91%) ended the program with the head aligned, the neck normal, the shoulders normal, without hyperkyphosis, lumbar spine and iliac crest normal, coxo-femoral joint normal, and knees and malleolus external aligned.

3.3.5. Functional Assessment

A qualitative DNS functional assessment was performed. In the "Seated Diaphragm test," the women had the ability to expand the abdominal wall and perform a symmetrical activation without elevating the ribs and shoulders and without losing the verticality of the spine. A similar situation was observed in the following evaluations.

Regarding the "Intra-abdominal Pressure test," in the initial evaluation, most participants (64%) presented a hyperextension of the lumbar area, as well as a weak activation of the abdominal wall with bulging and a slight diastasis of the rectus abdominis. In the following evaluation, only three women had the same condition. At the end of the program, all participants performed correct activation of the abdominal wall and stabilization of the lower back.

Regarding the "Quadruped Rockforward test," in the initial evaluation, most of the participants (82%) presented a cervical hyperextension, bringing the head to reclining, an unequal load of the palms of the hands, a scapular elevation, and anterior pelvic inclination. In subsequent evaluations, all participants were able to activate their abdominal, back, diaphragm, and pelvic muscles, straightening their spine.

Qualitative and quantitative FMS functional assessment was performed. Regarding the "Shoulder mobility test," in the initial evaluation, the wrists of most women were at a mean of 8.5 cm apart on the right and 12.8 cm apart on the left, changing to 6.3 and 9.4, respectively, and then to 6.3 and 8.3. They did not report any pain.

In the "Deep Squat test," six women (55%) were able to perform the movement without a plate, with the hips parallel, the tibia and trunk parallel, the knees aligned over the toes, symmetrical, without noticeable lumbar flexion, without the feet turning externally, without taking the heels off the ground and without pain; two performed the movement with pain. In the following evaluation, only one participant could not perform the movement correctly, and none of them had pain. A similar situation was observed in the final evaluation.

In the "Straight leg raise test," only one woman was unable to perform the correct movement. All the others were able to perform the movement, keeping the opposite hip neutral, the toes pointed upwards, the opposite knee in contact with the board, and without pain. A similar situation is in the following evaluations.

In the rotary stability test, most women performed the exercise with the spine parallel to the plate, hips parallel to the floor, knees and elbows aligned with the plate, the support ankle dorsiflexed, touching the elbow to the knee, and without pain, but only contralateral. At the end of the program, only one participant managed to do homolateral.

3.4. Satisfaction with the Program

As for the level of satisfaction with the program, 91% of the participants reported being very satisfied, while one reported being satisfied. All participants reported being very satisfied with the exercise professional. In total, 45% agreed that it is more motivating to exercise in a small group than in an individual setting, while 45% strongly agreed and one did not agree. All participants reported an improvement in their physical fitness, namely in terms of strength (81.8%), cardiorespiratory fitness (54.6%), flexibility (36.4%), posture (81.8%), body composition (72.7%), and balance and coordination (45.5%). 91% of women reported increased levels of physical activity, while one reported that she was equally active. All participants felt more energy for daily activities and less stress, reporting that they would recommend the training program and confirming their participation in a possible future postpartum period. There were no substantial differences in the opinions and level of satisfaction between online and in-person participants.

4. Discussion

The aim of this pilot study was to test the feasibility and analyze the effects of the "Active Mums" supervised postpartum exercise program on selected maternal health and fitness parameters.

At the psychological level, it is known that postpartum depression is experienced by approximately 20% of women; however, up to 50% of women experience high levels of depressive symptoms during this period [49]. When analyzing the depression results, we found that, initially, only two women scored levels that could be related to postpartum depression, a situation that, over the course of the program, disappeared. These results are in accordance with Carter et al. [8], who demonstrated that exercise is effective in reducing symptoms of depression in postpartum women.

The systematic review of Wilson et al. [50] concluded that there is a strong correlation between fatigue and depressive symptoms among women in the first two years after childbirth. However, our study did not confirm this outcome, as apparently, the levels of fatigue were higher than those on the depression scale, in addition to the fact that there were no significant decreases throughout the program, as we found on the scale of depression.

When analyzing the pain level of the participants through the RMDQ and the PGQ, we verified that in the first one, no differences were observed between the different evaluations and that the mean found was 1 in relation to the scores of functional capacities in the RMDQ, which can range up to 24. This situation led us to question the validity of the RMDQ in the immediate postpartum period. In the PGQ, although the participants did not appear to have a high degree of disability, some women reported pelvic pain in the initial evaluation, which decreased throughout the program until becoming scarce at the end of it. These results are in line with existing literature that has reported a positive association between exercise and decreased pain in the lumbar and pelvic region [50–52].

In the current literature, there is a correlation between postpartum depression and quality of life [53,54]. The results of the WHOQOL-Bref showed that most of the women participating in this study had a good quality of life right from the start and that it continued to improve throughout the program, as happened with the symptoms of depression, apparently supporting this correlation. Our results also support the findings of Yang and Chen [55] regarding the impact of an exercise program on fatigue.

Regarding the physical parameters, with regard to changes in body mass and BMI, we found that there was a decrease in values, with almost half of the participants returning to pre-pregnancy values. These results were not expected, according to Meyers and Hong [56], who concluded that exercise had no significant effect on weight loss in the short or long term in breastfeeding women. The same study [56] also reported findings about exercise and body composition, with exercise reducing body fat and preserving fat-free mass.

In this pilot study, there was a loss of fat mass of about 1.2% after 8 weeks and about 6.4% after 16 weeks of intervention, and an increase in muscle mass of 0.4% in the second evaluation and 0.9% after 16 weeks of intervention, which may explain the slight loss of body weight throughout the program. The outcome of the four participants who were still overweight at the end of the program was related to gestational weight gain, which was higher than the recommendations based on pre-pregnancy BMI, according to the Institute of Medicine. However, Dalfra et al. [57] suggested that appropriate gestational weight gain should be personalized considering the three obesity classes. As for the other anthropometric measures used, such as waist circumference, these showed a decrease by the end of the program. The waist-hip ratio shown before the start of the program was within healthy values, except for one participant, and this ratio remained unchanged until the end of the program.

All the changes already mentioned can be potentially explained by the exercise program implemented and the consequent increase in physical activity. Despite the fact that postpartum women are still recommended not to perform physical activity during the puerperal period, as confirmed by the participants of the program, the results of the level of physical activity using IPAQ were positive. Thus, the level of physical activity before starting the program was very low, limited to "Housework, House Maintenance, and Caring for Family." Regarding the mean time spent per week on physical activity, there was an exponential increase from that initial period until the end of the program in the sections "Housework, House Maintenance, and Caring for Family" and in the section "Recreation, Sport, and Leisure-time Physical Activity." "Sitting time" is an indicator of a sedentary lifestyle in IPAQ, having decreased throughout the program, both during the week and at the weekend. Thus, our participants progressively met the recommendations of the WHO [14] and ACOG [1], which include: postpartum women should start by doing small amounts of physical activity and gradually increase frequency, intensity, and duration over time; performing at least 150 min of moderate-intensity aerobic physical activity during the week, as well as incorporating a variety of aerobic and muscle-strengthening activities; adding gentle stretching may also be beneficial; and limiting the amount of time being sedentary. Therefore, our program seemed to positively influence the physical activity habits of these participants in their daily lives.

In terms of musculoskeletal parameters, postpartum postural analysis proves to be extremely important, as postural changes are frequent among puerperal women, both due to the gait biomechanical compensations during pregnancy [58–60] and potentially due to the tensions and overloads generated with the baby care and breastfeeding. In our study, before starting the program, all mothers presented a similar postural profile, with changes in pelvic tilt and lumbar and thoracic curvatures being the most noticeable, in line with the study by Gilleard et al. [61]. Biviá-Roig et al. [62], when analyzing the lumbar curvature in pregnant, non-pregnant, and postpartum women, hypothesized that it is the muscular responses, and not the curvatures, that are altered by pregnancy. This finding may encourage further research on whether the increase in tension in the pelvic muscles is due to the progressive displacement of the center of gravity during pregnancy and the abdominal weakness due to the diastasis of the rectus abdominis.

Postural compensations resulting from the dominance of the upper limb were also noted, such as a rise in the shoulder on that side in the frontal plane. During the program, strengthening and/or stretching exercises were used and combined, which made it possible to notice great postural differences at the end of the program. Before starting the program, such postural and muscular weaknesses were noticed when the DNS and FMS™ functional tests were applied. At the DNS level, before starting the program, the greatest fragility was noted in the abdominal muscles, while in the FMS™, weaknesses and asymmetries in movement were noted. After a qualitative analysis of the same, we concluded that there was a very positive progression in the functionality of the participants, mainly between the first and the second evaluations. Flexibility exercises can also improve postural stability, especially when combined with resistance exercises. In this line, these types of exercise were inserted into the program, and the respective assessments were made, leading to positive outcomes in flexibility and muscular resistance. This observation may encourage further research regarding the effectiveness of neuromotor exercise and the inclusion of this type of exercise in the guidelines for exercise prescriptions for postpartum women.

The functional state of the respiratory, cardiovascular, and musculoskeletal systems depends on cardiorespiratory fitness, which is related to health [15]. The cardiorespiratory fitness of the participants obtained very positive results, progressing from an "average" condition in the initial assessment to an almost "excellent" condition at the end of the exercise program, according to the ACSM's VO_{2max} classification [15]. In short, it was demonstrated that the fitness parameters showed an improvement, which is consistent with the available research regarding the importance of exercise programs for the postpartum period.

As for satisfaction, it was evident that the program met or exceeded the participants' expectations, emphasizing the importance of group sessions, the improvement of physical fitness, and the satisfaction with the exercise professional. The levels of satisfaction highlight the importance of the supervision of pre and postnatal exercise programs by qualified exercise professionals, preferably involved in multidisciplinary teams [63]. Supervision of exercise programs is recommended to ensure proper technique, provide confidence, and ensure the progression of appropriate levels of intensity and complexity. The exercise professional should provide regular feedback, positive reinforcement, and behavioral strategies to enhance adherence. The exercise professional should also provide the safest possible training and testing environment, prevent exercise-related emergencies, and be familiar with the safety and emergency procedures available at the fitness setting where the exercise program is delivered.

Regarding the feasibility of the postpartum exercise program both in person and online, it was demonstrated to be safe, tailored, motivating, and a contribution to improving the quality of life, functional capacity, and physical fitness of postpartum women, as shown by either qualitative or quantitative tools such as questionnaires and physical fitness tests performed. These results will encourage a study protocol with a larger sample in order to prove its effectiveness and incorporate this program into a routine healthcare setting.

The main limitation of this study is the small number of participants, preventing extrapolating conclusions, and the non-use of inferential statistics to support the observed trends. Moreover, it is recognized that there are various methodological and ethical constraints when working with this population, as well as a lack of validated tools. As for strengths, we highlight the real context approach used, that is, the implementation of an exercise intervention, combining the advantages of a program for a specific population with the effectiveness of controlled exercise and a progressive stimulus adapted to the level of physical fitness of each participant, guaranteeing their adherence. Due to the number of variables that are involved, this pilot test involves a complexity that can provide guidelines for the preparation of an intervention study protocol with a larger sample.

5. Conclusions

This pilot study allowed us to test a tailored supervised postpartum exercise program performed for 16 weeks. This postpartum exercise program was feasible, safe, and motivating and contributed to improving selected parameters of quality of life, physical activity, health, functional capacity, and physical fitness of postpartum women. These results provided guidance to develop a study protocol with a larger sample in order to prove its effectiveness, improve the guidelines for postpartum exercise, and incorporate this program into a routine healthcare setting.

Author Contributions: Conceptualization, R.S.-R. and A.S.; methodology, R.S.-R., C.B.-L., L.R., and A.S.; formal analysis, R.S.-R., C.B.-L., L.R., and A.S.; data curation, R.S.-R. and C.B.-L.; writing—original draft preparation, R.S.-R. and C.B.-L.; writing—review and editing, R.S.-R. and A.S.; supervision, R.S.-R. and A.S.; and funding acquisition, R.S.-R. and A.S. All authors have read and agreed to the published version of the manuscript.

Funding: This research received no external funding. This work was developed as part of the NEPPE project: 'The New Era of Pre- and Postnatal Exercise-training for instructors and trainers of various forms of physical activity in the field of online provision of exercise for pregnant and postpartum women,' financially supported by the Polish National Academic Exchange Agency (NAWA) within the SPINNAKER program (No. PPI/SPI/2020/1/00082/DEC/02), coordinated by Anna Szumilewicz. Rita Santos-Rocha was partly supported by a national grant through the FCT—Fundação para a Ciência e Tecnologia within the unit I&D 447 (UIDB/00447/2020)—CIPER. Rita Santos-Rocha, Liliana Ramos and Carla Brites-Lagos were partly supported by a national grant through Instituto Português do Desporto e Juventude—Programa Nacional de Desporto para Todos 2022—GRAVIDEZ ATIVA: Promoção da Atividade Física, Exercício e Desporto durante a Gravidez e Pós-parto (CP/216/DDT/2022).

Institutional Review Board Statement: This study is part of the study protocol that has been approved by the Ethics Committee of the Polytechnic Institute of Santarém, Portugal (approval number 9-2021-ESDRM).

Informed Consent Statement: Informed consent was obtained from all subjects involved in the study.

Data Availability Statement: The data presented in this study are available on request from the corresponding author. E-BOOK EXERCISE PRESCRIPTION IN THE POSTPARTUM PERIOD: [Portuguese; ISBN: 978-989-8768-49-0] Available at: https://www.researchgate.net/publication/372866386_PRE SCRICAO_DO_EXERCICIO_NO_PERIODO_POS-PARTO_Exercise_Prescription_in_the_Postpartu m_Portuguese (accessed on 21 September 2023). E-BOOK ACTIVE PREGNANCY GUIDE: [Portuguese; ISBN: 978-989-8768-27-8] Available at: https://www.researchgate.net/publication/3403157 48_GUIA_da_GRAVIDEZ_ATIVA_-_Atividade_Fisica_Exercicio_Fisico_Desporto_e_Saude_na_Grav idez_e_Pos-Parto_Active_Pregnancy_Guide_Physical_Activity_Exercise_Sport_and_Health_in_Pre gnancy_and_Post-partum_Port (accessed on 21 September 2023); [Spanish; ISBN: 978-989-8768-50-6] Available at: https://www.researchgate.net/publication/372907216_Guia_Embarazo_Activo_-_Ac tividad_Fisica_Nutricion_y_Sueno_Active_Pregnancy_Guide_Spanish (accessed on 21 September 2023); [English; ISBN: 978-989-8768-50-6] Available at: https://www.researchgate.net/publication /370874994_ACTIVE_PREGNANCY_GUIDE_-Physical_activity_nutrition_and_sleep (accessed on 21 September 2023). E-BOOK PROMOTION OF PHYSICAL ACTIVITY AND EXERCISE DURING PREGNANCY AND POSTPARTUM. HEALTH PROFESSIONALS' GUIDE: [Portuguese; ISBN: 978-989-8768-36-0] Available at: https://www.researchgate.net/publication/364808045_PROMOCAO_ DA_ATIVIDADE_FISICA_E_DO_EXERCICIO_DURANTE_A_GRAVIDEZ_E_O_POS-PARTO_Guia _para_Profissionais_de_Saude (accessed on 21 September 2023); [English; ISBN: 978-989-8768-42-1] Available at: https://www.researchgate.net/publication/364806085_PROMOTION_OF_PHYSICAL _ACTIVITY_AND_EXERCISE_DURING_PREGNANCY_AND_POSTPARTUM_-_Health_Profession als\T1\textquoteright_Guide (accessed on 21 September 2023).

Acknowledgments: IPDJ—Instituto Português do Desporto e da Juventude (Portuguese Institute of Sport and Youth) for their support in the production of the educational materials. Grant numbers: CP/216/DDT/2022, CP/536/DDT/2020, and CP/704/DDT/2019.

Conflicts of Interest: The authors declare no conflict of interest.

References

1. American College of Obstetricians and Gynecologists. Physical Activity and Exercise During Pregnancy and the Postpartum Period: ACOG Committee Opinion, Number 804. *Obstet. Gynecol.* **2020**, *135*, e178–e188. [CrossRef] [PubMed]
2. Bø, K.; Artal, R.; Barakat, R.; Brown, W.J.; Davies, G.A.L.; Dooley, M.; Evenson, K.R.; Haakstad, L.A.H.; Kayser, B.; Kinnunen, T.I.; et al. Exercise and pregnancy in recreational and elite athletes: 2016/17 evidence summary from the IOC Expert Group Meeting, Lausanne. Part 3-exercise in the postpartum period. *Br. J. Sports Med.* **2017**, *51*, 1516–1525. [CrossRef] [PubMed]
3. Ruegsegger, G.N.; Booth, F.W. Health Benefits of Exercise. *Cold Spring Harb. Perspect. Med.* **2018**, *8*, a029694. [CrossRef]
4. Dipietro, L.; Evenson, K.R.; Bloodgood, B.; Sprow, K.; Troiano, R.P.; Piercy, K.L.; Vaux-Bjerke, A.; Powell, K.E. 2018 Physical Activity Guidelines Advisory Committee. Benefits of Physical Activity during Pregnancy and Postpartum: An Umbrella Review. *Med. Sci. Sports Exerc.* **2019**, *51*, 1292–1302. [CrossRef]
5. Saligheh, M.; Hackett, D.; Boyce, P.; Cobley, S. Can exercise or physical activity help improve postnatal depression and weight loss? A systematic review. *Arch. Womens Ment. Health* **2017**, *20*, 595–611. [CrossRef] [PubMed]
6. Poyatos-León, R.; García-Hermoso, A.; Sanabria-Martínez, G.; Álvarez-Bueno, C.; Cavero-Redondo, I.; Martínez-Vizcaíno, V. Effects of exercise-based interventions on postpartum depression: A meta-analysis of randomized controlled trials. *Birth* **2017**, *44*, 200–208. [CrossRef] [PubMed]
7. Pritchett, R.V.; Daley, A.J.; Jolly, K. Does aerobic exercise reduce postpartum depressive symptoms? a systematic review and meta-analysis. *Br. J. Gen. Pract.* **2017**, *67*, e684–e691. [CrossRef] [PubMed]
8. Carter, T.; Bastounis, A.; Guo, B.; Jane Morrell, C. The effectiveness of exercise-based interventions for preventing or treating postpartum depression: A systematic review and meta-analysis. *Arch. Womens Ment. Health* **2019**, *22*, 37–53. [CrossRef]
9. Pentland, V.; Spilsbury, S.; Biswas, A.; Mottola, M.F.; Paplinskie, S.; Mitchell, M.S. Does Walking Reduce Postpartum Depressive Symptoms? A Systematic Review and Meta-Analysis of Randomized Controlled Trials. *J. Womens Health* **2022**, *31*, 555–563. [CrossRef]
10. Woodley, S.J.; Lawrenson, P.; Boyle, R.; Cody, J.D.; Mørkved, S.; Kernohan, A.; Hay-Smith, E.J.C. Pelvic floor muscle training for preventing and treating urinary and faecal incontinence in antenatal and postnatal women. *Cochrane Database Syst. Rev.* **2020**, *5*, CD007471. [CrossRef]
11. Tseng, P.C.; Puthussery, S.; Pappas, Y.; Gau, M.L. A systematic review of randomised controlled trials on the effectiveness of exercise programs on Lumbo Pelvic Pain among postnatal women. *BMC Pregnancy Childbirth* **2015**, *15*, 316. [CrossRef]
12. Liu, N.; Wang, J.; Chen, D.D.; Sun, W.J.; Li, P.; Zhang, W. Effects of exercise on pregnancy and postpartum fatigue: A systematic review and meta-analysis. *Eur. J. Obstet. Gynecol. Reprod. Biol.* **2020**, *253*, 285–295. [CrossRef]
13. U.S. Department of Health and Human Services. *Physical Activity Guidelines for Americans*, 2nd ed.; U.S. Department of Health and Human Services: Washington, DC, USA, 2018. Available online: https://health.gov/sites/default/files/2019-09/Physical_ Activity_Guidelines_2nd_edition.pdf#page=79 (accessed on 21 September 2023).

14. Bull, F.C.; Al-Ansari, S.S.; Biddle, S.; Borodulin, K.; Buman, M.P.; Cardon, G.; Carty, C.; Chaput, J.P.; Chastin, S.; Chou, R.; et al. World Health Organization 2020 guidelines on physical activity and sedentary behaviour. *Br. J. Sports Med.* **2020**, *54*, 1451–1462. [CrossRef]

15. American College of Sports Medicine. *ACSM's Guidelines for Exercise Testing and Prescription*, 11th ed.; Wolters Kluwer Health: Alphen aan den Rijn, The Netherlands, 2021; ISBN 978-1975150181.

16. Campos, M.D.S.B.; Buglia, S.; Colombo, C.S.S.S.; Buchler, R.D.D.; Brito, A.S.X.; Mizzaci, C.C.; Feitosa, R.H.F.; Leite, D.B.; Hossri, C.A.C.; Albuquerque, L.C.A.; et al. Position Statement on Exercise During Pregnancy and the Post-Partum Period—2021. *Arq. Bras. Cardiol.* **2021**, *117*, 160–180. (In English and Portuguese) [CrossRef] [PubMed]

17. Brown, W.J.; Hayman, M.; Haakstad, L.A.H.; Lamerton, T.; Mena, G.P.; Green, A.; Keating, S.E.; Gomes, G.A.O.; Coombes, J.S.; Mielke, G.I. Australian guidelines for physical activity in pregnancy and postpartum. *J. Sci. Med. Sport* **2022**, *25*, 511–519. [CrossRef] [PubMed]

18. Santos-Rocha, R.; Szumilewicz, A. Exercise Prescription and Adaptations in Early Postpartum. In *Exercise and Physical Activity during Pregnancy and Postpartum*; Santos-Rocha, R., Ed.; Springer: Cham, Switzerland, 2022. [CrossRef]

19. Brites-Lagos, C.; Maranhão, C.; Szumilewicz, A.; Santos-Rocha, R. Development and Validation of the Physical Exercise Program "Active Mums" for Postpartum Recovery. Qualitative Study with Application of the CReDECI-2 Guidelines. *BMC Pregnancy Childbirth* **2023**. [CrossRef]

20. The PAR-Q+ Physical Activity Readiness Questionnaire for Everyone 2023. Available online: http://eparmedx.com/wp-content/uploads/2022/12/ParQ-Plus-Jan-2023-Image-File.pdf (accessed on 21 September 2023).

21. Schwartz, J.; Oh, P.; Takito, M.Y.; Saunders, B.; Dolan, E.; Franchini, E.; Rhodes, R.E.; Bredin, S.S.D.; Coelho, J.P.; Dos Santos, P.; et al. Translation, Cultural Adaptation, and Reproducibility of the Physical Activity Readiness Questionnaire for Everyone (PAR-Q+): The Brazilian Portuguese Version. *Front. Cardiovasc. Med.* **2021**, *8*, 712696. [CrossRef]

22. Sember, V.; Meh, K.; Sorić, M.; Starc, G.; Rocha, P.; Jurak, G. Validity and Reliability of International Physical Activity Questionnaires for Adults across EU Countries: Systematic Review and Meta Analysis. *Int. J. Environ. Res. Public Health* **2020**, *17*, 7161. [CrossRef]

23. Campaniço, H.M.P.G. Validade Simultânea do Questionário Internacional de Actividade Física Através da Medição Objectiva da Actividade Física por Actigrafia Proporcional [Concurrent Validity of the International Physical Activity Questionnaire Through Objective Measurement of Physical Activity by Proportional Actigraphy]. Master's Thesis, Universidade de Lisboa, Lisboa, Portugal, 2016. (In Portuguese)

24. The WHOQOL Group. Development of the World Health Organization WHOQOL-BREF quality of life assessment. *Psychol. Med.* **1998**, *28*, 551–558. [CrossRef]

25. Vaz Serra, A.; Canavarro, M.C.; Simões, M.; Pereira, M.; Gameiro, S.; Quartilho, M.J.; Rijo, D.; Carona, C.; Paredes, T. Estudos psicométricos do instrumento de avaliação da qualidade de vida da Organização Mundial de Saúde (WHOQOL-Bref) para Português de Portugal. *Psiquiatr. Clínica* **2006**, *27*, 41–49. (In Portuguese)

26. Stuge, B.; Garratt, A.; Krogstad Jenssen, H.; Grotle, M. The pelvic girdle questionnaire: A condition-specific instrument for assessing activity limitations and symptoms in people with pelvic girdle pain. *Phys. Ther.* **2011**, *91*, 1096–1108. [CrossRef] [PubMed]

27. Silva, A.F.F. Adaptação Cultural e Linguística para a População Portuguesa do Instrumento de Medição "The Pelvic Girdle Questionnaire" [Cultural and Linguistic Adaptation for the Portuguese Population of the Measuring Instrument "The Pelvic Girdle Questionnaire"]. Master's Thesis, Santa Casa da Misericórdia de Lisboa, Escola Superior de Saúde do Alcoitão, Lisboa, Portugal, 2019. (In Portuguese)

28. Roland, M.; Morris, R. A study of the natural history of back pain. Part I: Development of a reliable and sensitive measure of disability in low-back pain. *Spine* **1983**, *8*, 141–144. [CrossRef] [PubMed]

29. Monteiro, J.; Faísca, L.; Nunes, O.; Hipólito, J. Questionário de incapacidade de Roland Morris: Adaptação e validação para a população portuguesa com lombalgia [Roland Morris Disability Questionnaire: Adaptation and validation for the Portuguese population with low back pain]. *Acta Médica Port.* **2010**, *23*, 761–766. (In Portuguese)

30. Michielsen, H.J.; De Vries, J.; Van Heck, G.L.; Van de Vijver, F.J.; Sijtsma, K. Examination of the Dimensionality of Fatigue: The Construction of the Fatigue Assessment Scale (FAS). *Eur. J. Psychol. Assess.* **2004**, *20*, 39.

31. Alves, B.A. Validação da Fatigue Assessment Scale para a População Portuguesa [Validation of the Fatigue Assessment Scale for the Portuguese Population]. Dissertação de Mestrado, Universidade Lusófona de Humanidades e Tecnologia, Escola de Psicologia e Ciências da Vida, Lisboa, Portugal, 2017. (In Portuguese)

32. Cox, J.L.; Holden, J.M.; Sagovsky, R. Detection of postnatal depression. Development of the 10-item Edinburgh Postnatal Depression Scale. *Br. J. Psychiatry* **1987**, *150*, 782–786. [CrossRef]

33. Areias, M.E.; Kumar, R.; Barros, H.; Figueiredo, E. Comparative incidence of depression in women and men, during pregnancy and after childbirth. Validation of the Edinburgh Postnatal Depression Scale in Portuguese mothers. *Br. J. Psychiatry* **1996**, *169*, 30–35. [CrossRef]

34. American College of Sports Medicine. *ACSM's Resources for the Exercise Physiologist*, 3rd ed.; Wolters Kluwer Health: Alphen aan den Rijn, The Netherlands, 2021; ISBN 978-1975153168.

35. Mahdieh, L.; Zolaktaf, V.; Karimi, M.T. Effects of dynamic neuromuscular stabilization (DNS) training on functional movements. *Hum. Mov. Sci.* **2020**, *70*, 102568. [CrossRef]

36. Cook, G.; Burton, L.; Hoogenboom, B. Pre-participation screening: The use of fundamental movements as an assessment of function—Part 1. *N. Am. J. Sports Phys. Ther.* **2006**, *1*, 62–72.

37. Cook, G.; Burton, L.; Hoogenboom, B. Pre-participation screening: The use of fundamental movements as an assessment of function—Part 2. *N. Am. J. Sports Phys. Ther.* **2006**, *1*, 132–139.

38. Cureton, K.J.; Sloniger, M.A.; O'Bannon, J.P.; Black, D.M.; McCormack, W.P. A generalized equation for prediction of VO2peak from 1-mile run/walk performance. *Med. Sci. Sports Exerc.* **1995**, *27*, 445–451. [CrossRef]

39. Pober, D.M.; Freedson, P.S.; Kline, G.M.; McInnis, K.J.; Rippe, J.M. Development and validation of a one-mile treadmill walk test to predict peak oxygen uptake in healthy adults ages 40 to 79 years. *Can. J. Appl. Physiol.* **2002**, *27*, 575–589. [CrossRef] [PubMed]

40. Rikli, R.E.; Jones, C.J. *Senior Fitness Test Manual*, 2nd ed.; Human Kinetics: Champaign, IL, USA, 2012; ISBN 978-1450411189.

41. Cuberek, R.; Machová, I.; Lipenská, M. Reliability of V sit-and-reach test used for flexibility self-assessment in females. *Acta Gymnica* **2013**, *43*, 35–39. [CrossRef]

42. Santos Rocha, R.; Maranhão, C.; Brites Lagos, C.; Ramos, L. *Prescrição do Exercício Físico no Período Pós-Parto [Exercise Prescription in the Postpartum Period]*; Escola Superior de Desporto de Rio Maior—Instituto Politécnico de Santarém: Rio Maior, Portugal, 2023; ISBN 978-989-8768-49-0. Available online: https://www.researchgate.net/publication/372866386_PRESCRICAO_DO_EXERCIC IO_NO_PERIODO_POS-PARTO_Exercise_Prescription_in_the_Postpartum_Portuguese (accessed on 21 September 2023). (In Portuguese)

43. Santos-Rocha, R. *Guia da Gravidez Ativa. Atividade Física, Exercício, Desporto e Saúde na Gravidez e Pós-Parto [Active Pregnancy Guide]*; Escola Superior de Desporto de Rio Maior—Instituto Politécnico de Santarém: Rio Maior, Portugal, 2020; ISBN 978-989-8768-26-1/978-989-8768-27-8 (ebook). Available online: https://www.researchgate.net/publication/340315748_GUIA_ da_GRAVIDEZ_ATIVA_-_Atividade_Fisica_Exercicio_Fisico_Desporto_e_Saude_na_Gravidez_e_Pos-Parto_Active_Pregnanc y_Guide_Physical_Activity_Exercise_Sport_and_Health_in_Pregnancy_and_Post-partum_Port (accessed on 21 September 2023). (In Portuguese)

44. Santos-Rocha, R.; Szumilewicz, A.; Wegrzyk, J.; Hyvärinen, M.; Silva, M.R.G.; Jorge, R.; Oviedo-Caro, M.Á. *Active Pregnancy Guide—Physical Activity, Nutrition, and Sleep*; Escola Superior de Desporto de Rio Maior—Instituto Politécnico de Santarém: Rio Maior, Portugal, 2023; ISBN 978-989-8768-50-6. Available online: https://www.researchgate.net/publication/370874994_ACTIV E_PREGNANCY_GUIDE_-Physical_activity_nutrition_and_sleep (accessed on 21 September 2023).

45. Santos-Rocha, R.; Silva, M.R.G.; Dias, H.; Jorge, R. *Promoção da Atividade Física e do Exercício Durante a Gravidez e o Pós-Parto—Guia para Profissionais de Saúde [Promotion of Physical Activity and Exercise during Pregnancy and Postpartum—Health Professionals' Guide]*; Instituto Politécnico de Santarém—Escola Superior de Desporto de Rio Maior: Rio Maior, Portugal, 2022; ISBN 978-989-8768-35- 3/978-989-8768-36-0 (ebook). Available online: https://www.researchgate.net/publication/364808045_PROMOCAO_DA_ATI VIDADE_FISICA_E_DO_EXERCICIO_DURANTE_A_GRAVIDEZ_E_O_POS-PARTO_Guia_para_Profissionais_de_Saude (accessed on 21 September 2023). (In Portuguese)

46. Santos-Rocha, R.; Szumilewicz, A.; Wegrzyk, J.; Hyvärinen, M.; De Labrusse, C.; Schlappy, F.; Silva, M.R.G.; Oviedo-Caro, M.Á. *Promotion of Physical Activity and Exercise during Pregnancy and Postpartum—Health Professionals' Guide*; Instituto Politécnico de Santarém—Escola Superior de Desporto de Rio Maior: Rio Maior, Portugal, 2022; ISBN 978-989-8768-42-1. Available online: https://www.researchgate.net/publication/364806085_PROMOTION_OF_PHYSICAL_ACTIVITY_AND_EXERCISE _DURING_PREGNANCY_AND_POSTPARTUM_-_Health_Professionals\T1\textquoteright_Guide (accessed on 21 September 2023).

47. Santos-Rocha, R.; Pajaujiene, S.; Szumilewicz, A. Active Pregnancy: Workshop on Promotion of Physical Activity in Pregnancy for Exercise Professionals. *J. Multidiscip. Healthc.* **2022**, *15*, 2077–2089. [CrossRef] [PubMed]

48. Haakstad, L.A.; Sanda, B.; Vistad, I.; Sagedal, L.R.; Seiler, H.L.; Torstveit, M.K. Evaluation of implementing a community-based exercise intervention during pregnancy. *Midwifery* **2017**, *46*, 45–51. [CrossRef]

49. Gavin, N.I.; Gaynes, B.N.; Lohr, K.N.; Meltzer-Brody, S.; Gartlehner, G.; Swinson, T. Perinatal depression: A systematic review of prevalence and incidence. *Obstet. Gynecol.* **2005**, *106 Pt 1*, 1071–1083. [CrossRef]

50. Teymuri, Z.; Hosseinifar, M.; Sirousi, M. The Effect of Stabilization Exercises on Pain, Disability, and Pelvic Floor Muscle Function in Postpartum Lumbopelvic Pain: A Randomized Controlled Trial. *Am. J. Phys. Med. Rehabil.* **2018**, *97*, 885–891. [CrossRef]

51. Saleh, M.S.M.; Botla, A.M.M.; Elbehary, N.A.M. Effect of core stability exercises on postpartum lumbopelvic pain: A randomized controlled trial. *J. Back Musculoskelet. Rehabil.* **2019**, *32*, 205–213. [CrossRef]

52. Ehsani, F.; Sahebi, N.; Shanbehzadeh, S.; Arab, A.M.; ShahAli, S. Stabilization exercise affects function of transverse abdominis and pelvic floor muscles in women with postpartum lumbo-pelvic pain: A double-blinded randomized clinical trial study. *Int. Urogynecol. J.* **2020**, *31*, 197–204. [CrossRef]

53. Sadat, Z.; Abedzadeh-Kalahroudi, M.; Kafaei Atrian, M.; Karimian, Z.; Sooki, Z. The Impact of Postpartum Depression on Quality of Life in Women After Child's Birth. *Iran. Red. Crescent. Med. J.* **2014**, *16*, e14995. [CrossRef]

54. Mousavi, F.; Shojaei, P. Postpartum Depression and Quality of Life: A Path Analysis. *Yale J. Biol. Med.* **2021**, *94*, 85–94.

55. Yang, C.L.; Chen, C.H. Effectiveness of aerobic gymnastic exercise on stress, fatigue, and sleep quality during postpartum: A pilot randomized controlled trial. *Int. J. Nurs. Stud.* **2018**, *77*, 1–7. [CrossRef] [PubMed]

56. Meyers, K.; Hong, M.Y. The physical effects of exercise in lactating women: A review. *J. Hum. Sport Exerc.* **2021**, *16*, 740–751. [CrossRef]

57. Dalfra', M.G.; Burlina, S.; Lapolla, A. Weight gain during pregnancy: A narrative review on the recent evidences. *Diabetes Res. Clin. Pract.* **2022**, *188*, 109913. [CrossRef] [PubMed]
58. Branco, M.; Santos-Rocha, R.; Vieira, F.; Aguiar, L.; Veloso, A.P. Three-Dimensional Kinetic Adaptations of Gait throughout Pregnancy and Postpartum. *Scientifica* **2015**, *2015*, 580374. [CrossRef] [PubMed]
59. Branco, M.A.; Santos-Rocha, R.; Vieira, F.; Aguiar, L.; Veloso, A.P. Three-dimensional kinematic adaptations of gait throughout pregnancy and post-partum. *Acta Bioeng. Biomech.* **2016**, *18*, 153–162.
60. Mei, Q.; Gu, Y.; Fernandez, J. Alterations of Pregnant Gait during Pregnancy and Post-Partum. *Sci Rep.* **2018**, *8*, 2217. [CrossRef]
61. Gilleard, W.L.; Crosbie, J.; Smith, R. Static trunk posture in sitting and standing during pregnancy and early postpartum. *Arch. Phys. Med. Rehabil.* **2002**, *83*, 1739–1744. [CrossRef]
62. Biviá-Roig, G.; Lisón, J.F.; Sánchez-Zuriaga, D. Changes in trunk posture and muscle responses in standing during pregnancy and postpartum. *PLoS ONE* **2018**, *13*, e0194853. [CrossRef]
63. Szumilewicz, A. Who and how should prescribe and conduct exercise programs for pregnant women? Recommendation based on the European educational standards for pregnancy and postnatal exercise specialists. *Dev. Period Med.* **2018**, *22*, 107–112. [CrossRef]

Article

Using the Health Belief Model to Assess the Physical Exercise Behaviors of International Students in South Korea during the Pandemic

Peng-Shuai Ma [1], Wi-Young So [2,*,†] and Hyongjun Choi [1,*,†]

[1] Department of Physical Education, Dankook University, Yongin 16890, Republic of Korea
[2] Sports Medicine Major, College of Humanities and Arts, Korea National University of Transportation, Chungju-si 27469, Republic of Korea
* Correspondence: wowso@ut.ac.kr (W.-Y.S.); chj2812@dankook.ac.kr (H.C.); Tel.: +82-43-841-5991 (W.-Y.S.); +82-31-8005-3859 (H.C.); Fax: +82-43-841-5990 (W.-Y.S.); +82-31-8021-7232 (H.C.)
† These authors contributed equally to this work.

Abstract: International students have the special status of being isolated in a foreign country during a pandemic. As Korea is a worldwide leader in education, it is important to understand the physical exercise behaviors of international students during this pandemic to assess the need for additional policies and support. The health belief model was used to score the physical exercise motivation and behaviors of international students in South Korea during the COVID-19 pandemic. In total, 315 valid questionnaires were obtained and analyzed for this study. The reliability and validity of the data were also assessed. For all variables, the values for combined reliability and the Cronbach's α were higher than 0.70. The following conclusions were drawn by comparing the differences between the measures. The results of the Kaiser–Meyer–Olkin and Bartlett tests were also higher than 0.70, confirming high reliability and validity. This study found a correlation between the health beliefs of international students and age, education, and accommodation. Consequently, international students with lower health belief scores should be encouraged to pay more attention to their personal health, participate in more physical exercise, strengthen their motivation to participate in physical exercise, and increase the frequency of their participation.

Keywords: COVID-19; health belief model; international students; physical exercise

Citation: Ma, P.-S.; So, W.-Y.; Choi, H. Using the Health Belief Model to Assess the Physical Exercise Behaviors of International Students in South Korea during the Pandemic. *Healthcare* **2023**, *11*, 469. https://doi.org/10.3390/healthcare11040469

Academic Editors: Rafael Oliveira and João Paulo Brito

Received: 13 January 2023
Revised: 31 January 2023
Accepted: 3 February 2023
Published: 6 February 2023

1. Introduction

As a top-ranking country among the Global Citizens for Human Rights' World's Best Education Systems [1], South Korea has been recognized worldwide for the comprehensiveness and quality of its education system. Hence, over the last decade, the number of international students in Korea has increased. In December 2021, this number reached 152,281, an increase of 44.9% over the total number in 2010 [2]. However, the coronavirus has had an impact on these numbers, as well as on the lives of these international students in Korea.

The coronavirus disease 2019 (COVID-19) was declared a global pandemic on 12 March 2020, by the World Health Organization [3]. Therefore, in the context of COVID-19, over the last few years the number of international students coming to South Korea has decreased. In 2020, the number decreased by 6470 compared with that in 2019, and the number in 2021 decreased by 1414 compared with that in 2020 [2]. With the end of the epidemic, the number of students studying abroad has increased by 14,611 compared to 2021 [2]. During this time, most countries imposed restrictions to reduce the spread of COVID-19; for example, these included restrictions on indoor and outdoor gatherings, business hours, quarantine periods, and international travel [4]. These restrictions have had many effects, including a reduction in physical activity. Due to the emergence of

COVID-19, international students in South Korea have fewer opportunities to go out to participate in sports activities. In such circumstances, students' physical health status may change. This is because social interaction behaviors that occur when going out to participate in physical exercise positively impact the mental health of international students. Thus, reduced physical exercise may lead to negative changes in the mental health of international students, which will also negatively impact their quality of life [5].

Physical exercise has been shown to be an important element in maintaining one's physical, as well as mental, health [5,6]. Physical activity and exercise might be key factors in helping the population better tolerate pandemic periods at both the mental and physical levels [7]. During the pandemic, participation in such activities changed, as physical exercise activities in most countries and regions around the world were affected [7,8]. This was particularly true in South Korea. During the pandemic, Korean citizens' participation in sports gradually shifted from outdoors to indoors, and from clustered multi-person sports to single-person non-clustered sports [8]. For an international student unfamiliar with the Korean lifestyle and the surrounding environment, physical exercise during the COVID-19 pandemic became difficult.

Regular physical exercise not only improves the body's immunity [5], but also effectively reduces disease fatality rates [6]. The health belief model (HBM), first proposed by American psychologist Rosenstock over six decades ago, assesses health-related attitudes and behaviors. It has been widely used, modified, and improved [9]. The model is a practical tool that can be applied to participation in physical exercise to improve our understanding of the underlying influential factors [10].

The HBM evaluates the health status of individuals in terms of their health beliefs, cues, or intentions, and behavioral constraints [11]. This research mainly focuses on five factors: perceived benefits of exercise, self-efficacy of physical evaluation, susceptibility to disease, severity of sensory illness and frailty, and paying attention to the results of physical evaluation. Although physical exercise is a good means of improving health, many people do not participate in it; thus, a better understanding of this phenomenon will enable us to promote better participation behaviors [11,12]. We use the HBM in our study to assess the health beliefs of international students and investigate how these beliefs correlate with physical activity during the pandemic.

Previous studies have found that health beliefs can help people take effective preventive measures. Therefore, establishing good health beliefs has a certain impact on physical exercise motivation and behavior. Physical exercise motivation is generally divided into physical health, appearance, entertainment, social communication, and other aspects [13]. However, groups with different educational levels and social backgrounds also have different health beliefs. People with higher educational backgrounds may have more advantages in the process of establishing personal health beliefs [14], as they have richer knowledge reserves and more life experience. Their personal health beliefs may be more comprehensive [13,15]. The environment may be a factor as well. In South Korea, international students can choose whether to live on campus or off campus, which also has an important impact on their lifestyles. Students living off campus may be more physically active than those living on campus [16,17].

At present, there are many studies on the physical exercise of college students in the context of COVID-19, but few focusing on the international student group [5,9,17–20]. We investigate the correlations between the health beliefs of international students, their education level, physical activity, and their living situation before and after COVID-19. Based on the importance of physical activity for a student's mental health, this study aims to provide a reference for students, academia, and policymakers regarding the motivation and willingness of international students to participate in physical exercise during the normalization of the pandemic. To that end, we use the HBM to understand students' awareness of their health and assess the impact of their scores on their participation in physical exercise in South Korea.

2. Materials and Methods

2.1. Data Collection

As international students at Dankook University, Yongin, South Korea, we created an online questionnaire through the website www.wjx.cn, accessed on 12 August 2021. Through our teachers and friends, we sent the questionnaire's link to the surrounding foreign students. To explore physical exercise behavior and changes in such behavior during the COVID-19 pandemic in the context of the HBM, on 3 March 2022, an online questionnaire was used to maintain social distancing, and students in the Seoul and Gyeonggi-do regions of South Korea were randomly selected to participate. A survey of 320 foreign students studying in South Korea used the HBM to assess their health behaviors and changes in their physical exercise behaviors. All 320 online questionnaires distributed were returned. Of these, 315 were valid, which led to an effective response rate of 98.40%. Since data sets on this study from the online questionnaire did not include private identifier information, such as telephone numbers, home addresses, and social security numbers, ethical approval was not required. Ultimately, the study was approved by the institutional review board of Dankook University, Yongin, South Korea, and ethical approval was waived.

2.2. Scale Design

There are relatively complete measurement scales in the extant research for the HBM [18]. This study transformed the HBM scale created by DAI Xia, deleted the inappropriate variables, and adapted them according to the specificity of the COVID-19 context. We based the measurement scale in our study on similar variables in previous research with adequate content credibility [12,16,18,21]. The questions were modified and adjusted accordingly, as shown in Table 1. All the items on the scale were measured against a five-point Likert scale, where 1 meant "strongly disagree" and 5 meant "strongly agree." At the same time, a score from 1 to 5 was assigned to the questionnaire responses on cognitive health belief patterns in this survey (only perceived benefits of exercise, self-efficacy of physical evaluation, susceptibility to disease, severity of sensory illness and frailty, and paying attention to the results of physical evaluation).

Table 1. The reliability and validity test results for each factor (n = 315).

Variables	AVE	CR	Cronbach's α	KMO & Bartlett
Perceived benefits of exercise	0.942	0.697	0.919	0.936
Self-efficacy of physical evaluation	0.518	0.808	0.721	0.945
Susceptibility to disease	0.868	0.687	0.811	0.919
Severity of sensory illness and frailty	0.745	0.921	0.892	0.938
Paying attention to the results of physical evaluation	0.828	0.935	0.910	0.921
Differences in motivation to participate in physical exercise before and after the emergence of COVID-19	0.611	0.916	0.885	0.926
Physical exercise behavior differences	0.680	0.922	0.802	0.833

Note: AVE, average variance extracted; CR, composite reliability; KMO, Kaiser–Meyer–Olkin.

2.3. Data Analysis

To study the HBM from the perspective of international students in South Korea and their physical exercise behaviors, we included the HBM's questions as well as questions regarding the differences in student motivation to participate in physical exercise and students' actual exercise behaviors before and after the pandemic. We also conducted several tests to ensure the reliability and validity of our variables. The data of the Internet survey were organized. SPSS 26.0's (IBM Corp., Armonk, NY, USA) Kaiser–Meyer–Olkin (KMO) Test was used to verify the validity of the results. In order to understand the importance attached to health, appearance, entertainment, learning skills, and social interactions under different educational backgrounds, the data were compared and sorted by analysis of variance and least square difference (ANOVA-LSD) post-hoc tests. The statistical significance was set at $p < 0.05$.

3. Results

3.1. Reliability and Validity

The combined reliability and Cronbach's α of most variables were higher than the recommended level of 0.70, and the KMO and Bartlett test results were higher than 0.70, thus assuring high validity. The reliability and validity test results are shown in Table 1.

To accurately understand the differences in each group's physical exercise behaviors and motivation in terms of education and other factors, we used the ANOVA-LSD to analyze the differences assuming a variance.

3.2. Sociodemographic Variable Analysis

Among the participants in this survey, international students from China accounted for the majority (69.84%), followed by those from Vietnam (22.54%). The number of international students from other countries and regions was relatively small (7.61%). Most of the international students were female (61.59%), between 23–25 years old (41.27%), and in either their third or fourth semester (52.06%). International students over 23 years old in at least their second semester can adapt to the surrounding environment of the school. Most international students (60.32%) chose more free off-campus accommodations. Details of the sociodemographic variables are shown in Table 2.

Table 2. Descriptive statistics for the respondents ($n = 315$).

Variables	Category	Frequency	Percentage (%)
Gender	Male	121	38.41
	Female	194	61.59
Age	18–22	52	16.51
	23–25	130	41.27
	26–30	97	30.79
	Over 30	35	11.11
Educational level	College	102	32.38
	Postgraduate	112	35.56
	PhD	101	32.06
Semester	1–2	69	21.90
	3–4	164	52.06
	5–6	60	19.05
	7–8	22	6.98
Living accommodation	On campus	125	39.68
	Off campus	190	60.32

3.3. Correlations among Age, Gender, and Health Beliefs

To understand the differences among international students in the context of the HBM across different age groups, we first tested the variables using Pearson correlation analysis in SPSS. The test results are shown in Table 3.

Among the international students, the benefits of activity significantly correlated with age ($p < 0.01$), education ($p < 0.01$), and lifestyle ($p < 0.01$). Both male and female students with increased health awareness perceived the benefits of exercise and enjoyed it more [22]. Our tests regarding differences in awareness based on health beliefs among international students in different age groups indicated that higher HBM scores correlated significantly with age, gender, educational background, and lifestyle among the international students surveyed. Age was closely related to personal life experience and perception of ontological health, and those with extensive experiences were more conscious of their health [23].

Table 3. Pearson correlation analysis of age, gender, and health beliefs ($n = 315$).

Variables	Contents	Sex	Age	Education	Living Situation
Perceived benefits of exercise	I think physical exercise can enhance physical health.	0.086	0.678 **	0.561 **	0.252 **
	I think physical activity can prevent disease and prolong life.	0.109	0.557 **	0.452 **	0.254 **
	I think physical exercise is good for the mind and body.	0.109	0.594 **	0.483 **	0.230 **
	I think physical exercise can keep me fit.	0.097	0.503 **	0.412 **	0.210 **
	I think physical activity improves quality of life.	0.064	0.776 **	0.640 **	0.277 **
	I think physical exercise is an important way to maintain physical health.	0.079	0.496 **	0.419 **	0.235 **
	I think enhancing physical health requires regular physical exercise.	0.130 *	0.551 **	0.469 **	0.267 **
Self-efficacy of physical evaluation	When my physique is in poor health, I feel a sense of urgency to exercise.	0.053	−0.041	0.039	0.098
	When my physique is not in good health, I will work out against the odds.	0.092	0.614 **	0.454 **	0.239 **
	When my physique is not healthy, I can overcome obstacles such as weather and the environment to train.	0.150 **	0.574 **	0.441 **	0.246 **
	Even when I get tired, I will follow a plan to exercise to enhance my physique and health.	0.051	0.487 **	0.383 **	0.177 **
Susceptibility to disease	People with poor physical health are more likely to be infected with COVID-19.	0.126 *	0.284 **	0.202 **	0.205 **
	Lack of exercise may lead to COVID-19 infection.	0.102	0.287 **	0.326 **	0.203 **
	Poor physical health made me worry about contracting COVID-19.	0.234 **	0.353 **	0.320 **	0.249 **
Perceived severity of illness and frailty	Serious investigations of perceived sickness and frailty—I am afraid of contracting COVID-19.	<0.001	−0.048	−0.020	−0.049
	Every time I get sick, I feel scared.	0.069	0.792 **	0.656 **	0.348 **
	I feel a serious threat to my physical health when I have an illness.	0.060	0.654 **	0.584 **	0.297 **
	Lack of exercise can lead to a serious illness.	0.141 *	0.478 **	0.516 **	0.292 **
Paying attention to the results of physical evaluation	Physical health evaluation is an important way for me to understand health.	0.091	0.489 **	0.506 **	0.283 **
	I am very concerned about changes in physical health information.	0.086	0.444 **	0.446 **	0.233 **
	The results of a physical health evaluation allow me to know more about my own health.	0.069	0.792 **	0.656 **	0.348 **

Note: * $p < 0.05$, ** $p < 0.01$.

Figure 1 presents the average scores of perceived health beliefs for groups. After re-organizing the HBM's score data from the questionnaire survey for different educational backgrounds, accommodations, and age groups, we found no significant correlations regarding comprehensive HBM scores for the vast majority of international students by gender. Those with higher average HBM scores were more educated, 26–30 years old, and lived off campus. Older international students with more education appeared to have increased safety awareness when living alone [18].

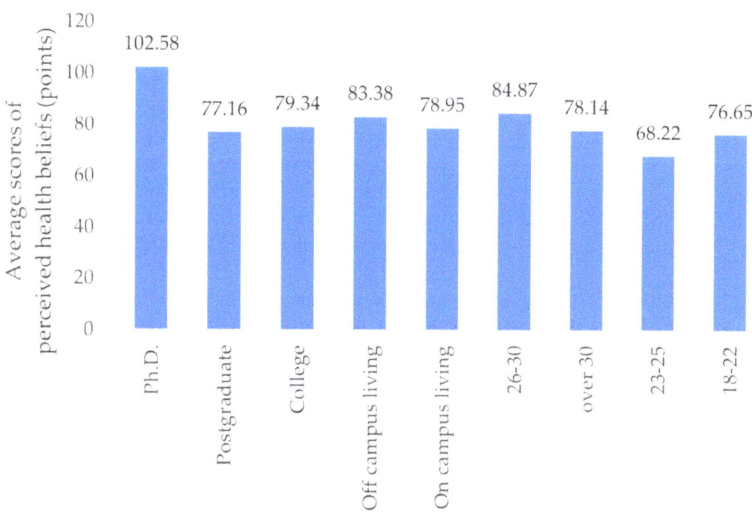

Figure 1. The average scores of perceived health beliefs for each group.

3.4. Correlations of Physical Exercise Motivation and Education

In Table 4, through the ANOVA-LSD analysis of academic qualifications, we found that following the pandemic outbreak, foreign undergraduate students had a stronger desire to improve their health, maintain their bodies, and enjoy sports, as well as a stronger need for social interaction, than did graduate and doctoral students. Thus, the sudden pandemic situation had a greater impact on the health motivation of international students with less education. Through the same analysis in terms of age difference, the motivation for physical exercise among 26–30-year-olds was lower than that among the other age groups. Although older international students had stronger personal health awareness, they perceived that the pandemic made it more difficult for them to get physical exercise. Motivation was less affected among the other age groups. In terms of living situation, students living off campus reflected a smaller difference in physical exercise motivation than did those living on campus. With the spread of the pandemic, more people chose to exercise at home. Students living off campus have more favorable conditions for such exercise.

Table 4. Differences in foreign students' physical exercise motivation based on academic qualifications.

Variables	Education	N	Mean	Standard Deviation	F	Least Significant Deviation
Health	Undergraduate	102	4.33	0.569	65.815 ***	1 > 2 > 3
	Masters	103	4.06	0.683		
	Ph.D.	110	3.43	0.515		
Appearance	Undergraduate	102	3.96	0.195	62.739 ***	1 > 2 > 3
	Masters	103	3.81	0.397		
	Ph.D.	110	3.39	0.490		
Entertainment	Undergraduate	102	3.86	0.446	40.888 ***	1 > 2 > 3
	Masters	103	3.76	0.474		
	Ph.D.	110	3.29	0.548		
Learning skills	Undergraduate	102	3.80	0.546	30.639 ***	1 > 2 > 3
	Masters	103	3.74	0.504		
	Ph.D.	110	3.28	0.544		
Social interaction	Undergraduate	102	3.75	0.624	27.439 ***	1 > 2 > 3
	Masters	103	3.71	0.517		
	Ph.D.	110	3.23	0.569		

Note: *** $p < 0.001$; tested using analysis of variance. 1: Undergraduate, 2: Masters, 3: Ph.D.

3.5. Influence of HBM Scores on International Students' Physical Exercise Motivation and Behavior

Figure 2 shows how the average HBM scores differ based on educational background, age, and living situation. Those with higher average scores were aged 26–30 years (84.87 points), had doctoral degrees (102.58 points), and had off-campus accommodations (83.38 points). Changes in motivation for training and exercise were also significantly different in the group of international students with higher health belief scores. Among them, students with doctoral degrees were more inclined to choose the "increase" option after the outbreak of the pandemic. The international students in the 26–30-year-old group chose the "unchanged" option, and the students living off campus chose the "increase" option.

In Table 5, the options from "Physical exercise behavior differences" questionnaire were assigned to the participants' corresponding scores of the questionnaire concerning the most exercise behaviors (except for the ninth question). Next, the average score was calculated for each group. In terms of physical exercise activities, the overall exercise intensity of the international students with doctoral degrees between the ages of 26–30 years was moderate, and their exercise time was generally 31–59 min about 5–6 times a week. Their exercise intensity did not change after the pandemic. However, in terms of physical exercise behavior, the average score for international students aged 26–30 years was 79.2 points, which was higher than that for those with doctoral degrees, at 59.69 points. The average score for international students living off campus was the lowest, at only 58.1 points. However, we also found that after the outbreak of the pandemic, international students living off campus changed their answers from "sweating a lot, vigorous but not lasting exercise (such as playing basketball, tennis, etc.)" to "shortness of breath, sweating a lot, intense and long-lasting exercise (such as long-distance running, playing badminton, swimming, etc.)", implying that their exercise intensity increased somewhat.

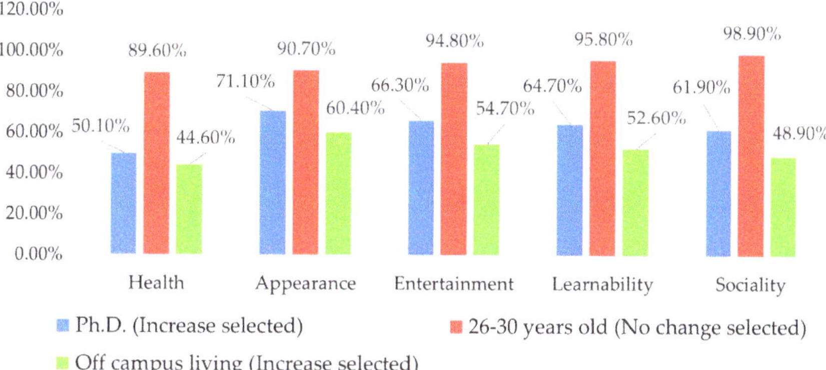

Figure 2. The exercise motivation intention factors with the highest average health belief scores.

Table 5. Summaries of physical exercise behaviors among those with strong health beliefs before and after the pandemic.

Variables	Exercise Intensity (B)	Exercise Intensity (A)	Exercise Time (B)	Exercise Time (A)	Exercise Frequency (B)	Exercise Frequency (A)	Range (B)	Range (A)
PhD.	High strength (52.2%)	High strength (52.0%)	31–59 min (70.0%)	31–59 min (64.4%)	5–6 times (65.5%)	5–6 times (56.6%)	501–1000 m (65.5%)	501–1000 m (53.3%)
26–30 years old	High strength (77.3%)	High strength (78.3%)	31–59 min (85.5%)	31–59 min (85%)	5–6 times (84.5%)	5–6 times (79.3%)	501–1000 m (81.4%)	501–1000 m (61.8%)
Off-campus living	Medium strength (33.5%)	High strength (34.0%)	31–59 min (65.4%)	31–59 min (62.7%)	5–6 times (73.9%)	5–6 times (69.6%)	501–1000 m (71.8%)	501–1000 m (50.5%)

4. Discussion

At present, there are many studies on college students' exercise within the context of the COVID-19 pandemic, but few studies focus on the international student group. International students faced a particularly difficult situation when the pandemic occurred because of their special status as foreigners. There are relatively few channels for understanding and interpreting the local pandemic prevention and control policies. Therefore, it is necessary to study international students' physical exercise behaviors during the pandemic [18,24].

This study investigated the HBM in relation to physical exercise motivation and behavior among 315 international students in South Korea within the context of the pandemic. The average health belief scores among international students with doctoral degrees, those aged 26–30 years, and those who lived in off-campus housing were higher than those of students with other educational backgrounds, in other age groups, and those who lived on campus. The motivation and behavior of international students with high average health belief scores did not change much before and after the pandemic. This shows that overseas students with higher academic qualifications, of a moderately mature age, and who live off campus in a more free-living environment have strong health beliefs. Their abilities to perceive and manage their bodies' health status are strong, and they have relatively regular physical exercise habits.

Previous studies have found that COVID-19 health beliefs can help people take effective preventive measures [25,26]; therefore, the establishment of good health beliefs has a certain impact on physical activity motivation and behavior. Physical exercise motivation is generally divided into physical health, appearance, entertainment, social communication, and other aspects [14,21,27]. Perceived exercise benefits allow participants to feel the physical benefits of exercise while performing it, so that people enjoy exercise. Physical fitness evaluation self-efficacy helps people to effectively recognize the relationship between physical health and exercise, so that exercise participants can reasonably arrange physical exercise according to their own health status. When someone is in poor physical health, there will also be concerns about and associations with physical illness. On the other hand, people's positive attitudes may also stimulate them to actively participate in sports. Since most people have no experience of being infected or sick with COVID-19, their ability to perceive the threat of the disease is poor. Therefore, their motivation to participate in preventive physical exercise will also be weakened; paying attention to the results of physical health evaluation is the fastest way to understand the physical health status of individuals, and it is also a more direct factor affecting physical exercise motivation and behavior.

However, groups with different educational levels and social backgrounds also have different health beliefs. People with higher educational backgrounds have more advantages in the process of establishing personal health beliefs [27,28], as well as richer knowledge reserves and life experience. This is consistent with the results of this study. This study also analyzed the correlations between the interviewees' educational background and age and found that for international students with more education and of a relatively advanced age, the establishment of the personal HBM is also more complete. People around the age of 30 have a strong awareness of health and physical exercise, and they have a relatively regular lifestyle. This is very helpful for maintaining physical and mental health [14,29,30]. In South Korea, students can choose whether to live in on-campus dormitories, and both on-campus and off-campus dormitories also have an important impact on students' lifestyles. In China, college students are not allowed to live off campus, so their lifestyles are relatively simple. South Korea's relatively free lifestyle brings more convenience to students' lives. For example, students can go to gyms to participate in physical exercise, and students living off campus are more active in physical activities than students living on campus [13,20]. In terms of intensity, frequency, and time of physical exercise, due to the impact of the pandemic, most people showed an overall downward trend. This has a stronger relationship with the inability to go out to participate in or enter a gym for physical exercise.

Within the context of the COVID-19 pandemic, the health beliefs of international students in South Korea have a certain influence on their physical exercise motivation and behavior. This, in turn, has a significant impact on physical exercise intensity, exercise frequency, exercise time, and activity range. This study did not investigate respondents' access to personal health information, so there are some limitations. In the future, we can learn about the specific sports and fitness status of international students by investigating how they obtain information about their health beliefs. In order to study the impact of the HBM on the health status of international students, international students should be assisted in establishing personal health beliefs and maintaining a good level of physical health.

5. Conclusions

This study surveyed 315 international students in Seoul and Gyeonggi-do, South Korea. Through the HBM, differences in physical exercise motivation and behavior before and after the pandemic were studied. Our conclusions are as follows: First, during the pandemic, physical exercise motivation among international students with higher health belief scores did not change significantly. This was closely related to the positive physical exercise habits formed through personal behaviors.

Second, the group health belief scores of foreign students with doctoral degrees, who were aged 26–30 years, and who lived off campus were higher than those of other groups.

The higher the health belief score, the higher the awareness and concern regarding the student's personal health status. Third, groups with higher health belief scores showed only small differences in terms of exercise intensity, frequency, time, and scope before and after the pandemic, reinforcing the benefits of strong health beliefs for establishing a strong health awareness and motivating regular physical activity. International students with lower health belief scores should be encouraged to pay more attention to their personal health, participate in more physical exercise, strengthen their motivation to participate in physical exercise, and increase the frequency of their participation.

Author Contributions: Conceptualization, P.-S.M., W.-Y.S. and H.C.; data curation, P.-S.M., W.-Y.S. and H.C.; formal analysis, P.-S.M., W.-Y.S. and H.C.; methodology, P.-S.M., W.-Y.S. and H.C.; project administration, P.-S.M., W.-Y.S. and H.C.; supervision, P.-S.M., W.-Y.S. and H.C.; writing—original draft preparation, P.-S.M., W.-Y.S. and H.C.; writing—review and editing, P.-S.M., W.-Y.S. and H.C. All authors have read and agreed to the published version of the manuscript.

Funding: This research received no external funding.

Institutional Review Board Statement: Ethical review and approval were waived for this study due to the questionnaire has been collected by online survey system and privacy agreement has been collected during the survey.

Informed Consent Statement: Informed consent was obtained from all patients involved in the study.

Data Availability Statement: The data presented in this study are available upon request from the authors. Some variables were restricted to preserve the anonymity of study participants.

Conflicts of Interest: The authors declare no conflict of interest.

References

1. Korea Ministry of Education. *Statistics of Foreign Students in Korea Domestic Higher Education Institutions in 2021*; Korea Ministry of Education: Sejong, Republic of Korea, 2022.
2. Available online: https://www.moe.go.kr/boardCnts/viewRenew.do?boardID=350&boardSeq=90123&lev=0&searchType=null&statusYN=W&page=1&s=moe&m=0309&opType=N (accessed on 22 December 2022).
3. Global Education Report 2022 World Best Education Systems. Available online: https://worldtop20.org/worldbesteducationsystem (accessed on 3 October 2022).
4. World Health Organization. WHO Announces COVID-19 Outbreak a Pandemic. World Health Organization. Available online: https://www.euro.who.int/en/health-topics/health-emergencies/coronavirus-covid-19/news/news/2020/3/who-announces-covid-19-outbreak-a-pandemic (accessed on 3 October 2022).
5. Sidebottom, C.; Ullevig, S.; Cheever, K.; Zhang, T. Effects of COVID-19 pandemic and quarantine period on physical activity and dietary habits of college-aged students. *Sport. Med. Health Sci.* **2021**, *3*, 228–235. [CrossRef] [PubMed]
6. Marques, A.; Santos, T.; Martins, J.; Matos, M.G.; Valeiro, M.G. The association between physical activity and chronic diseases in European adults. *Eur. J. Sport Sci.* **2018**, *18*, 140–149. [CrossRef] [PubMed]
7. Amatriain-Fernández, S.; Murillo-Rodríguez, E.S.; Gronwald, T.; Machado, S.; Budde, H. Benefits of physical activity and physical exercise in the time of pandemic. *Psychol. Trauma Theory Res. Pract. Policy* **2020**, *12*, S264. [CrossRef] [PubMed]
8. Hansen, B.H.; Dalene, K.E.; Ekelund, U.; Wang Fagerland, M.; Kolle, E.; Steene-Johannessen, J. Step by step: Association of device-measured daily steps with all-cause mortality-A prospective cohort Study. *Scand. J. Med. Sci. Sport.* **2020**, *30*, 1705–1711. [CrossRef] [PubMed]
9. Park, K.H.; Kim, A.R.; Yang, M.A.; Lim, S.J.; Park, J.H. Impact of the COVID-19 pandemic on the lifestyle, mental health, and quality of life of adults in South Korea. *PLoS ONE* **2021**, *16*, e0247970. [CrossRef]
10. Ekelund, U.; Tarp, J.; Steene-Johannessen, J.; Hansen, B.H.; Jefferis, B.; Fagerland, M.W.; Whincup, P.; Diaz, K.M.; Hooker, S.P.; Chernofsky, A.; et al. Dose-response associations between accelerometry measured physical activity and sedentary time and all cause mortality: Systematic review and harmonised meta-analysis. *Br. Med. J.* **2019**, *366*, l4570. [CrossRef]
11. Rosenstock, I.M. Historical Origins of the Health Belief Model. *Health Educ. Monogr.* **1974**, *2*, 328–335. [CrossRef]
12. Yu, Z.H.; Gong, Q.B.; Wu, J.T. Development of physical exercise health belief scale. *J. Wuhan Inst. Phys. Educ.* **2006**, *5*, 53–56.
13. Wanja, W.; Maik, B.; Johanna, S.; Julia, S. Too Bored for Sports? Adaptive and less-adaptive latent personality profiles for exercise behavior. *Psychol. Sport Exerc.* **2021**, *53*, 101851.
14. Younis, I.; Longsheng, C.; Zulfiqar, M.I.; Imran, M.; Shah, S.A.; Hussain, M.; Solangi, Y.A. Regional disparities in Preventive measures of COVID-19 pandemic in China. A study from international students' prior knowledge, perception and vulnerabilities. *Environ. Sci. Pollut. Res.* **2021**, *28*, 40355–40370. [CrossRef]
15. Zhang, D.; Zhang, X.; Ma, L. Review of Physical Exercise Motivation. *J. Chifeng Univ. (Nat. Sci. Ed.)* **2012**, *21*, 170–172.

16. Shin, S. Relationships between health promoting lifestyle, health belief about emerging infectious disease and hygiene behavior of college students. *J. Korea Converg. Soc.* **2019**, *10*, 285–293.
17. Taghrir, M.H.; Borazjani, R.; Shiraly, R. COVID-19 and Iranian medical students; a survey on their related-knowledge, preventive behaviors and risk perception. *Arch. Iran. Med.* **2020**, *23*, 249–254. [CrossRef]
18. Dai, X.; Yin, H.; Zhu, L. Compilation and preliminary application of the scale of physical fitness health belief of college students. *J. Beijing Sport Univ.* **2011**, *12*, 72–74.
19. Herby, J.; Jonung, L.; Hanke, S. A literature review and meta-analysis of the effects of lockdowns on COVID-19 mortality. *Stud. Appl. Econ.* **2022**, 200.
20. Strielkowski, W.; Wang, J. An introduction: COVID-19 pandemic and academic leadership. In Proceedings of the 6th International Conference on Social, Economic, and Academic Leadership (ICSEAL-6-2019), Prague, Czech, 13–14 December 2019; Atlantis Press: Amsterdam, The Netherlands, 2020; pp. 1–4.
21. Becker, M.H.; Maiman, L.A.; Kirscht, J.P.; Haefner, D.P.; Drachman, R.H. The Health Belief Model and prediction of dietary compliance: A field experiment. *J. Health Soc. Behav.* **1977**, *18*, 348–366. [CrossRef] [PubMed]
22. Irwin, J.D. Prevalence of university students' sufficient physical activity: A systematic review. *Percept. Mot. Ski.* **2004**, *98*, 927–943. [CrossRef] [PubMed]
23. Gadi, N.; Saleh, S.; Johnson, J.A.; Trinidade, A. The impact of the COVID-19 pandemic on the lifestyle and behaviours, mental health and education of students studying healthcare-related courses at a British university. *BMC Med. Educ.* **2022**, *22*, 115. [CrossRef] [PubMed]
24. Ammigan, R. Institutional Satisfaction and Recommendation: What Really Matters to International Students? *J. Int. Stud.* **2019**, *9*, 262–281. [CrossRef]
25. Ryu, S.; Ali, S.T.; Lim, J.S.; Chun, B.C. Estimation of the Excess COVID-19 Cases in Seoul, South Korea by the students arriving from China. *Int. J. Environ. Res. Public Health* **2020**, *17*, 3113. [CrossRef]
26. Aaltonen, S.; Rottensteiner, M.; Kaprio, J.; Kujala, U.M. Motives for physical activity among active and inactive persons in their mid-30s. *Scand. J. Med. Sci. Sport.* **2014**, *24*, 727–735. [CrossRef] [PubMed]
27. Sabiston, C.M.; Pila, E.; Vani, M.; Thogersen-Ntoumani, C. Body image, physical activity, and sport: A scoping review. *Psychol. Sport Exerc.* **2019**, *42*, 48–57. [CrossRef]
28. Elbe, A.M.; Lyhne, S.N.; Madsen, E.E.; Krustrup, P. Is regular physical activity a key to mental health? Commentary on "Association between physical exercise and mental health in 1.2 million individuals in the USA between 2011 and 2015: A cross-sectional study", by Chekroud et al., published in Lancet Psychiatry. *J. Sport Health Sci.* **2019**, *8*, 6–7. [PubMed]
29. Wang, C.; Tee, M.; Roy, A.E.; Fardin, M.A.; Srichokchatchawan, W.; Habib, H.A.; Tran, B.X.; Hussain, S.; Hoang, M.T.; Le, X.T.; et al. The impact of COVID-19 pandemic on physical and mental health of Asians: A study of seven middle-income countries in Asia. *PLoS ONE* **2021**, *16*, e0246824. [CrossRef]
30. Tison, G.H.; Avram, R.; Kuhar, P.; Abreau, S.; Marcus, G.M.; Pletcher, M.J.; Olgin, J.E. Worldwide Effect of COVID-19 on Physical Activity: A Descriptive Study. *Ann. Intern. Med.* **2020**, *173*, 767–770. [CrossRef] [PubMed]

Article

Additional Active Movements Are Not Required for Strength Gains in the Untrained during Short-Term Whole-Body Electromyostimulation Training

Holger Stephan *, Udo Frank Wehmeier, Tim Förster, Fabian Tomschi and Thomas Hilberg *

Department of Sports Medicine, University of Wuppertal, Moritzstraße 14, 42117 Wuppertal, Germany
* Correspondence: hstephan@uni-wuppertal.de (H.S.); hilberg@uni-wuppertal.de (T.H.)

Abstract: Recommendations for conventional strength training are well described, and the volume of research on whole-body electromyostimulation training (WB-EMS) is growing. The aim of the present study was to investigate whether active exercise movements during stimulation have a positive effect on strength gains. A total of 30 inactive subjects (28 completed the study) were randomly allocated into two training groups, the upper body group (UBG) and the lower body group (LBG). In the UBG ($n = 15$; age: 32 (25–36); body mass: 78.3 kg (53.1–114.3 kg)), WB-EMS was accompanied by exercise movements of the upper body and in the LBG ($n = 13$; age: 26 (20–35); body mass: 67.2 kg (47.4–100.3 kg)) by exercise movements of the lower body. Therefore, UBG served as a control when lower body strength was considered, and LBG served as a control when upper body strength was considered. Trunk exercises were performed under the same conditions in both groups. During the 20-min sessions, 12 repetitions were performed per exercise. In both groups, stimulation was performed with 350 µs wide square pulses at 85 Hz in biphasic mode, and stimulation intensity was 6–8 (scale 1–10). Isometric maximum strength was measured before and after the training (6 weeks set; one session/week) on 6 exercises for the upper body and 4 for the lower body. Isometric maximum strength was significantly higher after the EMS training in both groups in most test positions (UBG $p < 0.001$–0.031, r = 0.88–0.56; LBG $p = 0.001$–0.039, r = 0.88–0.57). Only for the left leg extension in the UBG ($p = 0.100$, r = 0.43) and for the biceps curl in the LBG ($p = 0.221$, r = 0.34) no changes were observed. Both groups showed similar absolute strength changes after EMS training. Body mass adjusted strength for the left arm pull increased more in the LBG group ($p = 0.040$, r = 0.39). Based on our results we conclude that concurring exercise movements during a short-term WB-EMS training period have no substantial influence on strength gains. People with health restrictions, beginners with no experience in strength training and people returning to training might be particularly suitable target groups, due to the low training effort. Supposedly, exercise movements become more relevant when initial adaptations to training are exhausted.

Keywords: WB-EMS; electric stimulation; strength training; strength test

Citation: Stephan, H.; Wehmeier, U.F.; Förster, T.; Tomschi, F.; Hilberg, T. Additional Active Movements Are Not Required for Strength Gains in the Untrained during Short-Term Whole-Body Electromyostimulation Training. *Healthcare* **2023**, *11*, 741. https://doi.org/10.3390/healthcare11050741

Academic Editors: Rafael Oliveira and João Paulo Brito

Received: 26 January 2023
Revised: 24 February 2023
Accepted: 27 February 2023
Published: 3 March 2023

1. Introduction

Whole-body electromyostimulation (WB-EMS) is a training method that can complement or to some extent replace traditional resistance training, as it can be used alone, superimposed, or combined (different training time points). Since several electrodes are used [1], different muscles can be stimulated at the same time [2]. Strength improvements can be achieved with both high-intensity resistance training and WB-EMS [3]. Previous studies have shown that WB-EMS is applicable in healthy people [4] and also in patients, e.g., in people suffering from Parkinson [5] or sarcopenic obesity [6]. In conventional resistance training, the one repetition maximum (1-RM) is used to describe the training intensity [7]. Since it represents the maximal voluntary contraction, a comparison between electromyostimulation (EMS) and normal contraction is possible [8]. A low-cost way to determine the intensity of strength training is to capture the perceived exertion using the Borg scale [9], which is also used in WB-EMS training [10,11]. In contrast to the 1-RM, where voluntary force production under external load is recorded, the perceived exertion reflects the internal load. For beginners in conventional strength training, at least 2 training sessions per week are recommended. Both multi-joint and single-joint exercises can be performed, using a variety of equipment and the own body weight. Per set, 8 to 12 repetitions should be completed at 60–70% of the repetition maximum [12]. To provide a safe and effective application of WB-EMS, guidelines recommend restricting the duration of one session to a maximum of 20 min. Moreover, the frequency should be limited to one session a week for at least the first eight weeks or a minimum interval of four days should be maintained thereafter [13]. Perceived exertion should be rated approximately as "hard" to "hard+" (lower during initial training) [13], corresponding to 5 to 6 on the Borg CR 10 scale [14]. Nevertheless, in some trials, the training frequencies were higher [2,15], and sometimes lower with one session a week [16,17] compared to the aforementioned recommendation after familiarization. The aggregated training stimulus consists of the number of sessions a week and the length of the training period. Usually, eight sessions or more have been conducted in strength related WB-EMS studies with healthy subjects [10,18]. Early strength improvements due to strength training can be attributed mainly to neural factors. From the third to fifth week on, strength development is mainly caused by hypertrophy [19]. Increases after very few sessions (as seen after three training sessions) are supposedly attributable to lower antagonist activity or motoric improvements of synergists [20]. Elgueta-Cancino and colleagues [21] elicited less inhibitory activity in the cortex, higher corticospinal excitability, and altered motor unit activation as assumed mechanisms of initial strength gain. Muscle growth and strength gain can also be achieved by compact training (eight weeks with three sessions a week) with neuromuscular electrical stimulation [22]. Similar to conventional strength training early strength gains owing to EMS-training are achieved without muscle growth [23].

The body of research on WB-EMS training is growing [24]. EMS can be superimposed on maximum or sub-maximum voluntary dynamic or isometric contractions or applied without any concomitant voluntary contraction. Nevertheless, little is known about the importance of active exercise movements during stimulation. Strength gains due to EMS with exercise movements were previously shown [25,26] and some authors addressed the impact of EMS superimposed on intense strength training [27,28]. To our knowledge, only Kemmler and colleagues [29] investigated the effects of smaller, WB-EMS accompanying movements. In this randomized controlled trial (RCT), participants trained once a week for 12 weeks. However, only older females with little muscle mass were included for the comparison between dynamic use (movements during stimulation) and passive use (only isometric contractions during stimulation) limiting the generalizability of the results obtained. Therefore, the present study aims to investigate whether active exercise movements during stimulation have a positive effect on strength gains of selected upper and lower body muscles in young healthy subjects of both sexes in training sessions using mobile, easily accessible fitness equipment, or the own body mass. We hypothesized that WB-EMS combined with concurrent exercise movements will result in higher strength gains than WB-EMS alone. Hence, this study was designed to clarify whether movement sequences are necessary for strength gains during WB-EMS or, whether the electrostimulation alone induces strength gains. The results might help fitness professionals and EMS-users to optimize recommendations for WB-EMS training depending on individual goals and requirements.

2. Materials and Methods

2.1. Subjects

The number of subjects to be included in the study was determined using an a priori sample size calculation for statistical comparison of the means of two unpaired groups (using the program GPower 3.1) based on the mean of the effect sizes (Δ strength leg extension: d = 1.67; Δ strength leg flexion: d = 0.79) reported by Kemmler and colleagues [29]. This study is similar to the present study. A predefined lower limit of statistical power of 80% and an α error probability of 0.05 were assumed. A dropout rate of 20% was further added. Based on the results of this calculation, a total of 30 subjects were initially recruited for participation. Subjects were included when being aged between 20 and 40 years and having abstained from physical activity for at least six months prior to the start of the study. Access was possible for both sexes. Subjects were excluded when acute injuries or physical complaints were reported or when contraindications as listed by Kemmler and colleagues [30] or Stöllberger and Finsterer [31] were present (e.g., epilepsy, bleeding disorders). No other exclusion criteria were defined (e.g., BMI, VO$_2$max). The study was conducted in accordance with the principles of the Declaration of Helsinki [32] and approved by the ethics committee of the University of Wuppertal (MS/BBL 200114 Wehmeier). All subjects signed a written consent to participate in the study.

2.2. Experimental Design

The procedure was based on a randomized controlled trial design (Figure 1). Subjects were randomly assigned to two training groups (with the program RandList 1.2), with the number of subjects in both groups being equal. In the upper body group (UBG), WB-EMS was accompanied by exercise movements of the upper body only and in the lower body group (LBG) by exercise movements of the lower body only. Therefore, the UBG served as a control when lower body strength is considered, and the LBG served as a control when upper body strength is considered. With this design, WB-EMS without exercises and WB-EMS with exercises could be compared. Intervention duration was set to six weeks, training frequency to one session/week, and the duration of the training session to 20 min. Before and after the training period, maximum force was determined during various exercises. Blinding of subjects was not possible because the intervention is identifiable. Blinding of the investigator was not applicable because the training instructions and the test instructions were given by the same person, a professional EMS trainer with a bachelor's degree in sports science. Subjects were asked to maintain their dietary habits and to keep their physical activity levels constant, which also meant avoiding additional physical activity. All interventions and measurements were conducted in an EMS studio (go!Orange—Studio für EMS, Remscheid, Germany).

2.3. WB-EMS Procedure

Both the UBG and the LBG received the same WB-EMS application (miha bodytec II; miha bodytec GmbH, Gersthofen, Germany) once a week. Subjects wore thin tight-fitting underwear. The vest with wetted electrodes was placed on the upper body and the wetted electrode bands on the arms, buttocks, and legs (miha bodytec). During the 20-min training, the upper and lower back, abdominal muscles, buttocks, muscles around the thigh, chest, and muscles around the upper arm were stimulated with 85 Hz of 350 μs wide rectangular pulses in biphasic mode. Both the duration of the pulse interval (stimulation on) and the pulse pause (stimulation off) were set to 4 s. The pulses were ramped up to the targeted intensity without delay (full intensity directly available) and similarly ramped down to zero (direct interruption of the stimulation) at the end of the stimulation phase. To maintain the same conditions, the stimulation intensity was adjusted to 6–8 on a scale of 1 (hardly noticeable) to 10 (painful) [33]. Regardless of group affiliation, muscles were voluntarily tensed during the stimulation episode.

2.4. Exercise Procedure

Both groups received WB-EMS and performed exercises meanwhile. (Supplementary Figures S1–S3). The UBG used upper body exercise movements (chest and upper back including shoulders and arms) and the LBG used lower body exercise movements (buttocks and thigh muscles including abductors and adductors). The UBG training consisted of rowing, butterfly reverse, latissimus pulls, pushups, butterfly, biceps curls, and triceps pulldowns. The LBG training consisted of squats, lunges, adductions, abductions, hip lifts, and leg raises. Both groups exercised the trunk (abdomen and lower back) with back extensions, crunches, and oblique crunches. Selected exercises were performed with additional fitness equipment (fitness tubes and elastic bands, each with varying resistance, and a Swiss Ball). During the first 1 to 2 sessions (depending on the training level), subjects maintained the position over the period of stimulation that they had taken at the onset of the stimulus. One set of 12 repetitions was performed per exercise, with each repetition beginning with the onset of the pulse. To maintain the same physical load level, i.e.,16 to 17 on the Borg RPE scale [33], the number of movements during an impulse interval could be increased up to three. If the training stimulus was not sufficient after the aforementioned customization, the originally targeted static exercise position should be maintained during the interval break. However, overexertion led to a backward correction. Another way to increase the intensity to the desired level was to increase the resistance either by giving an additional fitness tube or rubber band, or by using a version that offered more resistance.

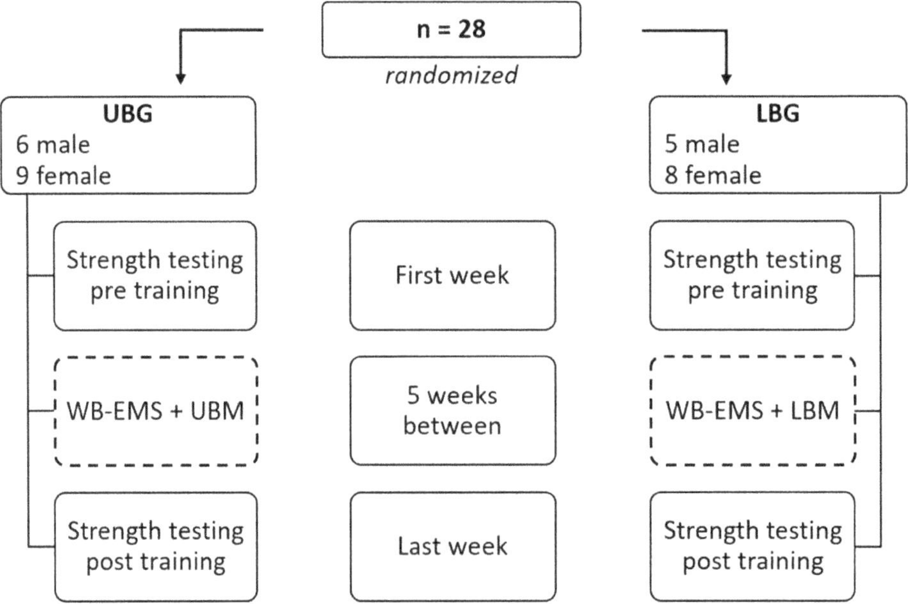

Figure 1. Overview of the study design and procedure. UBG: upper body group; LBG: lower body group; WB-EMS: whole-body electromyostimulation (dashed frame); UBM: upper body exercise movements; LBM: lower body exercise movements.

2.5. Isometric Strength Testing Procedure

Isometric maximum strength (N) was determined during 10 different exercises (arm adduction, arm pull, leg extension, and leg curl, each unilateral left and unilateral right, as well as during biceps curl and triceps pulldown, each bilateral) in standardized positions (Supplementary Figure S4) pre (initial measurement) and post (final measurement) intervention using a mobile device (KD 9363 including DMS measuring amplifier GVS-2; ME-measuring systems GmbH, Hennigsdorf, Germany), which was more practicable than the determination of the 1-RM. Reliability of the isometric maximum strength measurement method was verified by Runkel and colleagues for several test positions (triceps pulldown, biceps curl, arm pull, sit-up, leg curl, leg extension) in healthy subjects with a comparable body mass index [34] by a high interclass correlation coefficient (r = 0.764 to 0.934). At both time points, the tests were performed three times in each position. The pause was set to 10 seconds between individual tests. In each case, the maximum value was used for analysis. The whole testing procedure lasted approximately 20 min.

2.6. Statistical Analysis

Due to the presence of some discordant values (see box plots), skewed distribution in some cases (Shapiro–Wilk test), partial heterogeneity of error variances (Levene's test), and partial heterogeneity of covariances (Box test), nonparametric statistical tests were employed. The differences between the initial and the final maximum isometric strength were determined separately for each group using the Wilcoxon test. The initial and the final values were compared between the groups using the Mann–Whitney U test. Absolute differences were calculated by subtracting the initial values from the final values, and relative differences were calculated by dividing the final values by the initial values (the initial value was set to 100%). Group comparisons were performed using the Mann–Whitney U test for absolute and relative differences. The significance level was set to < 0.05. Two-tailed analyses were used. The results of the non-parametric tests were used to calculate the effect sizes [35]. A distinction was made between large effects (r $\geq$ 0.5), medium effects (< 0.5 to 0.3), and small effects (< 0.3 to 0.1) [36]. Statistics were calculated using SPSS (IBM SPSS Statistics for Windows, Version 28.0., IBM Corp., Armonk, NY, USA) and Excel (Microsoft Excel for Windows, 16.0., Microsoft Corp., Redmond, WA, USA). An intention-to-treat analysis was not possible due to dropouts occurring at baseline.

3. Results

Of the included subjects, 28 completed the study. The dropouts occurred due to personal reasons. The characteristics of the groups did not differ significantly from each other (Table 1) and the total training volume was similar in both groups. Most subjects (n = 9 in each group) completed five sessions and no adverse effects occurred. The body mass remained unchanged in both the UBG and the LBG (Table 1). Neither the initial nor the final values differed significantly between the two groups. Isometric maximum strength was significantly higher after EMS training in both groups, both in absolute terms (Table 2 UBG; Table 3 LBG) and body mass adjusted (N/kg), except for left leg extension in the UBG and biceps curl in the LBG. The changes in absolute strength were similar in both groups (Table 4). Body mass adjusted strength during left arm pull showed a higher increase in the LBG (Figure 2). In the other test positions, group affiliation made no difference (Figures 2–4). Furthermore, the LBG achieved a higher percentage strength gain in left arm pull, both absolute (Table 4) and body mass adjusted (UBG median 114.25% vs. LBG median 137.05%; p = 0.020; r = 0.44).

Table 1. Characteristics presented as medians (ranges) of the total collective (n = 28) and the two groups.

	Total (m 11; f 17)	UBG (m 6; f 9)	LBG (m 5; f 8)
Age [years]	28 (20–36)	32 (25–36)	26 (20–35)
Height [cm]	173.0 (159–186)	174.0 (159.0–186.0)	171.0 (160.0–186.0)
Body mass pre [kg]	74.1 (47.4–114.3)	78.3 (53.1–114.3)	67.2 (47.4–100.3)
Body mass post [kg]	74.4 (48.0–112.9)	78.2 (52.8–112.9)	68.0 (48.0–99.7)
BMI pre [kg/m^2]	25.33 (18.21–40.98)	25.68 (19.27–40.98)	23.88 (18.21–30.35)
BMI post [kg/m^2]	25.06 (18.08–38.57)	25.63 (19.16–38.57)	23.74 (18.08–30.65)
Number of sessions	5 (3–6)	5 (3–6)	5 (3–6)

m: male; f: female; UBG: upper body group; LBG: lower body group; BMI: body mass index.

Table 2. Initial and final median maximum strength values (ranges) of the upper body group (UBG).

Test Position	Strength (N) Initial	Strength (N) Final	Significance	Effect Size r
Arm adduction right	83.3 (44.8–143.0)	116.2 (42.6–178.3)	0.002 **	0.81
Arm adduction left	83.0 (44.8–124.4)	124.4 (54.2–196.0)	<0.001 **	0.88
Arm pull right	173.0 (117.0–293.7)	232.6 (124.7–331.6)	0.001 **	0.84
Arm pull left	196.0 (114.1–331.7)	242.2 (132.4–378.3)	0.006 **	0.70
Triceps pulldown	253.0 (142.7–461.2)	279.2 (149.6–510.4)	0.012 *	0.65
Biceps curl	308.5 (117.6–512.7)	331.7 (143.6–528.2)	0.008 **	0.69
Leg extension right	377.2 (196.7–697.4)	404.8 (237.4–766.0)	0.015 *	0.63
Leg extension left	373.4 (130.9–682.5)	403.3 (218.6–769.0)	0.100	0.43
Leg curl right	184.6 (40.9–296.1)	200.1 (71.2–447.5)	0.005 **	0.72
Leg curl left	185.5 (48.0–296.5)	186.9 (85.5–396.0)	0.031 *	0.56

$n = 15$; * significant difference $p < 0.05$; ** highly significant difference $p < 0.01$.

Table 3. Initial and final median maximum strength values (ranges) of the lower body group (LBG).

Test Position	Strength (N) Initial	Strength (N) Final	Significance	Effect Size r
Arm adduction right	61.7 (28.8–127.4)	84.9 (43.1–155.5)	0.007 **	0.75
Arm adduction left	56.2 (29.0–111.3)	85.1 (46.1–170.0)	0.002 **	0.84
Arm pull right	140.0 (80.9–281.7)	170.4 (145.0–362.2)	0.001 **	0.88
Arm pull left	131.0 (94.9–216.8)	167.9 (136.8–378.2)	0.001 **	0.88
Triceps pulldown	178.0 (125.0–474.0)	203.0 (140.2–474.2)	0.039 *	0.57
Biceps curl	215.3 (137.0–559.5)	212.7 (167.9–531.0)	0.221	0.34
Leg extension right	330.5 (218.2–725.0)	385.6 (281.0–787.8)	0.002 **	0.86
Leg extension left	304.0 (184.8–612.0)	348.5 (228.1–704.9)	0.001 **	0.88
Leg curl right	139.4 (104.6–268.7)	170.0 (140.0–311.9)	0.001 **	0.88
Leg curl left	127.0 (105.0–261.4)	166.2 (132.0–287.9)	0.002 **	0.86

$n = 13$; * significant difference $p < 0.05$; ** highly significant difference $p < 0.01$.

Table 4. Median differences (ranges) between final and initial maximum strength values in both groups (UBG and LBG).

Test Position	Δ UBG (N)	Δ UBG (%)	Δ LBG (N)	Δ LBG (%)
Arm adduction right	26.5 (−11.9–68.3)	137.01 (83.43–175.63)	12.6 (−13.9–44.3)	120.68 (86.47–222.92)
Arm adduction left	27.5 (5.6–83.3)	131.70 (104.61–183.05)	29.1 (−15.2–58.7)	152.74 (84.85–244.67)
Arm pull right	29.4 (−6.3–85.4)	114.74 (96.14–149.36)	32.5 (10.4–96.1)	128.46 (105.28–182.82)
Arm pull left	21.6 (−27.2–90.2)	115.07 (84.68–144.24)	41.4 (4.6–164.0)	131.67 (102.47–176.56) *
Triceps pulldown	11.7 (−9.0–89.5)	105.88 (97.20–161.89)	19.6 (−22.5–74.5)	110.17 (93.72–130.56)
Biceps curl	24.5 (−31.3–69.4)	107.45 (88.79–139.86)	21.2 (−45.1–40.7)	108.82 (88.02–122.55)
Leg extension right	52.0 (−90.7–141.2)	117.40 (86.99–125.25)	55.1 (−4.4–130.3)	116.67 (98.94–135.90)
Leg extension left	41.8 (−94.5–120.3)	110.05 (80.11–167.00)	56.0 (4.7–215.7)	121.68 (101.14–144.09)
Leg curl right	30.3 (−27.8–151.4)	117.45 (79.41–185.79)	37.1 (15.5–126.3)	124.87 (111.12–168.34)
Leg curl left	13.9 (−27.3–99.5)	107.42 (88.28–178.13)	39.7 (−1.7–70.7)	131.41 (98.79–160.84)

Δ: differences; UBG ($n = 15$): upper body group; LBG ($n = 13$): lower body group; * significant difference compared to UBG, $p = 0.020$ (effect size r = 0.44).

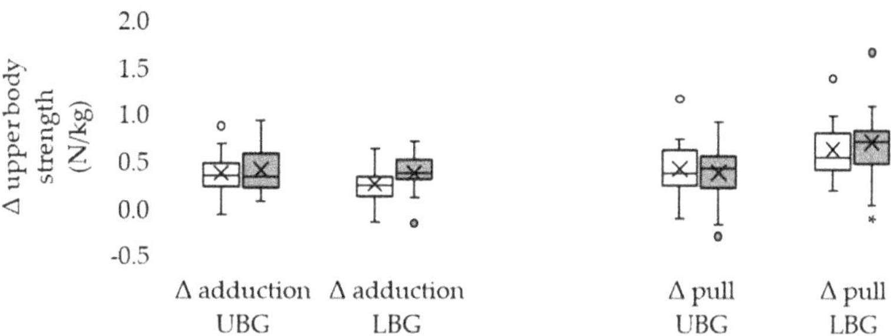

Figure 2. Differences between the final and the initial body weight adjusted maximum upper body strength (arm adduction and arm pull) in the upper body group and in the lower body group. Δ: differences; UBG (n = 15): upper body group; LBG (n = 13): lower body group; white box: right arm; grey box: left arm; circles represent discordant values; means are displayed by crosses and medians by crossbars; * significant difference between LBG and UBG, $p < 0.040$ (effect size r = 0.39).

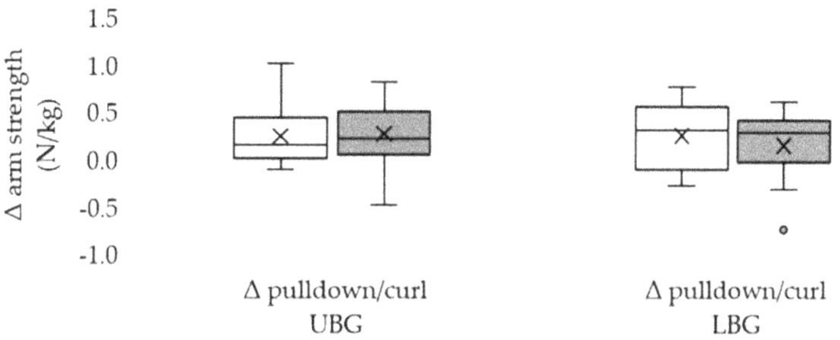

Figure 3. Differences between the final and the initial body weight adjusted maximum arm strength (biceps curl and triceps pulldown) in the upper body group and in the lower body group. Δ: differences; UBG (n = 15): upper body group; LBG (n = 13): lower body group; white box: triceps pulldown; grey box: biceps curl; the circle represents a discordant value; means are displayed by crosses and medians by crossbars; no significant differences occurred.

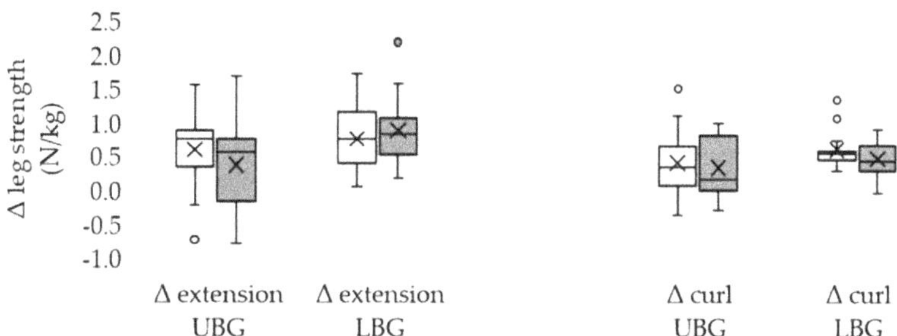

Figure 4. Differences between the final and the initial body weight adjusted maximum leg strength (leg extension and leg curl) in the upper body group and in the lower body group. Δ: difference; UBG (n = 15): upper body group; LBG (n = 13): lower body group; white box: right leg; grey box: left leg; circles represent discordant values; means are displayed by crosses and medians by crossbars; no significant differences occurred.

4. Discussion

4.1. Overview

Significant strength changes were observed in both groups after about five weeks training (one session per week). The percentage differences between the initial and final tests were higher than those found in the reliability analysis of the test device by Runkel and colleagues [34]. Therefore, the changes could be attributed to training. LBG training improved left arm pull strength more than UBG training. However, there were no group differences in the other exercises. Initial values between the two groups were not significantly different, but possibly at clinically relevant levels. If the higher initial values had been due to differences in training history, a lower ability to further increase strength would have be needed to be considered [37]. However, subjects should have abstained from intense physical activity for at least six months before starting the study.

4.2. Accompanying Voluntary Activity

Little is known about the effects of movements for strength gain during EMS. During local application, movements are usually avoided and isometric contractions are performed. Maffiuletti [38] summarized that there are no differences in strength increase between EMS and EMS superimposed on voluntary contractions. However, the conclusion is based on the results of isometric interventions. Although movements are thought to promote the activity of stimulated muscles [26], our results failed to show a consistent influence of active exercise movements on strength gains. Furthermore, strength gains from conventional resistance training depend, among others, on the range of motion used [39,40]. However, isometric contractions at multiple joint angles might cover at least in part the physiological range of motion. For EMS training, Maffiuletti [38] recommends changing the joint position and furthermore, changing the electrode positioning to increase recruitment. Admittedly, Kemmler and colleagues [29] demonstrated the benefit of movement during WB-EMS use, with participants exercising in supine position. In contrast, our participants performed exercises in different positions. Therefore, any movements of body parts that were not primarily intended for the exercises and possible differences in resistance to gravity might have influenced the results. Furthermore, it needs to be considered that additional fitness equipment (fitness tubes and elastic bands with different resistance as well as a Swiss Ball) was used for selected exercises. However, exercise movements using additional fitness equipment did not affect the results. In addition, both the UBG and LBG performed exercises for the trunk. Therefore, both groups received partially similar dynamic training stimuli (three exercises). Movements inevitably lead to changes in muscle length and shape (e.g., biceps muscle during curl). Hence, changes in the electrode contact were very likely to occur. Furthermore, training that aims to enhance endurance and strength at the same time, such as EMS superimposed on cycling [41,42], requires movements. However, stimulation intensity must be considered to ensure the range of motion [43].

4.3. Training Models and Adaptations

Supraspinal mechanisms appear to be responsible for the initial strength development through EMS training [23]. Bezerra and colleagues [44] showed increased strength after EMS superimposed onto maximum isometric quadriceps contractions, not only of the exercised leg but also of the unexercised leg, confirming neural contribution. The potential to use EMS for rapid strength gains was demonstrated by Deley and colleagues [45], who reported that maximum dynamic leg extension torque in prepubertal girls could be increased by up to 50.6% with three weekly isometric applications over a three-week period. According to Adams [46], atrophic patients as well as casualties are target groups for the use of EMS. After 5 to 6 weeks, a 10 to 15% enhancement of muscle function can be achieved, but three sessions a week are recommended. Several studies confirmed the impact of WB-EMS on strength [10,26]. However, to our knowledge, only Kemmler and colleagues [29] have studied the effects of exercise during WB-EMS to date. In most cases, the lower body was investigated. Von Stengel and Kemmler [25] showed that leg/hip strength can be improved with 1.5 WB-EMS training sessions (with unloaded, low effort exercises) per week over a 14 to 16 week period, regardless of age. Furthermore, strength gains due to unloaded WB-EMS were similar compared to a HIT training after 16 weeks with three sessions in two weeks [3]. An increase in strength was also observed after shorter training periods. For example, WB-EMS superimposed on jumps twice a week over seven weeks significantly improved leg strength in contrast to normal jump training [10,47,48]. In the study by Wirtz and colleagues [28], leg flexors strength increased only after combining stimulation of multiple body parts with loaded squats (100% 10 RM) twice per week and it was higher three weeks after the six-week training compared to the same training without stimulation. Dörmann and colleagues [18] showed significant improvements in leg strength after a four-week, eight-session WB-EMS training program that were similar to those seen in the control group, which performed the same training that included strength exercises, without additional stimulation, and in which intensification was accomplished using other training tools. However, not only leg muscles but also upper body muscles could benefit from dynamic WB-EMS. Reljic and colleagues [26] observed improvements throughout the entire body after a 12-week WB-EMS program with slight motions, consisting of two sessions per week. Our results suggest that even fewer training sessions are beneficial than previously described, whether or not exercise movements are performed during stimulation, which appears to be due to neural factors. Therefore, not only locally applied EMS training regimens have the potential to increase strength, but also WB-EMS training regimens without additional exercise movements.

4.4. Transferability

Benefits from WB-EMS can also be expected, for example, for patients suffering from sarcopenia, sarcopenic obesity, and low back pain [14]. It might be useful especially for beginners to start WB-EMS training with a five-week training period without additional exercise movements to improve basic strength before starting a more challenging exercise program. WB-EMS without additional exercise movements can be a first access to training when health conditions do not allow conventional exercises or when a lack of compliance exists. Relative to WB-EMS, local application appears to be superior in gaining strength [49]. However, the lack of focus on selected zones owing to stimulation of the entire body is a suggested explanation for the difference [14]. Therefore, only target muscles could be stimulated and not all available electrodes could be used, even if an electrode suit is worn, or zones could be stimulated in an individual order.

4.5. Limitations

We have shown that the effect of WB-EMS on strength gains is independent of the concomitant exercise movements. Nevertheless, some limitations need to be acknowledged. A test of core strength would have been useful, as both groups performed core strength exercises under the same conditions and a higher strength can be expected as observed in the study by Berger and colleagues [1], although they used a more extensive training program. Owing to two dropouts, the group sizes were slightly different, which affected the comparison. Furthermore, the strength gains of the dynamically trained muscles might have been underestimated, since only isometric strength was tested. It must also be mentioned that the increase in strength might have been influenced by deviations from the predefined number of training sessions. To evaluate the intensity of the movement sequences, an unstimulated group could have been used. Furthermore, an inactive group could have been used as a reference for the interventions. However, the study focused on the comparison between the EMS application without and the application with concurring exercise movements. When using WB-EMS training technology, the load parameters must be set with care to avoid unintended side effects, particularly during the first sessions of novices when adaptation to the load has not yet occurred in the form of the "repeated bout effect" [50].

5. Conclusions

WB-EMS training without accompanying movement exercises leads to substantial strength gains even during a short WB-EMS training period. At the beginning of WB-EMS training, electromyostimulation is more important for strength gains than active exercise movements. Therefore, future studies should examine the effects of exercise movements during long-term training periods, or consider individuals already adapted to WB-EMS training or strength training. The transferability of the results to a collective experienced with WB-EMS or strength training should be questioned, as movements (and maybe other approaches, e.g., additional mass or complicating tasks) may become more relevant when initial adaptations to training are exhausted. Since the training effort with WB-EMS is low, people with health restrictions, beginners without experience in strength training, and those returning to training might benefit from these results. These groups could refrain from exercise movements during the first WB-EMS training sessions and integrate them during the course of the subsequent training.

Supplementary Materials: The following supporting information can be downloaded at: https://www.mdpi.com/article/10.3390/healthcare11050741/s1, Figure S1: Exercise movements for the upper body performed by the upper body group (UBG); Figure S2: Exercise movements for the lower body performed by the lower body group (LBG); Figure S3: Exercises for the trunk performed by both groups; Figure S4: Isometric strength tests; Table S1: Data overview.

Author Contributions: Conceptualization, U.F.W. and T.F.; methodology, U.F.W.; validation, H.S., U.F.W. and T.F.; formal analysis, H.S., U.F.W., F.T. and T.H.; investigation, T.F.; resources, H.S., U.F.W., T.F. and T.H.; data curation, H.S., U.F.W. and T.F.; writing—original draft preparation, H.S.; writing—review and editing, H.S., U.F.W., T.F., F.T. and T.H.; visualization, H.S. and T.H.; supervision, U.F.W. and T.H.; project administration, U.F.W. and T.F. All authors have read and agreed to the published version of the manuscript.

Funding: This research received no external funding.

Institutional Review Board Statement: The study was conducted in accordance with the Declaration of Helsinki, and approved by the ethics committee of the University of Wuppertal (MS/BBL 200114 Wehmeier).

Informed Consent Statement: Informed consent was obtained from all subjects involved in the study.

Data Availability Statement: Data are available (Table S1).

Acknowledgments: We would like to thank go!ORANGE–Studio für EMS for providing the WB-EMS equipment and the studio. We acknowledge the support of the Open Access Publication Fund of the University of Wuppertal.

Conflicts of Interest: The authors declare no conflict of interest.

References

1. Berger, J.; Ludwig, O.; Becker, S.; Backfisch, M.; Kemmler, W.; Fröhlich, M. Effects of an Impulse Frequency Dependent 10-Week Whole-body Electromyostimulation Training Program on Specific Sport Performance Parameters. *J. Sports Sci. Med.* **2020**, *19*, 271–281.
2. Jee, Y.S. The efficacy and safety of whole-body electromyostimulation in applying to human body: Based from graded exercise test. *J. Exerc. Rehabil.* **2018**, *14*, 49–57. [CrossRef]
3. Kemmler, W.; Teschler, M.; Weissenfels, A.; Froehlich, M.; Kohl, M.; von Stengel, S. Whole-Body Electromyostimulation Versus High Intensity (Resistance Exercise) Training—Impact on Body Composition and Strength. *Dtsch. Z. Sportmed.* **2015**, *66*, 321–327. [CrossRef]
4. Pano-Rodriguez, A.; Beltran-Garrido, J.V.; Hernandez-Gonzalez, V.; Reverter-Masia, J. Effects of Whole-Body Electromyostimulation on Physical Fitness in Postmenopausal Women: A Randomized Controlled Trial. *Sensors* **2020**, *20*, 1482. [CrossRef] [PubMed]
5. Fiorilli, G.; Quinzi, F.; Buonsenso, A.; Casazza, G.; Manni, L.; Parisi, A.; Di Costanzo, A.; Calcagno, G.; Soligo, M.; di Cagno, A. A Single Session of Whole-Body Electromyostimulation Increases Muscle Strength, Endurance and proNGF in Early Parkinson Patients. *Int. J. Environ. Res. Public Health* **2021**, *18*, 5499. [CrossRef] [PubMed]

6. Kemmler, W.; Teschler, M.; Weissenfels, A.; Bebenek, M.; von Stengel, S.; Kohl, M.; Freiberger, E.; Goisser, S.; Jakob, F.; Sieber, C.; et al. Whole-body electromyostimulation to fight sarcopenic obesity in community-dwelling older women at risk. Resultsof the randomized controlled FORMOsA-sarcopenic obesity study. *Osteoporos. Int.* **2016**, *27*, 3261–3270. [CrossRef]

7. Kraemer, W.J.; Ratamess, N.A. Fundamentals of resistance training: Progression and exercise prescription. *Med. Sci. Sports Exerc.* **2004**, *36*, 674–688. [CrossRef]

8. Filipovic, A.; Kleinöder, H.; Dörmann, U.; Mester, J. Electromyostimulation–a systematic review of the influence of training regimens and stimulation parameters on effectiveness in electromyostimulation training of selected strength parameters. *J. Strength Cond. Res.* **2011**, *25*, 3218–3238. [CrossRef]

9. Tiggemann, C.L.; Korzenowski, A.L.; Brentano, M.A.; Tartaruga, M.P.; Alberton, C.L.; Kruel, L.F. Perceived exertion in different strength exercise loads in sedentary, active, and trained adults. *J. Strength Cond. Res.* **2010**, *24*, 2032–2041. [CrossRef]

10. Filipovic, A.; DeMarees, M.; Grau, M.; Hollinger, A.; Seeger, B.; Schiffer, T.; Bloch, W.; Gehlert, S. Superimposed Whole-Body Electrostimulation Augments Strength Adaptations and Type II Myofiber Growth in Soccer Players during a Competitive Season. *Front. Physiol.* **2019**, *10*, 1187. [CrossRef]

11. Filipovic, A.; Bizjak, D.; Tomschi, F.; Bloch, W.; Grau, M. Influence of Whole-Body Electrostimulation on the Deformability of Density-Separated Red Blood Cells in Soccer Players. *Front. Physiol.* **2019**, *10*, 548. [CrossRef] [PubMed]

12. American College of Sports Medicine. *ACSM's Guidelines for Exercise Testing and Prescription*, 11th ed.; Wolters Kluwer: Philadelphia, PA, USA, 2021.

13. Kemmler, W.; Froehlich, M.; von Stengel, S.; Kleinöder, H. Whole-Body Electromyostimulation—The Need for Common Sense! Rationale and Guideline for a Safe and Effective Training. *Dtsch. Z. Sportmed.* **2016**, *67*, 218–221. [CrossRef]

14. Kemmler, W.; Weissenfels, A.; Willert, S.; Shojaa, M.; von Stengel, S.; Filipovic, A.; Kleinöder, H.; Berger, J.; Fröhlich, M. Efficacy and Safety of Low Frequency Whole-Body Electromyostimulation (WB-EMS) to Improve Health-Related Outcomes in Non-athletic Adults. A Systematic Review. *Front. Physiol.* **2018**, *9*, 573. [CrossRef]

15. Micke, F.; Kleinöder, H.; Dörmann, U.; Wirtz, N.; Donath, L. Effects of an Eight-Week Superimposed Submaximal Dynamic Whole-Body Electromyostimulation Training on Strength and Power Parameters of the Leg Muscles: A Randomized Controlled Intervention Study. *Front. Physiol.* **2018**, *9*, 1719. [CrossRef] [PubMed]

16. Schuhbeck, E.; Birkenmaier, C.; Schulte-Göcking, H.; Pronnet, A.; Jansson, V.; Wegener, B. The Influence of WB-EMS-Training on the Performance of Ice Hockey Players of Different Competitive Status. *Front. Physiol.* **2019**, *10*, 1136. [CrossRef]

17. Ludwig, O.; Berger, J.; Schuh, T.; Backfisch, M.; Becker, S.; Fröhlich, M. Can A Superimposed Whole-Body Electromyostimulation Intervention Enhance the Effects of a 10-Week Athletic Strength Training in Youth Elite Soccer Players? *J. Sports Sci. Med.* **2020**, *19*, 535–546.

18. Dörmann, U.; Wirtz, N.; Micke, F.; Morat, M.; Kleinöder, H.; Donath, L. The Effects of Superimposed Whole-Body Electromyostimulation during Short-Term Strength Training on Physical Fitness in Physically Active Females: A Randomized Controlled Trial. *Front. Physiol.* **2019**, *10*, 728. [CrossRef]

19. Moritani, T.; deVries, H.A. Neural factors versus hypertrophy in the time course of muscle strength gain. *Am. J. Phys. Med.* **1979**, *58*, 115–130.

20. Coburn, J.W.; Housh, T.J.; Malek, M.H.; Weir, J.P.; Cramer, J.T.; Beck, T.W.; Johnson, G.O. Neuromuscular responses to three days of velocity-specific isokinetic training. *J. Strength Cond. Res.* **2006**, *20*, 892–898. [CrossRef]

21. Elgueta-Cancino, E.; Evans, E.; Martinez-Valdes, E.; Falla, D. The Effect of Resistance Training on Motor Unit Firing Properties: A Systematic Review and Meta-Analysis. *Front. Physiol.* **2022**, *13*, 817631. [CrossRef]

22. Natsume, T.; Ozaki, H.; Kakigi, R.; Kobayashi, H.; Naito, H. Effects of training intensity in electromyostimulation on human skeletal muscle. *Eur. J. Appl. Physiol.* **2018**, *118*, 1339–1347. [CrossRef] [PubMed]

23. Hortobágyi, T.; Maffiuletti, N.A. Neural adaptations to electrical stimulation strength training. *Eur. J. Appl. Physiol.* **2011**, *111*, 2439–2449. [CrossRef] [PubMed]

24. Rodrigues-Santana, L.; Adsuar, J.C.; Denche-Zamorano, Á.; Vega-Muñoz, A.; Salazar-Sepúlveda, G.; Contreras-Barraza, N.; Galán-Arroyo, C.; Louro, H. Bibliometric Analysis of Studies on Whole Body Electromyostimulation. *Biology* **2022**, *11*, 1205. [CrossRef]

25. von Stengel, S.; Kemmler, W. Trainability of leg strength by whole-body electromyostimulation during adult lifespan: A study with male cohorts. *Clin. Interv. Aging* **2018**, *13*, 2495–2502. [CrossRef]

26. Reljic, D.; Konturek, P.C.; Herrmann, H.J.; Neurath, M.F.; Zopf, Y. Effects of whole-body electromyostimulation exercise and caloric restriction on cardiometabolic risk profile and muscle strength in obese women with the metabolic syndrome: A pilot study. *J. Physiol. Pharmacol.* **2020**, *71*, 89–98. [CrossRef]

27. Evangelista, A.L.; Teixeira, C.V.S.; Barros, B.M.; de Azevedo, J.B.; Paunksnis, M.R.R.; Souza, C.R.; Wadhi, T.; Rica, R.L.; Braz, T.V.; Bocalini, D.S. Does whole-body electrical muscle stimulation combined with strength training promote morphofunctional alterations? *Clinics* **2019**, *74*, e1334. [CrossRef]

28. Wirtz, N.; Zinner, C.; Doermann, U.; Kleinoeder, H.; Mester, J. Effects of Loaded Squat Exercise with and without Application of Superimposed EMS on Physical Performance. *J. Sports Sci. Med.* **2016**, *15*, 26–33.

29. Kemmler, W.; Teschler, M.; von Stengel, S. Role of whole body-electromyostimulation—"A series of studies". *Osteologie* **2015**, *24*, 20–29. [CrossRef]

30. Kemmler, W.; Weissenfels, A.; Willert, S.; Fröhlich, M.; Ludwig, O.; Berger, J.; Zart, S.; Becker, S.; Backfisch, M.; Kleinöder, H.; et al. Recommended Contraindications for the Use of Non-Medical WB-Electromyostimulation. *Dtsch. Z. Sportmed.* **2019**, *70*, 278–282. [CrossRef]

31. Stöllberger, C.; Finsterer, J. Side effects of and contraindications for whole-body electro-myo-stimulation: A viewpoint. *BMJ Open Sport Exerc. Med.* **2019**, *5*, e000619. [CrossRef]

32. Association, W.M. World Medical Association Declaration of Helsinki: Ethical principles for medical research involving human subjects. *JAMA* **2013**, *310*, 2191–2194. [CrossRef]

33. Morishita, S.; Tsubaki, A.; Nakamura, M.; Nashimoto, S.; Fu, J.B.; Onishi, H. Rating of perceived exertion on resistance training in elderly subjects. *Expert. Rev. Cardiovasc. Ther.* **2019**, *17*, 135–142. [CrossRef] [PubMed]

34. Runkel, B.; Kappelhoff, M.; Hilberg, T. Complex strength performance in patients with haemophilia A. Method development and testing. *Hamostaseologie* **2015**, *35* (Suppl. S1), S12–S17. [CrossRef] [PubMed]

35. Tomczak, M.; Tomczak, E. The need to report effect size estimates revisited. An overview of some recommended measures of effect size. *Trends Sport Sci.* **2014**, *1*, 19–25.

36. Cohen, J. *Statistical Power Analysis for the Behavioral Sciences*; Erlbaum: Hillsdals, NJ, USA, 1988; Volume 2.

37. Hughes, D.C.; Ellefsen, S.; Baar, K. Adaptations to Endurance and Strength Training. *Cold Spring Harb. Perspect. Med.* **2018**, *8*, a029769. [CrossRef]

38. Maffiuletti, N.A. Physiological and methodological considerations for the use of neuromuscular electrical stimulation. *Eur. J. Appl. Physiol.* **2010**, *110*, 223–234. [CrossRef]

39. Martínez-Cava, A.; Hernández-Belmonte, A.; Courel-Ibáñez, J.; Morán-Navarro, R.; González-Badillo, J.J.; Pallarés, J.G. Bench Press at Full Range of Motion Produces Greater Neuromuscular Adaptations Than Partial Executions after Prolonged Resistance Training. *J. Strength Cond. Res.* **2022**, *36*, 10–15. [CrossRef]

40. Pallarés, J.G.; Cava, A.M.; Courel-Ibáñez, J.; González-Badillo, J.J.; Morán-Navarro, R. Full squat produces greater neuromuscular and functional adaptations and lower pain than partial squats after prolonged resistance training. *Eur. J. Sport Sci.* **2020**, *20*, 115–124. [CrossRef]

41. Omoto, M.; Matsuse, H.; Hashida, R.; Takano, Y.; Yamada, S.; Ohshima, H.; Tagawa, Y.; Shiba, N. Cycling Exercise with Electrical Stimulation of Antagonist Muscles Increases Plasma Growth Hormone and IL-6. *Tohoku J. Exp. Med.* **2015**, *237*, 209–217. [CrossRef]

42. Masayuki, O.; Matsuse, H.; Takano, Y.; Yamada, S.; Ohshima, H.; Tagawa, Y.; Shiba, N. Oxygen Uptake during Aerobic Cycling Exercise Simultaneously Combined with Neuromuscular Electrical Stimulation of Antagonists. *J. Nov. Physiother.* **2013**, *3*, 185. [CrossRef]

43. Paillard, T. Training Based on Electrical Stimulation Superimposed Onto Voluntary Contraction Would be Relevant Only as Part of Submaximal Contractions in Healthy Subjects. *Front. Physiol* **2018**, *9*, 1428. [CrossRef] [PubMed]

44. Bezerra, P.; Zhou, S.; Crowley, Z.; Brooks, L.; Hooper, A. Effects of unilateral electromyostimulation superimposed on voluntary training on strength and cross-sectional area. *Muscle Nerve* **2009**, *40*, 430–437. [CrossRef] [PubMed]

45. Deley, G.; Cometti, C.; Fatnassi, A.; Paizis, C.; Babault, N. Effects of combined electromyostimulation and gymnastics training in prepubertal girls. *J. Strength Cond. Res.* **2011**, *25*, 520–526. [CrossRef] [PubMed]

46. Adams, V. Electromyostimulation to fight atrophy and to build muscle: Facts and numbers. *J. Cachexia Sarcopenia Muscle* **2018**, *9*, 631–634. [CrossRef] [PubMed]

47. Filipovic, A.; Kleinöder, H.; Plück, D.; Hollmann, W.; Bloch, W.; Grau, M. Influence of Whole-Body Electrostimulation on Human Red Blood Cell Deformability. *J. Strength Cond. Res.* **2015**, *29*, 2570–2578. [CrossRef]

48. Filipovic, A.; Grau, M.; Kleinöder, H.; Zimmer, P.; Hollmann, W.; Bloch, W. Effects of a Whole-Body Electrostimulation Program on Strength, Sprinting, Jumping, and Kicking Capacity in Elite Soccer Players. *J. Sports Sci. Med.* **2016**, *15*, 639–648.

49. Filipovic, A.; Kleinöder, H.; Dörmann, U.; Mester, J. Electromyostimulation–a systematic review of the effects of different electromyostimulation methods on selected strength parameters in trained and elite athletes. *J. Strength Cond. Res.* **2012**, *26*, 2600–2614. [CrossRef]

50. Kemmler, W.; Teschler, M.; Bebenek, M.; von Stengel, S. [(Very) high Creatinkinase concentration after exertional whole-body electromyostimulation application: Health risks and longitudinal adaptations]. *Wien. Med. Wochenschr.* **2015**, *165*, 427–435. [CrossRef]

Article

The Effects of Combining Aerobic and Heavy Resistance Training on Body Composition, Muscle Hypertrophy, and Exercise Satisfaction in Physically Active Adults

Jerrican Tan [1,2], Oleksandr Krasilshchikov [3,*], Garry Kuan [2], Hairul Anuar Hashim [2], Monira I. Aldhahi [4], Sameer Badri Al-Mhanna [5] and Georgian Badicu [6]

[1] Fitness Innovations Malaysia Sendirian Berhad, Petaling Jaya 47820, Selangor, Malaysia; jerrican@student.usm.my

[2] School of Health Sciences, Universiti Sains Malaysia, Kubang Kerian 16150, Kelantan, Malaysia; garry@usm.my (G.K.); hairulkb@usm.my (H.A.H.)

[3] Faculty of Sports Science and Recreation, Universiti Teknologi MARA, Shah Alam 40450, Selangor, Malaysia

[4] Department of Rehabilitation Sciences, College of Health and Rehabilitation Sciences, Princess Nourah bint Abdulrahman University, P.O. Box 84428, Riyadh 11671, Saudi Arabia; mialdhahi@pnu.edu.sa

[5] Department of Physiology, School of Medical Sciences, Universiti Sains Malaysia, Kubang Kerian 16150, Kelantan, Malaysia; sameerbadri9@gmail.com

[6] Department of Physical Education and Special Motricity, Faculty of Physical Education and Mountain Sports, Transilvania University of Braşov, 500068 Braşov, Romania; georgian.badicu@unitbv.ro

* Correspondence: oleksandr@uitm.edu.my

Citation: Tan, J.; Krasilshchikov, O.; Kuan, G.; Hashim, H.A.; Aldhahi, M.I.; Al-Mhanna, S.B.; Badicu, G. The Effects of Combining Aerobic and Heavy Resistance Training on Body Composition, Muscle Hypertrophy, and Exercise Satisfaction in Physically Active Adults. *Healthcare* **2023**, *11*, 2443. https://doi.org/10.3390/healthcare11172443

Academic Editor: Herbert Löllgen

Received: 27 July 2023
Revised: 26 August 2023
Accepted: 29 August 2023
Published: 31 August 2023

Abstract: This study investigated the effects of combined aerobic and heavy resistance training on the variables of body composition, muscle hypertrophy, and exercise satisfaction in physically active adults in comparison with heavy resistance training only (predominantly designed for hypertrophy). Twenty-two healthy male adults between the ages of 18 and 35, who had limited previous experience with muscle resistance training, participated in the intervention program while maintaining their physical activity level. The participants were randomly allocated into two groups: the resistance training group (control group) and the combined training group (experimental group), which involved both resistance training and aerobic training. Aerobic training consisted of 30 min aerobic interval training sessions three times a week with a total of 8 min work bouts in each at 60–70% of heart rate reserve (HRR). The intervention training program lasted for eight weeks. Resistance training consisted of a 3-day muscle group split (2–3 exercises per muscle group, 8 sets per muscle group, 6–12 repetition maximum (RM). Upon completion, body composition, muscle hypertrophy, and exercise satisfaction were analyzed using the mixed-design ANOVA. Variables selected for this study as markers of body composition responded differently to the different interventions and time; however, some trends were not statistically significant. Overall, it is not possible to state unequivocally that one training modality was superior to another in the body composition cluster, for significant improvements were observed within the groups from pre- to post-interventions, but no significant differences were observed between the resistance training and combined training groups, while, both interventions showed improvement with time in some variables of muscle hypertrophy. Compared to baseline, the exercise satisfaction post-intervention improved within the groups. From pre- to post-testing, both resistance and combined training groups improved exercise satisfaction ($p < 0.05$ in both groups). However, there was no significant difference in exercise satisfaction observed between the resistance training and combined training groups after the training intervention ($p > 0.05$).

Keywords: heavy resistance training; aerobic training; body composition; muscle hypertrophy; exercise satisfaction; physically active adults

---

1. Introduction

Muscle hypertrophy results in an increase in the human metabolism rate [1]. The metabolic rate of muscle is estimated to be around 10 to 15 kcal/kg per day, which is equivalent to approximately 4.5 to 7.0 kcal/lbs per day [2]. Studies of physical activity and strength training interventions (lasting from 8 to 52 weeks) showed an increase in muscle mass of about 2.2 to 4.5 lbs [3,4]. This indicates that the increase of 4.5 pounds of muscle mass would increase the resting metabolic rate by about 50 kilocalories per day. Therefore, greater muscle mass results in a higher energy demand by muscle tissue during physical activity. Additionally, participating in aerobic endurance training leads to increased caloric expenditure, which aids in the reduction of body fat [5]. The incorporation of aerobic endurance training within a weight management plan has been identified as a significant factor in achieving optimal health-related outcomes [6,7].

Performing aerobic and strength training concurrently is an essential part of physical training aimed at improving not only health but athletic performance as well [8]. Beyond the health-related benefits of combining training modalities in one intervention program, researchers also addressed the fitness-related effects of such combinations. Among the attempts to increase strength component in view of fitness-related outcomes, a study on improving power by combining two strength training modalities within one intervention program proved the efficiency of complex training and its superiority over the traditional resistance training in Malaysia amateur weightlifters with at least 2 years of competitive weightlifting experience at the state level [9].

Aerobic endurance exercise is generally thought to have a limited effect on muscle hypertrophy. This is supported by research suggesting that aerobic exercise activates catabolic pathways, while anaerobic exercise stimulates anabolic pathways [10]. There are differences in the intracellular signaling response between the two types of exercises [11,12]. The findings led to the hypothesis of the AMP-activated protein kinase pathway (AMPK)—phosphatidylinositol 3-kinase (AKT) switch, suggesting that there is a discrepancy in the signaling responses produced by anaerobic and aerobic exercises, which may not complement each other to optimize muscular adaptations. AMPK signaling is for the catabolic pathway and is often associated with aerobic endurance exercises, while AKT signaling is for the anabolic pathway and is often associated with anaerobic or resistance training exercises [12]. However, this theory can be overly simplified and leads to misinterpretation. Multiple studies have shown increased mTOR (mammalian target of rapamycin) activation following aerobic endurance exercise [13]. At the same time, resistance training has consistently been found to increase the levels of AMPK [12]. Both aerobic exercise and resistance exercise have been found to be beneficial for increasing muscle mass, strength, and function. The question remains whether combining these two types of exercise yields superior results compared to performing either exercise modality alone.

Some recent studies [14,15] indicated no interference effect of aerobic training when combined with strength training. One study [14] concluded that concurrent training, regardless of the exercise order, can be a viable strategy to improve lower-body maximal strength and total lean mass comparably to resistance-only training with no reference to muscle hypertrophy. The same study suggests that conventional resistance training may facilitate increasing strength in concurrent training. No hypertrophy variables were involved in this study either.

There is limited research on the effects of combining aerobic and resistance exercise on muscle mass in adults. In one study [16], after 5 weeks of intervention, both the strength and strength plus endurance training groups experienced a significantly greater increase in strength (i.e., bench press, biceps curl) and arm cross-sectional area (i.e., the left and right arms) in comparison to the control (no exercise) group. However, there were no significant differences between the strength training group and the strength plus endurance training group. In a small cohort of untrained young males, strength plus endurance training did not impede strength gains or muscle hypertrophy when compared to strength training alone.

Understanding the effects of combined aerobic and resistance exercise on muscle mass could have important implications for developing beneficial programs that target the individual's optimal physiological benefit and satisfaction as well. Therefore, the aim of the current study is to investigate the effect of combined aerobic exercise and resistance exercise on muscle-mass-related variables in adults compared to performing resistance training alone. The objective of the study will be achieved through the following specific aims: to assess the effects of combined aerobic and heavy resistance training on body composition and muscle hypertrophy in physically active adults. In addition, the study will also assess the effect of these training regimens on exercise satisfaction in physically active adults.

2. Methods

2.1. Study Design and Participants

An experimental study design was conducted to investigate the effect of combining traditional muscle hypertrophy training with aerobic training on muscle mass and other related variables. The study was conducted in a workout studio in Kuala Lumpur, Malaysia, and the research protocol was approved by the Universiti Sains Malaysia Human Ethics Committee Protocol No. USM/JEPeM/19090542.

In this study, a convenience sampling method was applied, and the participants were randomly assigned to two groups: a control group and an experimental group. The control group performed traditional muscle hypertrophy training, which included exercises that targeted specific muscle groups using heavy weights and low repetitions. The experimental group combined traditional muscle hypertrophy training with aerobic training. Both groups underwent an 8-week training program, which consisted of three training sessions per week, with each session lasting approximately one hour. The subjects were monitored by trained instructors to ensure proper form and technique during each exercise session.

The research inclusion criteria in this study were healthy physically active male adults between the ages of 18 and 35. The age range was chosen to avoid hormonal factors that could affect hypertrophy, which typically begins to decline after age 35. Additionally, the study focused exclusively on male participants to reduce any potential confounding effects of gender differences in muscle mass and strength.

To ensure that subjects respond homogeneously to hypertrophy training, WHO general population BMI classification was followed during the subjects' recruitment; hence the participants were expected to have a BMI of between 18.5 and 24.9 kg/m^2.

The study included participants who were familiar with resistance training and previously practiced free-weight exercises but had not practiced muscle-hypertrophy-oriented resistance training before. Their history of involvement was movement and health oriented. Their previous training volumes (sets per muscle group) were low, not exceeding 1–2 sets per muscle group per week. This was done to ensure that participants had not previously developed substantial muscle mass or strength, which could confound the results of the study. However, the study did include physically active adults who were regularly engaged in health-related fitness activities and exercised at least 3 times per week for an average of 60 min over the past 6 months. Any participants reported to have sedentary behavior, or a history of cardiovascular disease were excluded from the study.

2.2. Sample Size Calculation

The sample size for this study was estimated using a statistical power analysis. The strength was set at 0.80 and the confidence level was set at 95%, which is a common level of significance in research studies. Based on these parameters, the sample size was calculated to be 20 participants. However, to account for possible dropouts during the intervention period (expected to not exceed 20%), 24 subjects were recruited for the study with each study group containing 12 participants.

2.3. Study Outcomes

Research variables' clusters in the study were body composition, muscle hypertrophy, and exercise satisfaction, which were measured as follows:

Body composition: Data were collected using a Bioimpedance Analysis Machine (Tanita, Japan). Percentage of fat mass, body mass index, fat mass (kg), and lean body mass (kg) were calculated. All the procedures were followed by the trained fitness practitioner as prescribed by the equipment manufacturer's instructions.

Muscle hypertrophy: Upper and lower limb girths, chest girth (mm), shoulder girth (mm), waist girth (mm), thigh girth (mm), and hip girth (mm) were measured. To ensure accuracy, standards from the International Society for the Advancement of Kinanthropometry (ISAK) were followed. All the measurements were conducted with body and limbs relaxed, with no residual effects from the previous/last resistance training session.

Exercise satisfaction scale: An 8-item version of the Physical Activity Enjoyment Scale (PACES) from [17] that provided a valid instrument for assessing enjoyment in physical activity was distributed among the participants. The questionnaire was duly validated for Malaysian participants [18].

Respondents were asked to rate "how you feel at the moment about the physical activity you have been doing" using a 7-point Likert scale (1 = un-pleasurable; 7 = pleasurable). Two items were reverse-coded. The sum of all the items forms a unidimensional measure of enjoyment. Higher values reflect greater levels of enjoyment.

2.4. Training Intervention

Intervention protocols included classic hypertrophy-oriented resistance training alone and a combination of hypertrophy-oriented resistance training with aerobic training. Both groups underwent orientation (familiarization) to learn exercise techniques, familiarize themselves with exercise intensity, and set realistic training expectations. A briefing on suggested meals was also arranged. In the control group, the training protocol included 3 sessions per week, 45–60 min per session (Tables 1–3).

Table 1. Session/Day 1. 45–60 min per session. Used in both experimental and control groups.

Exercise	Reps	Rest Interval	Sets
Squats	10–12 RM	60 s	4
Romanian Deadlifts	10–12 RM	60 s	4
Leg Press/Leg Extension	10–12 RM	60 s	4

Table 2. Session/Day 2. 45–60 min per session. Used in both experimental and control groups.

Exercise	Reps	Rest Interval	Sets
Bench Press	10–12 RM	60 s	4
Incline Chest Press	10–12 RM	60 s	4
Shoulder Press	10–12 RM	60 s	4
Upright Row	10–12 RM	60 s	4
Lying Triceps Extension	10–12 RM	60 s	3

Training exercises included the use of barbells and dumbbells. To ensure maximal safety for the participants, particularly those without previous hypertrophy/heavy weights experience, 1 RM testing and subsequent conversions were replaced by the practical weight selection approach, whereby to determine the load to be lifted for 10 RM, we were determined the load at which the participant could do 10 RM. Instructors involved in the study observed and ensured that the last repetition was coming 1–2 reps before muscle fatigue/failure.

Table 3. Session/Day 3 45–60 min per session. Used in both experimental and control groups.

Exercise	Reps	Rest Interval	Sets
Bent-over Row	10–12 RM	60 s	4
Lat Pull-down	10–12 RM	60 s	4
Seated Low Row	10–12 RM	60 s	4
Biceps Curl	10–12 RM	60 s	3

Progression (when necessary) was ensured through the application of the same practical approach on the weights lifted, through the adjustments of the weight to ensure that the same RM is maintained post-progression.

Experimental group (Resistance + Aerobic) training intervention protocol included 3 sessions of resistance training per week plus 3 sessions of aerobic training of 30 min total duration (Table 4).

Table 4. Interval Aerobic Training. 30–35 min total duration per session three times a week.

	Work Intensity	Work Duration	Recovery Intensity	Recovery Duration	Rounds
Warm-up	Up to 60% HRR	5 min			
Jog	60–70% HRR	2 min	50–60% HRR	4 min	4
Cool-down		5 min			

This protocol was used in the experimental training group only. Resistance training was the same as with control group sessions.

As the above shows, the research protocol used in the training intervention for the combined training (experimental) group was 90 min per week longer in duration than the intervention protocol for the resistance (control) training group.

Counting the work bouts per session, however (with each work bout per aerobic training session being 8 min), the workload difference between sessions in the two protocols was 24 min per week. The rest of the time was spent in warm-up, cool-down, and rest intervals (all at <60% of HRR).

2.5. Data Analysis

The normality distribution of the study parameter scores was examined using the histogram plot and the Kolmogorov–Smirnov test for inferential statistics, and the scores were found to be normally distributed ($p > 0.05$), requiring the use of a parametric test. The data are presented as mean and standard deviation for continuous variables and frequency and percentage for categorical variables. The mixed-design ANOVA was performed to determine the mean differences of the study parameters between group effects (control and intervention groups), within-group effects (across time), and the interaction effect (group × time). All statistical analyses were performed using the Statistical Product and Service Solution (SPSS) version 27. The level of statistical significance was set at a p-value of <0.05.

3. Results

A total of 24 participants took part in the study, 12 who were to receive resistance hypertrophy-focused training only, and 12 who were to receive combined resistance and aerobic training. Out of the original 24 participants recruited to the study, one participant from each group discontinued (due to the non-adherence to the intervention program) their participation in the intervention program, hence 22 (11 participants per group) were eventually analyzed statistically (Figure 1).

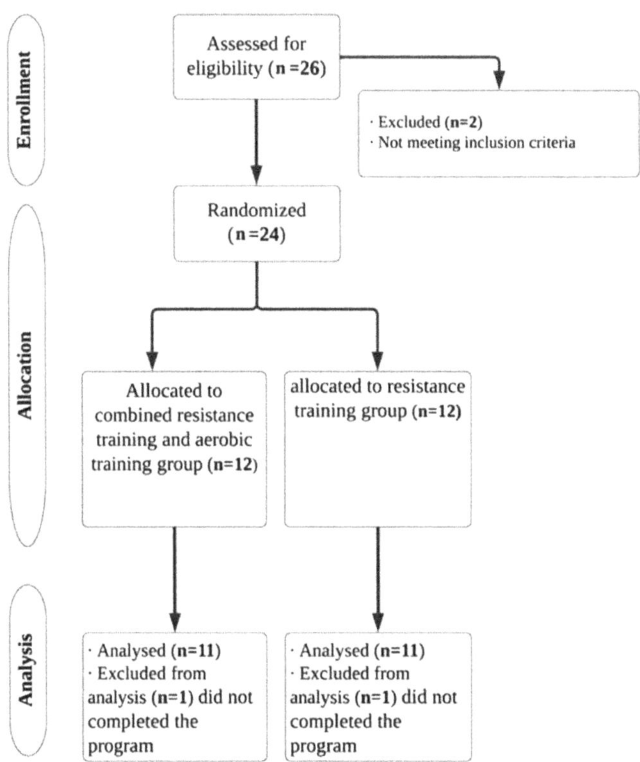

Figure 1. CONSORT flowchart of the study.

The participants had a mean age and mean height of 26.68 (SD = 4.34) and 172.00 (SD = 5.94), respectively.

3.1. Effects of Resistance Training & Combined Training on Body Composition

Analyzing the within-group effect, there were no significant differences observed in the resistance training group, whereas the combined group exhibited significant changes in some variables.

Weight loss was observed from the mid- to post-test (p = 0.006) period, signifying those participants lost weight due to the intervention training (Table 5).

Within-group changes were observed in body fat percentage, including a significant (p = 0.003) and (p = 0.016) reduction in the fat percentage from pre- to mid-intervention and pre- to post-intervention, respectively, in the combined training group (Table 5).

The dynamics of the fat mass data included within-group improvements in the fat mass, namely the significant (p = 0.007) and (p = 0.021) reduction in the fat mass from pre- to mid-intervention and pre- to post-intervention, respectively, in the combined training group (Table 5) with no significant changes observed in the resistance training group.

Lean body mass fitted into a similar dynamic with significant improvements observed from pre- to mid-intervention (p = 0.001) and from pre- to post-intervention (p = 0.015) assessments (Table 5).

No significant interactions between groups were observed in the body composition variables at any measurement point in this study.

Table 5. Resistance Group and Combined Group within-group effect in body composition variables (Time effect).

Variable	Pre-Intervention		*p*-Value	Mid-Intervention		*p*-Value	Post-Intervention		*p*-Value
	Mean ± SD	MD (95% CI)		Mean ± SD	MD (95% CI)		Mean ± SD	MD (95% CI)	
Body Weight (kg)									
Resistance	71.6 ± 12.2	−0.42 (−1.49, 0.65)	0.42	72.1 ± 13.3	0.17 (−0.60, 0.94)	0.64	71.9 ± 13.2	−0.25 (−1.85, 1.35)	0.75
Combined	83.1 ± 13.2	0.1 (−1.1, 1.4)	0.8	82.9 ± 12.8	1.3 (0.4, 2.2)	0.006 *	81.6 ± 11.6	1.5 (−0.3, 3.3)	0.1
Body Fat (%)									
Resistance	21.5 ± 5.5	0.62 (−0.38, 1.61)	0.2	20.9 ± 5.8	0.12 (−0.93, 1.17)	0.81	20.7 ± 5.99	0.74 (−0.92, 2.39)	0.36
Combined	23.8 ± 8.8	1.8 (0.7, 3.0)	0.003 *	21.9 ± 9.1	0.5 (−0.6, 1.8)	0.32	21.3 ± 7.7	2.46 (0.5, 4.4)	0.016 *
Fat Mass (kg)									
Resistance	15.8 ± 6.2	0.20 (−0.76, 1.16)	0.67	15.6 ± 6.99	0.09 (−1.02, 1.21)	0.86	15.4 ± 7.4	0.29 (−1.54, 2.12)	0.74
Combined	19.3 ± 8.9	1.6 (0.5, 2.7)	0.007 *	19.0 ± 9.7	0.9 (−0.3, 2.2)	0.14	18.1 ± 8.1	2.59 (0.4, 4.7)	0.021 *
Lean Body Mass (kg)									
Resistance	55.8 ± 7.2	−0.63 (−1.29, 0.04)	0.06	56.4 ± 6.98	0.09 (−0.51, 0.70)	0.75	56.3 ± 6.6	−0.54 (−1.25, 0.17)	0.13
Combined	62.4 ± 6.1	−1.4 (−2.2, −0.6)	0.001 *	63.9 ± 6.2	0.3 (−0.3, 1.1)	0.26	63.5 ± 5.7	−1.1 (−1.9, −0.2)	0.01 *

* Denotes significant differences. MD: Mean deviation; SD: Standard deviation, CI: Confidence interval

3.2. Effects of Resistance Training and Combined Training on Muscle Hypertrophy

Among the muscle hypertrophy variables, chest girth improved from pre- to post-intervention testing for resistance and combined training (*p* = 0.029 and *p* = 0.004, respectively) (Table 6).

Table 6. Resistance Group and Combined Group within-group effect in muscle hypertrophy variables (Time effect).

Variable	Group	Pre-Intervention		Post-Intervention		*p*-Values
		Mean ± SD	95% CI	Mean ± SD	95% CI	
Chest Girth	Resistance	96.18 ± 8.10	91.70, 100.66	98.00 ± 8.60	93.14, 102.87	0.029 *
(cm)	Combined	101.73 ± 5.98	97.25, 106.21	104.00 ± 6.77	99.36, 109.09	0.004 *
Shoulder Girth	Resistance	112.91 ± 7.97	108.41, 117.41	115.82 ± 7.72	110.81, 120.83	0.002 *
(cm)	Combined	121.09 ± 6.23	116.59, 125.59	122.00 ± 8.21	116.99, 127.01	0.284
Waist Girth	Resistance	81.14 ± 8.00	76.08, 86.19	82.55 ± 8.04	77.80, 87.29	0.122
(cm)	Combined	88.64 ± 8.06	83.58, 93.69	88.14 ± 7.01	83.39, 92.88	0.573
Thigh Girth	Resistance	60.96 ± 9.79	55.97, 65.94	61.36 ± 9.54	56.38, 66.35	0.477
(cm)	Combined	62.64 ± 5.48	57.65, 67.63	64.46 ± 5.87	59.47, 69.44	0.004 *
Hip Girth	Resistance	97.18 ± 8.71	92.17, 102.19	98.64 ± 8.97	93.52, 103.75	0.009 *
(cm)	Combined	103.82 ± 7.14	98.81, 108.83	103.46 ± 7.20	98.34, 108.57	0.477

* *p* ≤ 0.05.

Shoulder girth and hip girth improved significantly only in the resistance training group (*p* = 0.002 and *p* = 0.009 respectively) with no improvements in the combined group (Table 6).

Thigh girth, on the contrary, improved in the combined training group (*p* = 0.004) with no improvement of this variable observed in the resistance training group (Table 6).

No significant interactions between groups were observed in the muscle hypertrophy variables at any measurement point in this study.

3.3. Effects of Resistance Training and Combined Training on Exercise Satisfaction

Exercise satisfaction improved from the pre-test to post-test periods within the groups. Both resistance training and combined training groups showed improved satisfaction from pre- to post-testing (Table 7). There was, however, no significant difference observed between the groups (resistance training versus combined training) before the intervention and after training.

Table 7. Resistance Group and Combined Group within-group effect in exercise satisfaction (Time effect).

Variable	Group	Pre-Intervention	Post-Intervention	p-Values
		Mean $\pm$ SD (95% CI)	Mean $\pm$ SD (95% CI)	
Exercise satisfaction	Resistance	37.46 $\pm$ 9.94 (31.94, 42.97)	48.91 $\pm$ 6.43 (45.38, 52.44)	<0.001 *
	Combined	37.36 $\pm$ 7.40 (31.85, 42.88)	50.36 $\pm$ 4.65 (46.84, 53.89)	<0.001 *

* $p \leq 0.05$.

4. Discussion

Among the body composition variables, significant weight reductions observed in the combined group could be possibly related to the higher caloric expenditures during training with the said group, which were facilitated by the additional aerobic training component. The first 4 weeks of training possibly served as a cumulation time for the adaptations, which materialized and became noticeable later, between weeks 4 and 8. Previous research has demonstrated that a combination of aerobic and resistance exercise is quite beneficial in helping individuals lose weight [19–21].

At the same time, neither of the training programs proved more beneficial than the other in body weight loss/gain. It is possible that 8 weeks of training was not long enough to widen the observed changes and to produce significant differences between the groups in terms of body weight. Had the intervention period been longer, such differences could have become significant. The compliance, however, could be compromised by longer intervention and may have led to more participants discontinuing training toward the completion of the intervention. Previous studies used a longer intervention duration than 12 weeks [21], yielding mixed results that were significant in some studies [22] but not in others [23]. In the present study, we were targeting the best balance between the optimal intervention duration and the best possible training program adherence. So optimal outcomes could be achieved with minimum possible training program non-compliance.

Comparing the effects of resistance training and combined training on body fat, the latter brought some noticeable improvements; however, the difference between groups did not reach the desired level of significance. Earlier studies found that the combined group lost more body fat than the resistance training group [21,24,25].

In the current study, body fat improved more in the first 4 weeks of the intervention in the combined training group, whereas body weight reduced significantly in the second part of the intervention in the combined training group. Regardless, by linking these two variables, it becomes obvious that the observed significant weight reduction in the combined training group was due to significant reduction in body fat.

There was a significant mean difference in fat mass with time ($p = 0.044$). Some statistically significant improvements were observed within the combined group from pre- to mid-testing ($p = 0.007$) and from pre- to post-testing ($p = 0.021$).

There was, however, no significant mean difference in fat mass between the resistance training group and the combined training group. However, compared to past studies [21,26,27], the results reveal that the combined group reduced more body fat than the resistance training group.

Lean body mass improved from the pre-test to mid-test ($p = 0.001$) periods and from the pre-test to post-test ($p = 0.015$) periods within the combined training group. Resistance training alone did not facilitate any improvements from pre-training to mid-training and from mid-training to post-training; however, some close to statistical significance levels of lean body mass were observed from pre- to mid-tests in the resistance training group. Hence, adding aerobic components to resistance training significantly influenced the improvement in lean body mass, which is in line with earlier studies [28].

Among the muscle hypertrophy variables, there were mixed effects and improvements for both the resistance training and combined training groups. Namely, shoulder and hip girth were improved by resistance training, whereas thigh girth improved in the combined group only and chest girth improved as the result of training in both groups.

The observed improvements in chest girth can be attributed to various mechanisms. Some improvements can be related to the possible chest expansion facilitated using exercises like barbell pullovers, wide grip bench presses, and others in resistance training programs [29]. In our study, however, we cannot detect this mechanism since no increase in pectoralis mass was seen as such. The second mechanism can be related to the increased volumes of aerobic training in the combined group, leading to a possible increase in chest expansion [30,31]. That too, however, cannot be confirmed within the framework of the current study and using current research measurement instruments.

Resistance training resulted in increased shoulder girth, demonstrating the benefits of such a training modality. Traditional resistance training may be more efficient if the specific objective is to improve muscle hypertrophy as opposed to more functional kinds of exercise that just result in increases in shoulder girth [32,33].

In common with past research [21,34,35], the findings show that combined training increases thigh girth more than resistance training.

In terms of hip girth, the resistance training group showed a greater improvement, and this training modality looks more effective than combined training for this variable (although statistically insignificant). As mentioned earlier, a previous study reported that combining the two different training strategies may compromise muscle hypertrophy [34].

In the exercise satisfaction analysis, there was no significant mean difference in exercise satisfaction between the resistance training group and the combined training group. Similarly, although the means of exercise satisfaction were slightly higher at the end of the training intervention in the combined training group, the differences were not statistically significant.

Exercise satisfaction improving because of various training modalities is well documented [36,37]. The results of this study confirm an increase in exercise satisfaction in the resistance training group, indirectly suggesting that the training program was enjoyable as such and that training outcomes were achieved. However, adding the aerobic component to the resistance training didn't alter the attitude to training and satisfaction of the exercisers. The program remained satisfying and enjoyable, with training objectives being achieved as well.

Limitations of the Study

Although the study was carefully designed and executed, it is important to acknowledge some limitations that could affect the generalizability of the study results. One such limitation of the study is the relatively small sample size. Although the sample size was carefully calculated using a statistical power analysis, a larger sample size could have increased the statistical power of the study and reduced the possibility of type II errors. Additionally, a larger sample size may have increased the generalizability of the study results to other populations.

Another limitation of the study is that objective methods to measure muscle mass directly were not used to assess muscle hypertrophy. Instead, the study used indirect measures of muscle mass, such as circumferences, which are prone to measurement error. Although the study used standardized techniques for taking measurements, objective measures such as MRI or DEXA scans could have possibly provided more accurate and reliable measurements of muscle hypertrophy.

Despite these limitations, the study provides valuable insights into the effects of combined aerobic and resistance exercise on body composition, muscle hypertrophy, and exercise satisfaction.

5. Conclusions

Combining heavy resistance training with aerobic training positively influenced the variables of body composition in physically active males, leading to reduced body weight, body fat, and fat mass, and increased lean body mass.

Adding aerobic components to muscle-hypertrophy-focused training impacted the hypertrophy variables too. It led to thigh girth and chest girth improvements in the combined training group, proving that aerobic training does not oppose the effects of hypertrophy training and instead can facilitate these if resistance and aerobic training are combined in the right proportions.

The combination of resistance and aerobic training improves exercise satisfaction, as does resistance training alone. Hence, such a combination may be useful in solidifying exercise adherence in prolonged exercise interventions, making them more variable and satisfying.

According to the within-group effects analysis, there were no significant changes elicited by the resistance training in the control group in any of the body composition variables assessed in this study. Therefore, eight weeks of resistance hypertrophy-focused training were not sufficient to enforce the significant changes in the body composition domain. Among the muscle hypertrophy markers, resistance training resulted in significant improvements in chest, shoulder, and hip girths. Resistance training also improved the control group's level of exercise satisfaction.

Author Contributions: Conceptualization, J.T. and O.K.; Methodology, O.K. and H.A.H.; Validation, G.K. and S.B.A.-M.; Formal analysis, G.B.; Investigation, J.T.; Data curation, G.K.; Writing—original draft, J.T.; Writing—review & editing, O.K., M.I.A. and S.B.A.-M.; Supervision, O.K., G.K. and H.A.H.; Funding acquisition, M.I.A. All authors have read and agreed to the published version of the manuscript.

Funding: This research was funded by Princess Nourah bint Abdulrahman University Researchers Supporting Project number (PNURSP2023R 286), Princess Nourah bint Abdulrahman University, Riyadh, Saudi Arabia.

Institutional Review Board Statement: The study was conducted in a workout studio in Kuala Lumpur, Malaysia, and the research protocol was approved by the Universiti Sains Malaysia Human Ethics Committee Protocol No. USM/JEPeM/19090542.

Informed Consent Statement: Informed consent was obtained from all subjects involved in the study.

Data Availability Statement: Data will be available on request.

Acknowledgments: We would like to thank Princess Nourah bint Abdulrahman University for supporting this project through Princess Nourah bint Abdulrahman University Researchers Supporting Project number (PNURSP2023R 286), Princess Nourah bint Abdulrahman University, Riyadh, Saudi Arabia.

Conflicts of Interest: The authors declare no conflict of interest.

References

1. Martel, G.F.; Roth, S.M.; Ivey, F.M.; Lemmer, J.T.; Tracy, B.L.; Hurlbut, D.E.; Hopkins, W.G.; Fyfe, J.J.; Phillips, S.M.; Bishop, D.J.; et al. Age and sex affect human muscle fibre adaptations to heavy-resistance strength training. *Exp. Physiol.* **2006**, *91*, 457–464. [CrossRef] [PubMed]
2. Elia, M.; Stubbs, R.; Henry, C. Differences in fat, carbohydrate, and protein metabolism between lean and obese subjects undergoing total starvation. *Obes. Res.* **1999**, *7*, 597–604. [CrossRef]

3. Donnelly, J.E.; Lambourne, K. Classroom-based physical activity, cognition, and academic achievement. *Prev. Med.* **2011**, *52*, S36–S42. [CrossRef]

4. Thomas, M.; Burns, S. Increasing Lean Mass and Strength: A Comparison of High Frequency Strength Training to Lower Frequency Strength Training. *Int. J. Exerc. Sci.* **2016**, *9*, 159–167.

5. Garber, C.E.; Blissmer, B.; Deschenes, M.R.; Franklin, B.A.; Lamonte, M.J.; Lee, I.-M.; Nieman, D.C.; Swain, D.P. American College of Sports Medicine position stand. Quantity and quality of exercise for developing and maintaining cardiorespiratory, musculoskeletal, and neuromotor fitness in apparently healthy adults: Guidance for prescribing exercise. *Med. Sci. Sports Exerc.* **2011**, *43*, 1334–1359. [CrossRef] [PubMed]

6. Shaw, B.; Shaw, I.; Krasilshchikov, O. Effects of aerobic and concurrent aerobic and resistance training on effort- or muscle-dependant spirometry parameters in chronic obstructive pulmonary disease (COPD): A meta-analysis. *Asian J. Exerc. Sports Sci.* **2008**, *5*, 63–71.

7. Shaw, I.; Shaw, B.; Krasilshchikov, O. Comparison of aerobic and combined aerobic and resistance training on low-density lipoprotein cholesterol concentrations in men. *Cardiovasc. J. Afr.* **2009**, *20*, 290–295.

8. Schumann, M.; Feuerbacher, J.F.; Sünkeler, M.; Freitag, N.; Rønnestad, B.R.; Doma, K.; Lundberg, T.R. Compatibility of Concurrent Aerobic and Strength Training for Skeletal Muscle Size and Function: An Updated Systematic Review and Meta-Analysis. *Sports Med.* **2021**, *52*, 601–612. [CrossRef]

9. Kuan, G.; Wan Firdaus Krasilshchikov, O. Effects of using complex training method on muscular power among competitive male weightlifters. *Malays. J. Sport Sci. Recreat. (MJSSR)* **2018**, *14*, 11–23.

10. Hargreaves, M.; Spriet, L.L. Skeletal muscle energy metabolism during exercise. *Nat. Metab.* **2020**, *2*, 817–828. [CrossRef]

11. Atherton, P.J.; Babraj, J.A.; Smith, K.; Singh, J.; Rennie, M.J.; Wackerhage, H. Selective activation of AMPK-PGC-1α or PKB-TSC2-mTOR signaling can explain specific adaptive responses to endurance or resistance training-like electrical muscle stimulation. *FASEB J.* **2005**, *19*, 1–23.

12. Coffey, V.G.; Hawley, J.A. Concurrent exercise training: Do opposites distract? *J. Physiol.* **2017**, *595*, 2883–2896.

13. Stepto, N.K.; Benziane, B.; Wadley, G.D.; Chibalin, A.V.; Canny, B.J.; Eynon, N.; McConell, G.K. Short-term intensified cycle training alters acute and chronic responses of PGC1α and Cytochrome C oxidase IV to exercise in human skeletal muscle. *PLoS ONE* **2012**, *7*, e53080.

14. Lee, M.J.-C.; Ballantyne, J.K.; Chagolla, J.; Hopkins, W.G.; Fyfe, J.J.; Phillips, S.M.; Bishop, D.J.; Bartlett, J.D. Order of same-day concurrent training influences some indices of power development, but not strength, lean mass, or aerobic fitness in healthy, moderately-active men after 9 weeks of training. *PLoS ONE* **2020**, *15*, e0233134.

15. Timmins, R.G.; Shamim, B.; Tofari, P.J.; Hickey, J.T.; Camera, D.M. Differences in lower limb strength and structure after 12 weeks of resistance, endurance, and concurrent training. *Int. J. Sports Physiol. Perform.* **2020**, *15*, 1223–1230.

16. Antonio, J. The Effects of Upper Body Strength Training Versus Concurrent Strength + Endurance Training on Indices of Strength, Aerobic Power, and Muscle Cross-Sectional Area. *J. Exerc. Physiol. Online* **2023**, *26*, 183–191.

17. Mullen, S.P.; Olson, E.A.; Phillips, S.M.; Szabo, A.N.; Wójcicki, T.R.; Mailey, E.L.; Gothe, N.P.; Fanning, J.T.; Kramer, A.F.; McAuley, E. Measuring enjoyment of physical activity in older adults: Invariance of the physical activity enjoyment scale (paces) across groups and time. *Int. J. Behav. Nutr. Phys. Act.* **2011**, *8*, 103. [CrossRef]

18. Hashim, H.A.; Grove, J.R.; Whipp, P. Testing a model of physical education enjoyment and physical activity among high school students. *ICHPER-SD J. Res. Health* **2008**, *3*, 95.

19. Kaneda, K.; Sato, D.; Wakabayashi, H.; Hanai, A.; Nomura, T. A comparison of the effects of different water exercise programs on balance ability in elderly people. *J. Aging Phys. Act.* **2008**, *16*, 381–392.

20. Khan, A.A. Resistance Training: Benefits on Health & Fitness. *Int. J. Fitness Health Phys. Educ. Iron Games* **2016**, *3*, 60–65.

21. Pöchmüller, M.; Schwingshackl, L.; Colombani, P.C.; Hoffmann, G. A systematic review and meta-analysis of carbohydrate benefits associated with randomized controlled competition-based performance trials. *J. Int. Soc. Sports Nutr.* **2016**, *13*, 27. [PubMed]

22. Ibañez, J.; Izquierdo, M.; ARGuelles, I.; Forga, L.; Larrión, J.L.; García-Unciti, M.; Idoate, F.; Gorostiaga, E.M. Twice-weekly progressive resistance training decreases abdominal fat and improves insulin sensitivity in older men with type 2 diabetes. *Diabetes Care* **2005**, *28*, 662–667. [PubMed]

23. Ferreira, M.L.; Sherrington, C.; Smith, K.; Carswell, P.; Bell, R.; Bell, M.; Nascimento, D.P.; Pereira, L.S.M.; Vardon, P. Physical activity improves strength, balance and endurance in adults aged 40–65 years: A systematic review. *J. Physiother.* **2012**, *58*, 145–156. [PubMed]

24. Jago, R.; Drews, K.L.; McMurray, R.G.; Thompson, D.; Volpe, S.L.; Moe, E.L.; Jakicic, J.M.; Pham, T.H.; Bruecker, S.; Blackshear, T.B.; et al. Fatness, fitness, and cardiometabolic risk factors among sixth-grade youth. *Med. Sci. Sports Exerc.* **2010**, *42*, 1502.

25. Worthey, C.D. *The Effects of Two Different Resistance Training Protocols with Similar Volume on Muscular Strength, Muscle Thickness, and Fat-Free Mass*; Illinois State University: Normal, IL, USA, 2016.

26. Miller, S.; Wolfe, R.R. The danger of weight loss in the elderly. *J. Nutr. Health Aging* **2008**, *12*, 487–491.

27. Miller, T.; Mull, S.; Aragon, A.A.; Krieger, J.; Schoenfeld, B.J. Resistance training combined with diet decreases body fat while preserving lean mass independent of resting metabolic rate: A randomized trial. *Int. J. Sport. Nutr. Exerc. Metab.* **2018**, *28*, 46–54.

28. Voisin, S.; Almén, M.S.; Zheleznyakova, G.Y.; Lundberg, L.; Zarei, S.; Castillo, S.; Eriksson, F.E.; Nilsson, E.K.; Blüher, M.; Böttcher, Y.; et al. Many obesity-associated SNPs strongly associate with DNA methylation changes at proximal promoters and enhancers. *Genome Med.* **2015**, *7*, 103.

29. Osawa, K.; Miyoshi, T.; Koyama, Y.; Sato, S.; Akagi, N.; Morimitsu, Y.; Kubo, M.; Sugiyama, H.; Nakamura, K.; Morita, H. Differential association of visceral adipose tissue with coronary plaque characteristics in patients with and without diabetes mellitus. *Cardiovasc. Diabetol.* **2014**, *13*, 1–11.

30. Cribb, P.J.; Williams, A.D.; Stathis, C.; Carey, M.F.; Hayes, A. Effects of whey isolate, creatine and resistance training on muscle hypertrophy. *Med. Sci. Sports Exerc.* **2007**, *39*, 298–307.

31. DeFreitas, J.M.; Beck, T.W.; Stock, M.S.; Dillon, M.A.; Kasishke, P.R. An examination of the time course of training-induced skeletal muscle hypertrophy. *Eur. J. Appl. Physiol.* **2011**, *111*, 2785–2790.

32. Abdel-Aziz Habib, H. Effect of functional resistance drills with Elastic bands on some of physical and biomechanical variables and kicking accuracy in soccer. *Assiut J. Sport. Sci. Arts* **2018**, *2018*, 312–332.

33. Weiss, T.; Kreitinger, J.; Wilde, H.; Wiora, C.; Steege, M.; Dalleck, L.; Janot, J. Effect of functional resistance training on muscular fitness outcomes in young adults. *J. Exerc. Sci. Fit.* **2010**, *8*, 113–122.

34. Karavirta, L.; Häkkinen, A.; Sillanpää, E.; García-López, D.; Kauhanen, A.; Haapasaari, A.; Alen, M.; Pakarinen, A.; Kraemer, W.J.; Izquierdo, M. Effects of combined endurance and strength training on muscle strength, power and hypertrophy in 40–67-year-old men. *Scand. J. Med. Sci. Sports* **2011**, *21*, 402–411.

35. Ozaki, H.; Sakamaki, M.; Yasuda, T.; Fujita, S.; Ogasawara, R.; Sugaya, M.; Nakajima, T.; Abe, T. Increases in thigh muscle volume and strength by walk training with leg blood flow reduction in older participants. *J. Gerontol. Ser. A Biomed. Sci. Med. Sci.* **2011**, *66*, 257–263.

36. Collins, K.A.; Fos, L.B.; Ross, L.M.; Slentz, C.A.; Davis, P.G.; Willis, L.H.; Piner, L.W.; Bateman, L.A.; Houmard, J.A.; Kraus, W.E. Aerobic, resistance, and combination training on health-related quality of life: The STRRIDE-AT/RT randomized trial. *Front. Sports Act. Living* **2021**, *2*, 620300. [PubMed]

37. Srismith, D.; Dierkes, K.; Zipfel, S.; Thiel, A.; Sudeck, G.; Giel, K.E.; Behrens, S.C. Physical activity improves body image of sedentary adults. Exploring the roles of interoception and affective response. *Curr. Psychol.* **2022**. [CrossRef]

MDPI

Article

Amateur Female Athletes Perform the Running Split of a Triathlon Race at Higher Relative Intensity than the Male Athletes: A Cross-Sectional Study

Guilherme Corrêa De Araújo Moury Fernandes [1], José G. G. Barbosa Junior [2], Aldo Seffrin [2], Lavínia Vivan [2], Claudio A. B. de Lira [3], Rodrigo L. Vancini [4], Katja Weiss [5], Beat Knechtle [6,*] and Marilia S. Andrade [2]

[1] Sports Medicine Residency Program, Department of Orthopedics and Traumatology, Federal University of São Paulo, São Paulo 04021-001, São Paulo, Brazil
[2] Department of Physiology, Federal University of São Paulo, São Paulo 04021-001, São Paulo, Brazil
[3] Human and Exercise Physiology Division, Faculty of Physical Education and Dance, Federal University of Goiás, Goiânia 74690-900, Goiás, Brazil
[4] Center for Physical Education and Sports, Federal University of Espírito Santo, Vitória 29075-210, Espirito Santo, Brazil
[5] Institute of Primary Care, University of Zurich, 8091 Zurich, Switzerland
[6] Medbase St. Gallen Am Vadianplatz, 9001 St. Gallen, Switzerland
* Correspondence: beat.knechtle@hispeed.ch

Citation: De Araújo Moury Fernandes, G.C.; Barbosa Junior, J.G.G.; Seffrin, A.; Vivan, L.; de Lira, C.A.B.; Vancini, R.L.; Weiss, K.; Knechtle, B.; Andrade, M.S. Amateur Female Athletes Perform the Running Split of a Triathlon Race at Higher Relative Intensity than the Male Athletes: A Cross-Sectional Study. *Healthcare* **2023**, *11*, 418. https://doi.org/10.3390/healthcare11030418

Academic Editors: João Paulo Brito and Rafael Oliveira

Received: 17 December 2022
Revised: 27 January 2023
Accepted: 31 January 2023
Published: 1 February 2023

Abstract: Maximal oxygen uptake ($\dot{V}O_2max$), ventilatory threshold (VT) and respiratory compensation point (RCP) can be used to monitor the training intensity and the race strategy, and the elucidation of the specificities existing between the sexes can be interesting for coaches and athletes. The aim of the study was to compare ventilatory threshold (VT), respiratory compensation point (RCP), and the percentage of the maximal aerobic speed (MAS) that can be maintained in a triathlon race between sexes. Forty-one triathletes (22 men and 19 women), 42.1 ± 8.4 (26 to 60) years old, that raced the same Olympic triathlon underwent a cardiorespiratory maximal treadmill test to assess their VT, RPC, and MAS, and race speed. The maximal oxygen uptake ($\dot{V}O_2max$) (54.0 ± 5.1 vs. 49.8 ± 7.7 mL/kg/min, $p < 0.001$) and MAS (17 ± 2 vs. 15 ± 2 km/h, $p = 0.001$) were significantly higher in male than in female athletes. Conversely, there were no sex differences according to the percentage of $\dot{V}O_2max$ reached at VT (74.4 ± 4.9 vs. $76.1 \pm 5.4\%$, $p = 0.298$) and RCP (89.9 ± 3.6 vs. $90.6 \pm 4.0\%$, $p = 0.560$). The mean speed during the race did not differ between sexes (12.1 ± 1.7 km/h and 11.7 ± 1.8 km/h, $p = 0.506$, respectively). Finally, men performed the running split at a lower percentage of speed at RCP than women (84.0 ± 8.7 vs. $91.2 \pm 7.0\%$, respectively, $p = 0.005$). Therefore, male and female athletes accomplished the running split in an Olympic triathlon distance at distinct relative intensities, as female athletes run at a higher RCP percentage.

Keywords: ventilatory threshold; $\dot{V}O_2max$; performance; respiratory compensation point; women

1. Introduction

Over the last few decades, there has been a notable increase in the number of female athletes [1], reaching its maximum at the 2020 Tokyo Olympics, with over 48% of the 11,300 athletes being women. Thus, the 2020 Tokyo Olympics were the most equal in terms of sex distribution. Specifically in the triathlon event, the participation of women has increased considerably during the last decades reaching between 25% and 40% of the total field [2].

In the triathlon, as in any other endurance sport, the main physiological determinants of performance are the maximum ability to absorb or utilize oxygen ($\dot{V}O_2max$), the fraction of $\dot{V}O_2max$ that might be sustained for long periods of time, and running economy [3]. Among these factors, $\dot{V}O_2max$ is certainly one of the most studied variables [4]. There is a relative consensus in the literature that the $\dot{V}O_2max$ is limited by the ability of the cardiorespiratory system to deliver oxygen to the exercising muscles, and not by the muscular capacity of extraction [3]. Previous studies have investigated sex differences in $\dot{V}O_2max$ between sedentary, amateur, and elite athletes, and all of them agree that female athletes present lower values than their male counterparts [5,6]. These differences are commonly attributed to different factors, such as higher body fat, lower red cell mass and hemoglobin levels, and lower end-diastolic, end-systolic, stroke volume, and cardiac output in women [7–9].

Other physiological variables that are commonly used as predictors of endurance performance are ventilatory thresholds [10]. During an incremental exercise test performed with the concomitant measurement of $\dot{V}O_2$, carbon dioxide production ($\dot{V}CO_2$), and minute ventilation, two distinct thresholds were identified [11]. The first, commonly called the ventilatory threshold (VT), is defined as the highest sustained intensity of exercise for which the measurement of oxygen uptake can account for the entire energy requirement. The second, called the respiratory compensation point (RCP), is characterized by the highest exercise intensity in which the body can buffer hydrogen ions production, preventing their accumulation [11,12]. There is relative consensus in the literature that the limiting factors for VT and RCP are peripheral conditions (extraction capacity) [5]. Considering that women present a greater proportional area of type I muscle fibers and preserve more glycogen than males, as they use more fatty acids, it is possible that female athletes present different VT and RCP than males. Although some recent studies have shown that the percentage of $\dot{V}O_2max$ at VT and RCP varies significantly between the sexes [13–15], the actual literature data are conflicting [5].

Considering that the speed of runs associated with $\dot{V}O_2max$, VT, and RCP can be used to monitor the training intensity and the race strategy [15], the elucidation of the specificities existing between the sexes can be interesting for coaches and athletes. Therefore, the aim of this study was to compare male and female athletes according to VT, RCP, and the percentage of the maximal aerobic speed that athletes can run in a triathlon race.

2. Materials and Methods

2.1. Participants and Study Design

Participant recruitment was carried out through direct contact with triathlon trainers and social media. Forty-one athletes participated in the study (22 male and 19 female). The male athletes were 41.4 ± 9.2 years old, 74.1 ± 6.9 kg and 174.2 ± 7.0 cm, and the female athletes were 42.8 ± 7.2 years old, 58.7 ± 6.6 kg and 163.9 ± 4.6 cm. The age was not significantly different between sex groups ($p = 0.609$) but male athletes were significantly heavier ($p < 0.001$) and taller ($p < 0.001$) than the female athletes. The inclusion criteria to take part in the study included having at least 3 years triathlon training experience, and being enrolled in the 32nd International Olympic Triathlon Santos, Brazil (Olympic distance). The exclusion criteria were not finishing the competition or having any medical condition that prevented the cardiorespiratory maximum treadmill test.

2.2. Experimental Procedures

All athletes underwent a cardiorespiratory incremental maximal test to determine $\dot{V}O_2max$, VT, and RCP, in addition to measure the maximal aerobic speed (MAS). All tests were conducted less than a month before the triathlon race competition between 20 January and 10 February 2022.

The cardiorespiratory incremental maximal test was performed using a treadmill (Inbrasport, ATL, Porto Alegre, Brazil) and computer-based metabolic analyzer (Quark, Rome, Italy). Before each test, the metabolic analyzer was calibrated according to the manufacturer's guidelines. All the participants were instructed to avoid vigorous exercises the day before the test, and to avoid consuming stimulating beverages on the day of the test, such as coffee or tea. Additionally, the participants completed the Physical Activity

Readiness Questionnaire (PAR-Q) [16]. Participants who did not answer "YES" to any of the PAR-Q questions and did not present any contraindication to participation were enrolled in the study. No-one answered "YES" to any question of the PAR-Q questions.

The test started with a 3 min warm up at 8 km/h, increased by 1 km/h in each minute until volitional exhaustion. The treadmill was programmed to have a 1% inclination to simulate the difficulties of open-air running [17]. Each test lasted between 8 and 12 min. Heart rate was monitored during the whole test using a heart rate monitor (Ambit 2S, Suunto, Finland), and perceived effort was rated according to the Borg scale (Noble et al., 1983).

The $\dot{V}O_2$, $\dot{V}CO_2$, O_2 end-tidal pressure (PET O_2), CO_2 end-tidal pressure (PET CO_2), and minute ventilation ($\dot{V}_E$), were measured breath-by-breath, and all data were averaged over 20 s for analysis. VT and RCP were identified through the O_2 and CO_2 ventilatory equivalents and end-tidal pressures [18] by two independent investigators. In the case of discordance about the VT or RCP, a third investigator was consulted to identify these variables, so that, in all cases, there was agreement regarding the VT and RCP between at least two investigators. The speed at VT and RCP, in addition to the $\dot{V}O_2max$ percentage at VT and RCP have been presented. The $\dot{V}O_2max$ was determined as the stabilization of O_2 consumption (increase lower than 2.1 mL $\cdot$ kg^{-1}. min^{-1}), even after increasing treadmill speed [19]. The maximal aerobic speed reached during the test was defined as the minimal speed eliciting $\dot{V}O_2max$ [20]. All the tests were collected by qualified professionals trained and experienced in the method.

2.3. Statistical Analysis

Data are presented as the mean and standard deviation. Descriptive analysis was conducted to evaluate the distribution of the variables. All variables presented normal distribution and homogeneous variability according to the Shapiro–Wilk and Levene tests, respectively. Student's t-test was used to compare the mean values. The SPSS version 21.0 (SPSS, Inc., Chicago, IL, USA) was employed to perform the analysis. The G*Power version 3.1.9.2 (Franz, Universität Kiel, Germany) was used to determine the sample size and analyze the test power level. A sample size calculation on the velocity maintained during the running split, using data from a pilot study (n = 12), which are 11.0 ± 1.0 km/h for male athletes and 10.2 ± 0.9 km/h for female athletes, showed that 19 athletes in each group (male or female) were needed to detect a relevant difference with 80% power and a significance level of 5%. For power level calculation, a t-test family was selected, and mean values, standard deviations, and effect sizes (Cohen d) were included in the calculation. The power of the test varies from 0 to 1. Usually, researchers use 0.80 as the power level of the test [21]. The measurement of the effect size for differences between sexes were determined by calculating the mean difference between the two sexes, and then dividing the result by the pooled standard deviation. Calculating effect sizes, the magnitude of any change was judged according to the following criteria: $d < 0.2$ considered no effect, $0.2 \leq d < 0.5$ considered a "small" effect size; $0.5 \leq d < 0.8$ represented a "medium" effect size; and $d \geq 0.8$ a "large" effect size [22]. The level of significance was set at $p < 0.05$.

3. Results

- The absolute and relative to body mass values for $\dot{V}O_2max$ and the maximal aerobic speed were significantly higher for male than for female athletes. Conversely, there were no sex differences in the percentage of $\dot{V}O_2max$ reached at VT and RCP. However, the speeds at VT and RCP were higher in male athletes (Table 1).
- Despite the sex differences according to $\dot{V}O_2max$, VT, and RCP, the mean speed maintained during the running split of an Olympic triathlon race was not different between sexes ($p = 0.506$, $d = 0.23$). The running speed during the race was situated between the speeds associated with VT and RCP for both sexes. However, female athletes performed the running split at a higher percentage of the speed at the RCP than male athletes (Table 1).

Table 1. Comparison between sexes of the descriptive characteristics of athletes.

	Male (n = 22)	Female (n = 19)	*p*-Value	Power (1-ß)	Effect Size (*d*)	CI for Effect Size
$\dot{V}O_2max$ (L/min)	4.00 ± 0.52	2.89 ± 0.40	<0.001	1.00	2.39	1.3 to 3.4
$\dot{V}O_2max$ (mL/kg/min)	54.0 ± 5.1	49.8 ± 7.7	0.047	0.64	0.64	0.1 to 1.2
MAS (km/h)	17 ± 2	15 ± 2	0.001	0.93	1.00	0.6 to 1.4
%$\dot{V}O_2$ max at VT	74.4 ± 4.9	76.1 ± 5.4	0.298	0.26	0.32	−0.2 to 0.8
Speed at VT (km/h)	11.8 ± 1.1	10.7 ± 1.5	0.021	0.83	0.83	0.3 to 1.4
%$\dot{V}O_2$ max at RCP	89.9 ± 3.6	90.6 ± 4.0	0.560	0.14	0.18	−0.3 to 0.7
Speed at RCP (km/h)	14.3 ± 1.2	12.8 ± 1.6	0.001	0.95	1.06	0.6 to 1.5
Mean speed in running split (km/h)	12.1 ± 1.7	11.7 ± 1.8	0.506	0.18	0.23	−0.4 to 0.8
% RCP maintained during the running split	84.0 ± 8.7	91.2 ± 7.0	0.005	0.88	0.91	0.4 to 1.4

Data are presented as mean ± standard deviation. $\dot{V}O_2max$: maximal oxygen uptake; MAS: maximal aerobic speed; VT: ventilatory threshold; RCP: respiratory compensation point; CI: confidence interval.

4. Discussion

The main findings of the present study were the following: (i) female and male amateur athletes presented similar VT and RCP related to $\dot{V}O_2max$; (ii) male athletes presented higher speed at VT and RCP than female athletes; (iii) the mean running speed maintained during the running split of a triathlon race did not differ between sexes; and (iv) female athletes maintained the running split at a higher relative intensity considering the percentage of RCP, than male athletes.

The running split represents approximately 25% of the overall race time of a triathlon race, although it has been demonstrated to be the main determinant in Olympic distance triathlon overall performance [23,24] for male and female athletes. Considering the importance of the running split for triathlon performance, several previous studies have investigated the determinant factors for this split time [25,26]. However, little is known about sex-related differences.

According to $\dot{V}O_2max$, which has been defined as the highest rate at which oxygen can be taken up and utilized by the body during severe exercise [3], the results of the present study corroborate previous findings that female athletes present lower $\dot{V}O_2$ max than male athletes [27]. In the same direction, Puccinelli et al. [13] also showed higher $\dot{V}O_2max$ values for male than for female triathletes (59.9 ± 6.3 and 49.5 ± 7.8 mL/kg/min, respectively). The $\dot{V}O_2max$ is limited by central cardiovascular factors [3]; therefore, the smaller cardiac volume, cardiac output, and the lower hematocrit level, are possibly associated factors to the lower $\dot{V}O_2max$ for female athletes [28].

Furthermore, the present results showed no sex differences in %$\dot{V}O_2max$ at VT and RCP. The data on sex differences in VT and RCP are conflicting [5,13]; VT and RCP are limited by peripheral conditions (e.g., muscle adaptations to aerobic exercises [3,29], such as increased capillary density, the increased mitochondrial content of muscle, and higher mitochondrial enzyme levels (i.e., succinate dehydrogenase, NADH dehydrogenase, and NADH-cytochrome c reductase)). This limitation can be explained because to have the necessary energy to muscular contraction, ATP is converted to ADP and Pi, and the high levels of ADP and Pi drive metabolic reactions to resynthesize new ATP molecules to continue the muscular contraction [3]. With few mitochondria in a muscle cell, ADP levels should increase substantially to reach the ADP demand via aerobic pathways, and high ADP levels also have a stimulatory effect on phosphofructose kinase (PFK), stimulating glycolysis via and generating a high demand for carbohydrates as an energetic substrate [3]. In a muscle with higher mitochondrial content, the ADP level increases less for the same $\dot{V}O_2$ demand; therefore, there is less PFK stimulus and less carbohydrate turnover, increasing the possibility of using fat as an energetic substrate, resulting in later lactate formation [3,29].

Considering that women show a greater proportional area of type I muscle fibers [5], higher VT and RCP could be expected in female athletes. Indeed, Puccinelli et al. [13] demonstrated that female amateur athletes presented with higher VT than male athletes. In addition, Puccinelli et al. [13] selected only amateur triathletes; only 18 women and 39 men were included in the study. Considering sex differences in the willingness to participate in certain types of research [5], we cannot exclude the hypothesis that only the best-trained female athletes volunteered to participate in the study; therefore, the different levels of training could be a reason for the different $\%\dot{V}O_2max$ at VT and RCP, and not the sex metabolic differences. In a recent review, Besson et al. [5] reported that there were no sex differences according to $\%\dot{V}O_2max$ at ventilatory thresholds. As expected, despite the lack of sex differences according to the $\%\dot{V}O_2max$ at ventilatory thresholds, as the male athletes presented significantly higher $\dot{V}O_2max$ and MAS, the speed at the ventilatory thresholds was higher for men than for women.

Furthermore, it is also generally accepted that carbohydrate feeding during prolonged exercises (longer than 2 h) can maintain the possibility of carbohydrate oxidation and the prevention of hypoglycemia [30]. Beyond that, a good level of hydration during prolonged exercise is of fundamental importance, once it has been demonstrated that dehydration steadily increased body temperature, heart rate, and oxygen uptake for the same speed of running, worsening the physical performance [31].

The percentage of running speed at RCP that was sustained during the running split of the triathlon race was significantly higher in the female group. This was an interesting finding. As women were able to sustain a higher percentage speed at RCP, the average speed maintained during the race did not differ between the sexes, and consequently the race time. There are some possible explanations for why a woman can maintain a higher percentage of RCP speed than a man during a 10 km run. As stated previously, female athletes showed a greater proportional area of type I fibers. In addition, women can spare more carbohydrates during exercise, as they are more capable in using fatty acids as an energetic substrate during aerobic exercises, as documented by Tarnopolsky et al. [32], who found lower respiratory exchange ratio values were found for women (0.87×0.94). Moreover, despite the lack of evidence, women appear to have some advantage in the biomechanics of running, especially due to shorter stride length and higher stride frequency, and consequently, a lower duty factor [33]. Women have also been shown to have a different neuromotor strategy, which could reduce their fatigability during endurance exercises. It was suggested that experienced ultra-endurance-trail women were more resistant to fatigue after ultra-trail running, as they had a lower decrease in maximal voluntary torque changes in the knee extensors and plantar flexors [34]. These differences make women more resistant to fatigue during long-term aerobic exercises [5].

In the present study, the swimming and cycling intensities of male and female athletes were not measured. It is possible that the athletes performed the first division (swimming and cycling) at different relative intensities, which could have affected the results of the running split. However, this does not invalidate the result that men and women perform running splits at different relative intensities. In the present study, the level of hydration of the athletes and carbohydrate feeding were not evaluated. Considering that a possible dehydration or hypoglycemia of the athletes can affect sports performance, this can be considered a limitation of the present study. The authors suggest that future studies should be developed to assess the relative intensity in which men and women perform the three triathlon splits.

5. Conclusions

Despite having no difference according to VT and RCP, male and female athletes performed the running split in an Olympic triathlon distance at different relative effort intensities, and female athletes ran at a higher relative intensity (i.e., a higher RCP percentage). These findings can be useful for coaches and athletes to consider sex differences when designing the strategy for a triathlon race, which may be different for each sex.

Author Contributions: Conceptualization, A.S. and M.S.A.; methodology, M.S.A.; validation, M.S.A. and C.A.B.d.L.; formal analysis, J.G.G.B.J. and G.C.D.A.M.F.; investigation, A.S.; resources, L.V. and M.S.A.; data curation, R.L.V. and B.K.; writing—original draft preparation, G.C.D.A.M.F. and A.S.; writing—review and editing, M.S.A., C.A.B.d.L., K.W. and B.K.; visualization, B.K.; supervision and project administration, M.S.A. All authors have read and agreed to the published version of the manuscript.

Funding: This work was supported by the Fundação de Amparo à Pesquisa do Estado de São Paulo (FAPESP) (grant number: 2021/08114-1).

Institutional Review Board Statement: All experimental procedures were approved by the Human Research Ethics Committee of the Federal University of São Paulo and conformed to the principles outlined in the Declaration of Helsinki (approval number 0973/2021). This study was guided by ethical standards and national and international laws. All athletes signed the consent form after receiving instructions regarding the possible risks and benefits and were granted privacy, confidentiality, and anonymity rights. The participants were free to stop participating at any stage of the experiment without giving reasons for their decision.

Informed Consent Statement: Informed consent was obtained from all subjects involved in the study.

Data Availability Statement: Data supporting the study results can be provided followed by a request sent to the corresponding author's e-mail.

Acknowledgments: We thank all volunteers who took the time to contribute to this research and the Olympic Training and Research Center (Centro Olímpico de Treinamento e Pesquisa, COTP, São Paulo, Brazil).

Conflicts of Interest: The authors declare no conflict of interest.

References

1. Santos, A.C.; Turner, T.J.; Bycura, D.K. Current and Future Trends in Strength and Conditioning for Female Athletes. *Int. J. Environ. Res. Public Health* **2022**, *19*, 2687. [CrossRef] [PubMed]
2. Lepers, R. Sex Difference in Triathlon Performance. *Front. Physiol.* **2019**, *10*, 973. [CrossRef] [PubMed]
3. Bassett, D.R.; Howley, E.T. Limiting Factors for Maximum Oxygen Uptake and Determinants of Endurance Performance. *Med. Sci. Sport. Exerc.* **2000**, *32*, 70–84. [CrossRef] [PubMed]
4. Basset, D.R., Jr. Scientific contributions of A. V. Hill: Exercise physiology pioneer. *J. Appl. Physiol.* **2002**, *93*, 1567–1582. [CrossRef]
5. Besson, T.; Macchi, R.; Rossi, J.; Morio, C.Y.M.; Kunimasa, Y.; Nicol, C.; Vercruyssen, F.; Millet, G.Y. Sex Differences in Endurance Running. *Sport. Med.* **2022**, *52*, 1235–1257. [CrossRef]
6. Millet, G.P.; Bentley, D.J. The Physiological Responses to Running after Cycling in Elite Junior and Senior Triathletes. *Int. J. Sport. Med.* **2004**, *25*, 191–197. [CrossRef]
7. Schmidt, W.; Prommer, N. Impact of Alterations in Total Hemoglobin Mass on VO2max. *Exerc. Sport Sci. Rev.* **2010**, *38*, 68–75. [CrossRef]
8. Mier, C.M.; Domenick, M.A.; Turner, N.S.; Wilmore, J.H. Changes in Stroke Volume and Maximal Aerobic Capacity with Increased Blood Volume in Men and Women. *J. Appl. Physiol.* **1996**, *80*, 1180–1186. [CrossRef]
9. Rutkowski, D.R.; Barton, G.P.; François, C.J.; Aggarwal, N.; Roldán-Alzate, A. Sex Differences in Cardiac Flow Dynamics of Healthy Volunteers. *Radiol. Cardiothorac. Imaging* **2020**, *2*, e190058. [CrossRef]
10. Bossi, A.H.; Lima, P.; de Lima, J.P.; Hopker, J. Laboratory Predictors of Uphill Cycling Performance in Trained Cyclists. *J. Sport. Sci.* **2017**, *35*, 1364–1371. [CrossRef]
11. Whipp, B.J.; Davis, J.A.; Wasserman, K. Ventilatory Control of the "isocapnic Buffering" Region in Rapidly-Incremental Exercise. *Respir. Physiol.* **1989**, *76*, 357–367. [CrossRef] [PubMed]
12. Beaver, W.L.; Wasserman, K.; Whipp, B.J. Bicarbonate Buffering of Lactic Acid Generated during Exercise. *J. Appl. Physiol.* **1986**, *60*, 472–478. [CrossRef] [PubMed]
13. Puccinelli, P.J.; de Lira, C.A.B.; Vancini, R.L.; Nikolaidis, P.T.; Knechtle, B.; Rosemann, T.; Andrade, M.S. The Performance, Physiology and Morphology of Female and Male Olympic-Distance Triathletes. *Healthcare* **2022**, *10*, 797. [CrossRef]
14. Iannetta, D.; Inglis, E.C.; Mattu, A.T.; Fontana, F.Y.; Pogliaghi, S.; Keir, D.A.; Murias, J.M. A Critical Evaluation of Current Methods for Exercise Prescription in Women and Men. *Med. Sci. Sport. Exerc.* **2020**, *52*, 466–473. [CrossRef] [PubMed]
15. Engel, F.A.; Ackermann, A.; Chtourou, H.; Sperlich, B. High-Intensity Interval Training Performed by Young Athletes: A Systematic Review and Meta-Analysis. *Front. Physiol.* **2018**, *9*, 1012. [CrossRef]
16. Schwartz, J.; Oh, P.; Takito, M.Y.; Saunders, B.; Dolan, E.; Franchini, E.; Rhodes, R.E.; Bredin, S.S.D.; Coelho, J.P.; dos Santos, P.; et al. Translation, Cultural Adaptation, and Reproducibility of the Physical Activity Readiness Questionnaire for Everyone (PAR-Q+): The Brazilian Portuguese Version. *Front. Cardiovasc. Med.* **2021**, *8*, 2696. [CrossRef]

17. Jones, A.M.; Doust, J.H. A 1% Treadmill Grade Most Accurately Reflects the Energetic Cost of Outdoor Running. *J. Sport. Sci.* **1996**, *14*, 321–327. [CrossRef]
18. Whipp, B.J.; Ward, S.A.; Wasserman, K. Respiratory Markers of the Anaerobic Threshold. *Adv. Cardiol.* **1986**, *35*, 47–64.
19. Howley, E.; Basset, D.; Welch, H. Criteria for Maximal Oxygen Uptake: Review and Commentary. *Med. Sci. Sport. Exerc.* **1995**, *27*, 1292–1301. [CrossRef]
20. di Prampero, P.E. Energetics of Muscular Exercise. *Rev. Physiol. Biochem. Pharmacol.* **1981**, *89*, 143–222. [CrossRef]
21. Banerjee, A.; Chitnis, U.B.; Jadhav, S.L.; Bhawalkar, J.S.; Chaudhury, S. Hypothesis Testing, Type I and Type II Errors. *Ind. Psychiatry J.* **2009**, *18*, 127. [CrossRef] [PubMed]
22. Faul, F.; Erdfelder, E.; Lang, A.G.; Buchner, A. G*Power 3: A Flexible Statistical Power Analysis Program for the Social, Behavioral, and Biomedical Sciences. *Behav. Res. Methods* **2007**, *39*, 175–191. [CrossRef] [PubMed]
23. Etxebarria, N.; Wright, J.; Jeacocke, H.; Mesquida, C.; Pyne, D.B. Running Your Best Triathlon Race. *Int. J. Sport. Physiol. Perform.* **2021**, *16*, 744–747. [CrossRef] [PubMed]
24. Ofoghi, B.; Zeleznikow, J.; Macmahon, C.; Rehula, J.; Dwyer, D.B. Performance Analysis and Prediction in Triathlon. *J. Sport. Sci.* **2016**, *34*, 607–612. [CrossRef]
25. Hausswirth, C.; le Meur, Y.; Bieuzen, F.; Brisswalter, J.; Bernard, T. Pacing Strategy during the Initial Phase of the Run in Triathlon: Influence on Overall Performance. *Eur. J. Appl. Physiol.* **2010**, *108*, 1115–1123. [CrossRef]
26. Puccinelli, P.J.; Lima, G.H.O.; Pesquero, J.B.; de Lira, C.A.B.; Vancini, R.L.; Nikolaids, P.T.; Knechtle, B.; Andrade, M.S. Previous Experience, Aerobic Capacity and Body Composition Are the Best Predictors for Olympic Distance Triathlon Performance: Predictors in Amateur Triathlon. *Physiol. Behav.* **2020**, *225*, 113110. [CrossRef]
27. Murphy, W.G. The Sex Difference in Haemoglobin Levels in Adults—Mechanisms, Causes, and Consequences. *Blood Rev.* **2014**, *28*, 41–47. [CrossRef]
28. Howden, E.J.; Perhonen, M.; Peshock, R.M.; Zhang, R.; Arbab-Zadeh, A.; Adams-Huet, B.; Levine, B.D. Females Have a Blunted Cardiovascular Response to One Year of Intensive Supervised Endurance Training. *J. Appl. Physiol.* **2015**, *119*, 37–46. [CrossRef]
29. Holloszy, J.O.C. Adaptations of Skeletal Muscle to Endurance Exercise and Their Metabolic Consequences. *J. Appl. Physiol. Respir. Environ. Exerc. Physiol.* **1984**, *56*, 831–838. [CrossRef]
30. Jeukendrup, A.E. Carbohydrate Intake during Exercise and Performance. *Nutrition* **2004**, *20*, 669–677. [CrossRef]
31. Périard, J.D.; Eijsvogels, T.M.H.; Daanen, H.A.M. Exercise under Heat Stress: Thermoregulation, Hydration, Performance Implications, and Mitigation Strategies. *Physiol. Rev.* **2021**, *101*, 1873–1979. [CrossRef] [PubMed]
32. Tarnopolsky, L.J.; MacDougall, J.D.; Atkinson, S.A.; Tarnopolsky, M.A.; Sutton, J.R. Gender Differences in Substrate for Endurance Exercise. *J. Appl. Physiol.* **1990**, *68*, 302–308. [CrossRef] [PubMed]
33. Nelson, R.C.; Brooks, C.M.; Pike, N.L. Biomechanical Comparison of Male and Female Distance Runners. *Ann. N. Y. Acad. Sci.* **1977**, *301*, 793–807. [CrossRef] [PubMed]
34. Temesi, J.; Arnal, P.J.; Rupp, T.; Féasson, L.; Cartier, R.; Gergelé, L., Verges, S.; Martin, V.; Millet, G.Y. Are Females More Resistant to Extreme Neuromuscular Fatigue? *Med. Sci. Sport. Exerc.* **2015**, *47*, 1372–1382. [CrossRef] [PubMed]

 *healthcare*

Article

Validity of On-Line Supervised Fitness Tests in People with Low Back Pain

Ana Myriam Lavín-Pérez [1,2], Juan Luis León-Llamas [3,*], Francisco José Salas Costilla [4], Daniel Collado-Mateo [1], Raúl López de las Heras [4], Pablo Gasque Celma [4,5,*] and Santos Villafaina [3,6]

[1] Centre for Sport Studies, Rey Juan Carlos University, 28943 Fuenlabrada, Spain
[2] GO fitLAB, Ingesport, 28003 Madrid, Spain
[3] Physical Activity and Quality of Life Research Group (AFYCAV), Facultad de Ciencias del Deporte, Universidad de Extremadura, 10003 Cáceres, Spain
[4] Sports Medicine Service, Alcobendas City Council, 28100 Alcobendas, Spain
[5] Department of Physical Education, Sport and Human Motricity, Autónoma Univesity, Ciudad Universitaria de Cantoblanco, 28049 Madrid, Spain
[6] Departamento de Desporto e Saúde, Escola de Saúde e Desenvolvimento Humano, Universidade de Évora, 7004-516 Évora, Portugal
* Correspondence: leonllamas@unex.es (J.L.L.-L.); pgasque@aytoalcobendas.org (P.G.C.)

Abstract: This study aimed to investigate the concurrent validity between online evaluations (OEs) and face-to-face evaluations (IPEs) of a Senior Fitness Test and two balance tests in people with low back pain (LBP). Forty participants of 58.48 (9.87) years were included. The 30 s chair stand-up, arm curl, 2 min step, chair-sit and reach, back scratch, 8 foot up-and-go, sharpened Romberg, and one-legged stance tests were administrated using both OE and IPE methods. The results indicated no significant differences ($p > 0.05$) between the two methods except in the 8-foot up-and-go test ($p = 0.007$). Considering the ICC values and Bland-Altman plots, excellent agreement was found for the chair-sit and reach test, moderate agreement for the arm-curl and 8-foot up-and-go tests, and good agreement for the other tests. Strong correlations ($p < 0.001$) were observed in all variables except for the arm-curl and 8-foot up-and-go tests, where moderate correlations were found ($p < 0.05$). These results support the validity of OEs and IPEs in all tests, except for the arm-curl and 8-foot up-and-go tests, where lower ICC values and moderate correlations were found. However, it is important to consider the range of fluctuation of the ICC and the significant values obtained through correlations.

Keywords: exercise; physical fitness; pain; patient outcome assessment; online intervention

Citation: Lavín-Pérez, A.M.; León-Llamas, J.L.; Salas Costilla, F.J.; Collado-Mateo, D.; López de las Heras, R.; Gasque Celma, P.; Villafaina, S. Validity of On-Line Supervised Fitness Tests in People with Low Back Pain. *Healthcare* **2023**, *11*, 1019. https://doi.org/10.3390/healthcare11071019

Academic Editors: Young-Chang Arai, Rafael Oliveira and João Paulo Brito

Received: 20 February 2023
Revised: 28 March 2023
Accepted: 31 March 2023
Published: 3 April 2023

1. Introduction

Low back pain (LBP) is the main contributor to the overall burden of musculoskeletal conditions (568 million prevalent cases worldwide, responsible for 47% of global years of healthy life lost due to disability, YLDs), and neck pain is the fourth (223 million people; 22 million YLDs) [1]. Moreover, LBP is a global health problem [2]. Moreover, this is a global health problem [2] with a growing prevalence from 377.5 million people in 1990 to 577 million in 2017 and that entailed a health cost in the US of $56.5 and $62.3 thousand billion annually, respectively, an increase of 112% since 1990 [3].

LBP has a multifactorial and complex pathology that can cause significant limitations to the activities of daily living [4]. This condition normally produces a decline in the physical activity level, caused by kinesiophobia or pain-catastrophizing [5]. Consequently, sedentary behavior could affect patients' physical fitness, reducing physical function and strength [6], which may worsen their prognosis. In addition, LBP prevalence increases with age, and it is related to a higher body mass index, less physical activity, and lower self-perceived health [7]. Thus, exercise could be a useful tool, as long as there is no medical contradiction and it is individualized, to prevent and rehabilitate LBP, improving patients' quality of life and the ability to perform activities of daily living [8]. Accordingly, to design an individualized and effective physical exercise intervention, it is necessary to perform physical fitness tests [9]. In addition, to have as complete a picture as possible of the person's clinical condition, it would be advisable to take into account parallel measures, such as the level of pain through specific assessments [10] or the risk of developing this LBP and the predictors of this condition [11].

Regarding physical fitness evaluations in people with LBP, the American College of Sports Medicine (ACSM) recommends the evaluation of cardiorespiratory fitness, strength, and flexibility [12]. The Senior Fitness Test Battery (SFT) [13] consists of six tests that assess the strength of lower and upper limbs (arm curl and chair stand tests), the flexibility of the upper and lower limbs (back scratch and chair sit and reach), cardiorespiratory fitness (the 6 min walking test or the 2 min step test can be used), and agility (8-foot up-and-go test). Thus, the Senior Fitness test battery (SFT) [13] is a good option to evaluate people with LBP. However, previous studies suggested that balance could also be evaluated in people with LBP due to an increased center of pressure sway compared with that in the healthy controls [14,15]. In this regard, previous studies have included complex and challenging balance tasks since they are more sensitive in detecting balance impairments in people with LBP [16,17]. Among these challenging balance tests, the tandem Romberg test and the one-legged stance test have been proposed as challenging for the LBP population [18].

Considering the current health situation due to COVID-19 [19] and the growth of telemedicine, taking advantage of the emergence of new technologies and internet access, physical fitness assessments have considered new ways of monitoring and controlling the participants´ functionality and physical fitness condition. Previous studies have started to address this need by assessing patients with multiple sclerosis [20], chronic respiratory disease [21], cardiac conditions [22], and cancer [23] and even assessments of veterans during the COVID-19 pandemic using remote physical fitness evaluations [24]. Among these articles, two were systematic reviews that analyzed the test used remotely [21,22]. However, Hwang, Fan, Bowe, Louis, Bertram, Morris, and Adsett [22] analyzed the most used physical fitness test remotely used in cardiac conditions (in this case the 6 min walking test) without analyzing the validity of the remote physical fitness test. In the same line, Holland, Malaguti, Hoffman, Lahham, Burge, Dowman, May, Bondarenko, Graco, and Tikellis [21], after performing a systematic review of articles that included at-home, remote, and face-to-face evaluations of people with pulmonary diseases, concluded that physical fitness tests are rarely conducted remotely.

Nevertheless, there are studies that have conducted remote assessments of physical fitness. Among these studies, Blair, Blair, Harding, Herman, Boyce, Demark-Wahnefried, Davis, Kinney, and Pankratz [23] participants received a toolkit and instructions for setting up the timed-up-and-go and the 30 s chair stand test. However, data are still not available since it is a protocol study. Differently, the study of Ogawa, Harris, Dufour, Morey, and Bean [24] determined the inter-rater reliability of three physical fitness tests performed via a telehealth visit (30 s arm curl test, 30 s chair stand test, 2 min step test) among community-dwelling older veterans. However, these tests were focused on mobility or strength, without addressing other relevant capacities for activities of daily living, such as flexibility, aerobic capacity, or balance. Furthermore, the online evaluations conducted in these previous studies have been employed by using isolated tests without a comparison and validation between online and face-to-face evaluations. In contrast, Winters-Stone, et al. [25] reported the validity of the chair stand test and the 4 m usual walk test, showing acceptable validity when remotely performed for older adults with cancer. In the same line, Hoenemeyer, et al. [26] reported the reliability and validity of the SFT (including the sit and reach test, the 30-chair stand test, the back scratch test, the 8-foot up-and-go test and go test, the timed 8 min walk test, and the 2 min step test) in cancer survivors and supportive partners. Conclusions highlighted that remote physical fitness assessments are reliable, valid, acceptable, and safe in these populations.

Thus, due to the lack of evidence regarding remote physical fitness assessments for one of the most prevalent conditions [1], such as LBP, as well as the need to compare remote vs. face-to-face physical fitness assessments, this study aimed to evaluate the validity of a novel online adaptation of the six functional tests included in the SFT (30 s chair stand-up test, arm curl test, 2 min step test, chair-sit and reach test, back scratch test, and 8-foot up-and-go test) and two balance tests (the sharpened Romberg test and the one-legged stance test). In this regard, the hypothesis was that the validity between the OE and IPE will range from a good to excellent level of agreement based on the results of a previous study [26].

2. Materials and Methods

2.1. Study Design and Participants

Considering the sample size calculation performed with the G*Power software 3.1.9.4 (Kiel University, Kiel, Germany), a minimum of 13 people was needed to achieve a 99% power to detect a significant correlation with an alpha of 0.001. For this purpose, regarding the different statistical tests presented in G*Power, the bivariate normal model was selected to correlate two variables analyzed in a former study and also included in the current study. Data of OEs and IPEs of the 8-foot up-and-go test provided by Hoenemeyer, Cole, Oster, Pekmezi, Pye, and Demark-Wahnefried [26] were used. Lastly, a convenience sample of 40 people (26 females and 14 males) with LBP were recruited in a Sport Medicine Service (Alcobendas, Spain) and enrolled in the study before June of 2021 (Figure 1). Since in this Sport Medicine Service, more than 13 people voluntarily decided to participate in this validity study, all of them were included for eligibility. Participants randomly performed a face-to-face (IPE) and an online evaluation (OE) in order to assess the validity of the SFT and two balance tests (sharpened Romberg test and one-legged stance test).

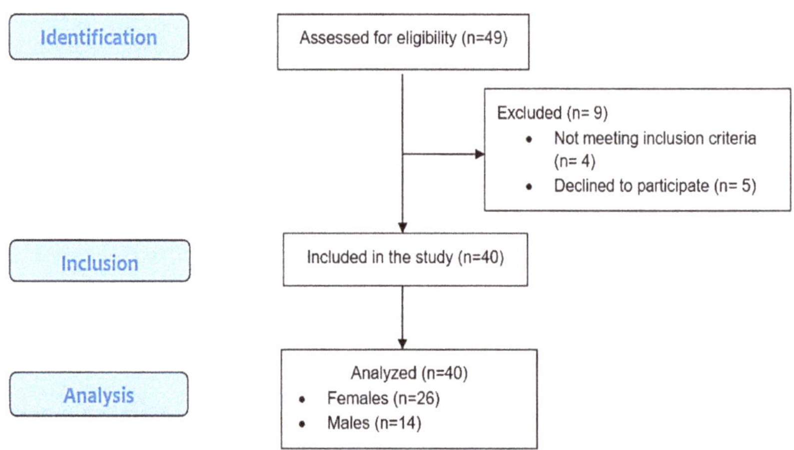

Figure 1. Flow diagram.

The specialist of the Sport Medicine Service recruited the patients from the LBP rehabilitation programs of the institution. In this regard, those patients who wanted to participate and fulfilled the following inclusion criteria were incorporated into the study: (a) people diagnosed with LBP, (b) people aged between 45 to 75-years-old, and (c) people without a medical contraindication for exercise. This age range was selected because it had the highest prevalence of low back pain in the sports medicine service of the city council. Moreover, the following exclusion criteria were considered: (a) people with other musculoskeletal or chronic diseases, (b) people without a phone, tablet, or computer with a camera and internet connection, (c) people injured or operated on in the last month, and

(d) people with mental disorders unable to understand or execute physical indications. Concretely, the included participants had persistent pain beyond 3 months of symptoms [27] and one or more of the following diagnoses: herniated disc(s), scoliosis, spinal stenosis, and spinal osteoarthritis.

All the procedures were approved by the Research Ethics Committee of the University of Extremadura (approval reference: 41/2021) according to the ethical standards of the Declaration of Helsinki. The study was conducted from December 2020 to July 2021. Before starting the study, all patients were informed of the procedures and signed an informed consent form.

2.2. Procedures and Assessments

The validity of the SFT and two balance tests were carried out by randomizing the order of the evaluations according to the modality (face-to-face or online evaluations). For this process, a series of random numbers was generated using a randomization tool available on the web (https://www.randomizer.org/ accessed on 8 January 2022) that allowed the order of the evaluations to be established for each participant. Both evaluation sessions were developed under the supervision of a physical activity and sports science professional. Each kind of evaluation was performed by the same evaluator and in the same order starting with the non-fatiguing test, then the muscular endurance tests, and ending with the submaximal aerobic capacity test [28]. The two evaluation procedures were spaced by a 2-week period.

The SFT [13] has been chosen as it is widely used in clinical and research settings and has the ability to be performed easily and cheaply in homes or clinics without extensive technical expertise. The six physical fitness tests included in the SFT [13] were as follows: 30 s chair stand-up test, arm curl test, 2 min step test, chair-sit and reach test, back scratch test and 8-foot up-and-go test. Furthermore, since the balance evaluation of the challenging balance test is recommended for people with LBP, two validated balance tests (sharpened Romberg test and one-legged stance test) were included [29,30]. Two balance tests were evaluated since the one-legged stance test is more complex than the sharpened Romberg test. By introducing two levels of difficulty, more complete information about participants' balance was obtained.

However, the arm curl test, chair-sit and reach test, and back scratch were adapted to be self-administered (see Table 1). Before the evaluations, independently of the assessment modalities (OE or IPE), the professional explained to participants how to perform the tests. In this regard, the main differences between the types of evaluation were as follows: (1) the professional and participant setting, and (2) the way to collect participants' results. In the IPE, the professional was next to the participants, with the material selected and placed, making the corresponding explanations and noting down the results that he measured. Whereas in the OE, the professional and the participant were connected by a video call. Following the professional instructions, the participant placed and prepared the necessary material, and after performing the test, the participants registered the results. During the tests, the professional indicated to the participant to position the camera in front of him or her at hip level and far enough away to be able to see the participant completely.

Table 1. Physical and measurement adaptations in the face-to-face and online evaluations for test self-administration.

Original Measure	Adaptions for the Face-to-Face Evaluation	Adaptions for the Online Evaluation
30 s Chair Stand-Up Test [13]	Same as original	Participants controlled the stopwatch by starting the test at second 10 and finishing it at second 40. Repetitions were counted by participants.
Arm Curl Test [13]	The weight was added to a sturdy cloth or plastic bag until 2.300 kg and 3.600 kg was reached (e.g., packets of rice, nuts, yogurt, etc.).	
		Participants controlled the stopwatch by starting the test at second 10 and finishing it at second 40. Repetitions were counted by participants.
2-min Step Test in Place [13]	Same as original	Repetitions were counted by participants. Time was controlled by each participant.
Chair-sit and Reach Test [13]	Same as original	A light object, such as a ruler, was held with both hands by the participants. Participants positioned the end of the object touching the toe tips and slid their hands elongating the trunk. The score was obtained by measuring the distance between the fingertip and the end of the object.
Back Scratch Test [13]	Same as original	A light object, such as a ruler, was held with the upper hand until the end touched the fingertip of the lower hand. The upper hand slid in the object as low as possible. The score was obtained by measuring the distance between the fingertip and the end of the object.
8-Foot Up-and-go Test [13]	Same as original	Participants started and stopped their stopwatch holding it during the test.
Sharpened Romberg Test [31]	Same as original	Participants controlled the stopwatch by pressing start and stop finishing the test when 60 s was reached or balance was lost.
One-Legged Stance Test [32]	Same as original	Participants controlled the stopwatch by pressing start and stop finishing the test when 60 s was reached or balance was lost.

2.3. Demographic Information and Quality of Life

Information regarding age, sex, body mass, and height parameters was collected. The Body Mass Index (BMI) calculation (kg/m^2) was conducted through bodyweight divided by squared height [BMI = weight (kg)/height (m^2)]. Furthermore, the physical and psychosocial states were registered using the EuroQol-5D (EQ-5D-3L) questionnaire. The questionnaire was composed of 5 dimensions concerning mobility, self-care, activities of daily life, pain, and anxiety/depression [33]. Participants had three levels of scoring options ranging from 1 to 3, being option 1 "I have no problems doing my usual activities" and option three "I am unable to do my usual activities". In addition, the questionnaire included a Visual Analogue Scale (VAS) in order to self-report the current health state. The VAS is rated from 0 to 100, where higher scores reflect a better health state [33]. The Spanish EQ-5D (3L) version was used [34]. This questionnaire has been employed before to assess the quality of life of chronic back pain patients [7].

2.3.1. The 30 s Chair Stand-Up Test

The participant started the test sitting in the middle of a chair approximately 43 cm from the floor, with the back or legs of the chair against the wall to avoid possible movements [13]. The test began with the participant standing with his or her back straight, feet flat on the floor, and arms crossed over the chest. In the IPE, at the signal "go", the participant had to stand up and sit down from the chair as many times as possible for 30 s [13]. In the OE, the participants, with a stopwatch, started the test at second 10 and finished it at second 40. The final score was the number of times the participant managed to stand up. The test aimed to assess the strength-endurance of the lower limbs. The performance was measured equally in both test modalities. In the IPE, the professional was the one who counted the repetitions, whereas in the OE, the participant was the one who did it.

2.3.2. Arm Curl Test

The participant began seated in the same chair as in the previous test, without armrests, with their back straight, feet flat on the floor, and the body part of the dominant arm with which he/she performed the test at the end of the chair [13]. The weight selected for the test was 2.300 kg for women and 3.600 kg for men [13]. For this purpose, in both evaluation modalities, the participants added contents to a sturdy cloth or plastic bag until the required weight was reached (e.g., packets of rice, nuts, yogurt, etc.). Although participants were well instructed to place the number of packages needed to reach the weight set in the arm curl test evaluation, it is possible that there may have been slight variations in the total weight incorporated. In the IPE, the test started with the participant holding the weight with his dominant hand and the elbow in extension. To complete the exercise, the participant should flex the elbow and return to full extension of the elbow as many times as possible within 30 s. In the OE, the participants started in the same position, but, with a stopwatch, they started the test at second 10 and finished it at second 40. The performance was measured equally in both test modalities. In the IPE, the professional noted the repetitions performed by the participants, while in the OE the participants counted their own repetitions. The objective was to assess upper limb strength.

2.3.3. 2 Min Step Test in Place

At the signal "go", the participant started to walk in place by raising the knees until the knees reached the midpoint between the knee and the iliac crest [13]. The number of times the participant raised the knee of the dominant leg for 2 min was recorded [13]. The objective was to assess aerobic endurance. The test was carried out equally in both evaluation modalities, although in the IPE, the professional reported participants' repetitions, whereas in the OE, the participants counted their own repetitions. Moreover, the 2 min was controlled by the professional in the IPE and by the participant in the OE.

2.3.4. Chair-Sit and Reach Test

The participant was seated on the edge of the chair used in the previous tests and stretched the preferred leg, and the other leg was kept flexed in front of the hip. In the OE, he/she grasped the end of a light, rigid object, e.g., a pen or a ruler, and held it with both hands, fingers outstretched and palms facing each other. The arms remained outstretched with hands together. The participant elongated the trunk by bending the hips with the intention of touching the toe of the outstretched leg. Participants were instructed to stretch the leg parallel to the flexed leg to avoid any hip abduction movements that could interfere with the correct assessment of truck flexibility. In this position, the participant stretched out his arms to their full length holding the object with its major axis parallel to the leg until the end of the object touched the toe. Once this was achieved, the participant simultaneously slid their hands in the object as close to the foot as possible. After stretching as much as possible, the participant was instructed to keep one hand at the maximum point reached. Then, with the other hand, using a ruler or a meter, each participant measured the distance between the end of the object and the fingertips of the hand. In contrast, in the IPE, the original test of the Senior Fitness Test battery was followed [13], so that the participant performed the same procedure as in the OE but without holding an object and keeping his or her hands together and at the same level during the whole test [13]. Therefore, the professional measured the distance reached between the fingertips.

The scoring of both modalities followed the same instructions. If they passed with the fingers, the score had a positive sign (+X cm), if the fingers touched the tip of the toe the scoring was 0 cm, and if they did not reach it, a negative sign before the number was added (−X cm). The objective was to assess the flexibility of the lower body.

2.3.5. Back Scratch Test

In both types of evaluations, the participant was standing with the hand of his/her dominant arm on the same shoulder intending to touch his/her back, from above. In the OE, the dominant hand, outstretched, held an object (e.g., a pen), while the other arm was placed behind the back around the waist with the palm of the hand pointing outwards (in line with the object of the dominant hand). In this way, the end of the object tried to touch the tip of the middle finger of the down hand by adjusting the length of the object left free from the upper hand. As in the previous test, the professional indicated to the participants to keep one hand at the maximum point reached in the object and measured the distance from the tip of the finger of the upper hand and the end of the object. Unlike that, in the IPE, the upper hand of the patient kept its fingers directed downwards and completely stretched trying to reach the fingertip of the other hand, without employing an object [13]. In this case, the professional measured the distance between the tips of the middle fingers. The objective was to assess upper limb flexibility.

The scoring of both modalities followed the same instructions. If they passed, the score had a positive sign (+X cm), if the fingertips of both hands touched each other, the scoring was 0 cm, and if they did not reach it, a negative sign before the number was added (−X cm). The objective was to assess the flexibility of the upper body.

2.3.6. Eight-Foot Up-and-Go Test

Before starting, a chair was placed against the wall, and a cone (or visible mark) was set at a distance of about 2.44 m in a straight line from the back of the chair [13]. At the signal "go", the participant got up from the chair, without leaning on it, walked to the cone, went around it, and sat back in the chair [13]. The time elapsed between the "go" signal and the moment when the participant sat down on the chair was recorded utilizing a stopwatch. In the IPE, the professional was responsible for controlling the stopwatch, whereas in the OE, the participants started and stopped their stopwatch. In this modality, they were instructed to hold the stopwatch during the test to press the buttons as fast as possible. The aim was to assess agility and dynamic balance.

2.3.7. Sharpened Romberg Test

The test was carried out in a tandem position [30,31]. The participant placed the heel of one foot in contact with the toe of his or her other foot. The time reached in this position was recorded, the test ended when the participant lost his/her balance or when 60 s on the position was reached. In the IPE, the professional started and stopped the stopwatch when the balance was lost or at 60 s, and in the OE modality, the patient controlled it by holding the stopwatch [30,31].

2.3.8. One-Legged Stance Test

The participant made the test with the dominant leg and stood with hands on hips. At the signal "go", he/she lifted their feet off the ground and placed this leg with the knee flexed in front of him/her. They had to keep their balance for as long as possible. The time that he/she held on without touching the ground with the knee flexed was recorded. The maximum time of the test was 60 s. In the IPE, the professional was responsible for controlling the stopwatch, whereas in the OE modality, the participant started and stropped their own stopwatch by holding it in the hand. The aim was to assess static balance [29,32]. This balance test was proposed since challenging balance task are more sensitive in detecting balance impairments in people with LBP [16,17] than easier ones.

2.4. Statistical Analytics

The Statistical Package for Social Science (version 25.0; SPSS, Inc., IBM Corp, Armonk, New York, NY, USA) was used to conduct the statistical analyses. Data are presented as the mean and standard deviation (SD) from each evaluation modality and from the difference between modality mean results. The explained statistical analysis procedures were performed based on the global sample size, as well as by dividing the participants according to their sex. The alpha level of significance was set at <0.05.

Furthermore, parametric and non-parametric tests were conducted based on the Shapiro–Wilk results. Validity is a multicomponent concept that assesses to what extent a tool measures what it is intended to measure [35]. Criterion validity is an estimate of the extent to which a measure agrees with a gold standard, and it is possible to evaluate by statistically testing a new measurement technique against an independent criterion or standard (concurrent validity) [36]. For this reason, to explore the concurrent validity, the performance differences between the evaluation modalities (IPE and OE), the one sample *t*-test or one-sample Wilcoxon signed rank test was used. In the same vein, the intraclass correlation coefficient (ICC) of two-way mixed effects, consistency, and single rater/measurement [37] were calculated to assess the level of agreement between OE and the IPE results. ICC values less than 0.50 are indicative of poor agreement, values between 0.5 and 0.75 indicate moderate agreement, values between 0.75 and 0.90 indicate good agreement, and values greater than 0.90 indicate excellent agreement [38]. Moreover, Bland-Altman plots were performed to illustrate the agreement between the OE and IPE scores [39] to provide the limits of agreement and bias, following the procedure proposed by Giavarina [40]. The software Graphpad Prism 8.0 was used to create the graphs. Lastly, the correlations between the tests of the two measurement methods were analyzed, and Cohen's recommendations [41] were followed to interpret the correlation coefficient. A score $\geq$0.5 was strong, it was moderate if the score was between 0.5 and 0.35, and it was poor if the score was $\leq$0.35.

3. Results

A total of 26 women and 14 men (65% and 35% of the total sample size, respectively) participated in this study. Their age ranged from 45 to 72 years. Regarding BMI results, 53.84% of the women had a normal weight, 38.46% were overweight, and 7.69% were obese, whereas 35.71% of the men had a normal weight, 50% were overweight, and 14.28% were obese.

Participants' quality of life (EQ-5D-3L coefficient) reached high scorings close to 1 (full healthy), the maximum achievable score. Concerning the EQ-5D Visual Analogue Scale, participants also perceived their health to be apparently elevated since it was close to 100. Analyzing the included dimensions, the highest quality of life problem was obtained in pain perception. More specific information about participants' characteristics is shown in Table 2.

Table 3 shows participants' descriptive data and the agreement in the OE and IPE of the total sample size. Comparisons between OE and IPE did not show significant differences between OE and IPE modalities in all tests ($p > 0.05$), except in the 8-foot up-and-go (p-value = 0.007). ICC values ranged from moderate to excellent agreement regarding the lowest and highest score (0.55 to 0.93). A moderate agreement was found for the arm-curl test and the 8-foot up-and-go test. In addition, a good agreement was found for the 30 s chair stand-up test, 2 min step-test in place, back scratch test, sharpened Romberg test, and the one-legged stance test. Moreover, excellent agreement was found in the chair-sit and reach test. Considering the Bland-Altman plots of the scores obtained by OE and IPE of the tests of the total sample size (see Figure 2), the graphs depict excellent agreement between OE and IPE with only reduced scores of observations falling outside the limits of agreement for the different test performed. Moreover, there was not significant bias since the line of equality was within the confidence interval of the mean difference. Finally, all variables showed strong correlations, except for the arm curl test and 8-foot up-and-go test, where moderate correlations were found. However, all variables showed significant relationships.

Table 2. Participants' physical and quality-of-life characteristics.

Participants' Characteristics	Total Sample Size	Women (*n* = 26)	Men (*n* = 14)
	Mean (SD) [range of variability]	Mean (SD) [range of variability]	Mean (SD) [range of variability]
Age	58.48 (9.87) [45–72]	58.46 (9.23) [45–72]	58.50 (10.14) [47–69]
Height (cm)	166.00 (9.72) [145–184]	164.38 (10.27) [145–176]	169.00 (8.09) [165–184]
Body mass (Kg)	71.10 (13.08) [53–107]	69.92 (14.52) [53–88]	73.29 (10.01) [65–107]
BMI	25.69 (3.37) [18.78–35.34]	25.74 (3.82) [18.78–31.55]	25.59 (2.42) [21.50–35.34]
EuroQol-5D (coefficient)	0.85 (0.14) [0.32–1]	0.87 (0.12) [0.74–1]	0.81 (0.18) [0.32–1]
EuroQol-5D (VAS)	76.25 (10.49) [60–100]	76.73 (11.13) [60–100]	69.87 (9.50) [60–90]
Mobility	1.08 (0.27) [1–2]	0.87 (0.12) [1–2]	1.21 (0.43) [1–2]
Self-Care	1.03 (0.16) [1–2]	1.04 (0.20) [1–2]	1.00 (0.00) [1]
Activities Of Daily Life	1.20 (0.41) [1–2]	1.19 (0.40) [1–2]	1.21 (0.43) [1–2]
Pain	1.63 (0.54) [1–2]	1.62 (0.50) [1–2]	1.64 (0.63) [1–2]
Anxiety/Depression	1.05 (0.22) [1–2]	1.00 (0.00) [1]	1.14 (0.36) [1–2]

SD: standard deviation, EuroQol-5D: European Quality of Life-5 Dimensions, VAS: visual analogue scale.

Table 3. Agreement under online and face-to-face evaluations of total sample size (N = 40).

Variable	Online Evaluation Mean (SD)	Face-to-Face Evaluation Mean (SD)	*p*-Value	ICC (95% CI)	Correlation Coefficient
30 s Chair Stand-Up test	19.00 (3.76)	18.18 (3.91)	0.108	0.79 (0.610–0.891)	0.66 ***
Arm-curl test	21.38 (4.42)	20.70 (4.28)	0.385	0.55 (0.143–0.760)	0.38 *
2 min step-test in place	107.50 (23.12)	110.10 (23.13)	0.306	0.87 (0.748–0.930)	0.76 ***
Chair-sit and reach test	−1.46 (9.59)	−0.93 (10.50)	0.528	0.93 (0.859–0.961)	0.86 ***
Back scratch test	−4.89 (8.53)	−2.68 (9.25)	0.305	0.79 (0.596–0.887)	0.64 ***
8-foot up-and-go test	4.89 (0.85)	4.45 (1.06)	0.007	0.64 (0.313–0.808)	0.48 **
Sharpened Romberg test	59.00 (4.41)	58.63 (3.92)	0.334	0.89 (0.792–0.942)	0.66 †††
One-legged stance test	50.57 (16.34)	47.88 (19.44)	0.363	0.83 (0.675–0.909)	0.75 †††

SD: standard deviation, ICC: intraclass correlation coefficient. Correlation Coefficient. Paired *t*-test or Wilcoxon tests were conducted depending on the distribution (variables with a *p*-value lower than 0.05 in the Shapiro–Wilk test were considered for a non-parametric analysis). *** *p*-value < 0.001; ** *p*-value < 0.01; * *p*-value < 0.05 based on Pearson's correlation coefficient. ††† *p*-value < 0.001; based on Spearman's Rho correlation coefficient.

Agreement data divided by sex is reported in the Supplementary Materials (See Supplementary Table S1).

Figure 2 shows the Bland-Altman plots of the scores obtained by OE and IPE of the tests of the total sample size. Moreover, in the Supplementary Materials, the Bland-Altman plots for women and men in the OE and IPE are reported (See Supplementary Figures S1 and S2). The plots of the total sample size (Figure 2) and the men and women analysis (Supplementary Figures S1 and S2) depicted excellent agreement between the OE and IPE with only reduced scores of observations falling outside of the limits of agreement for the different tests performed.

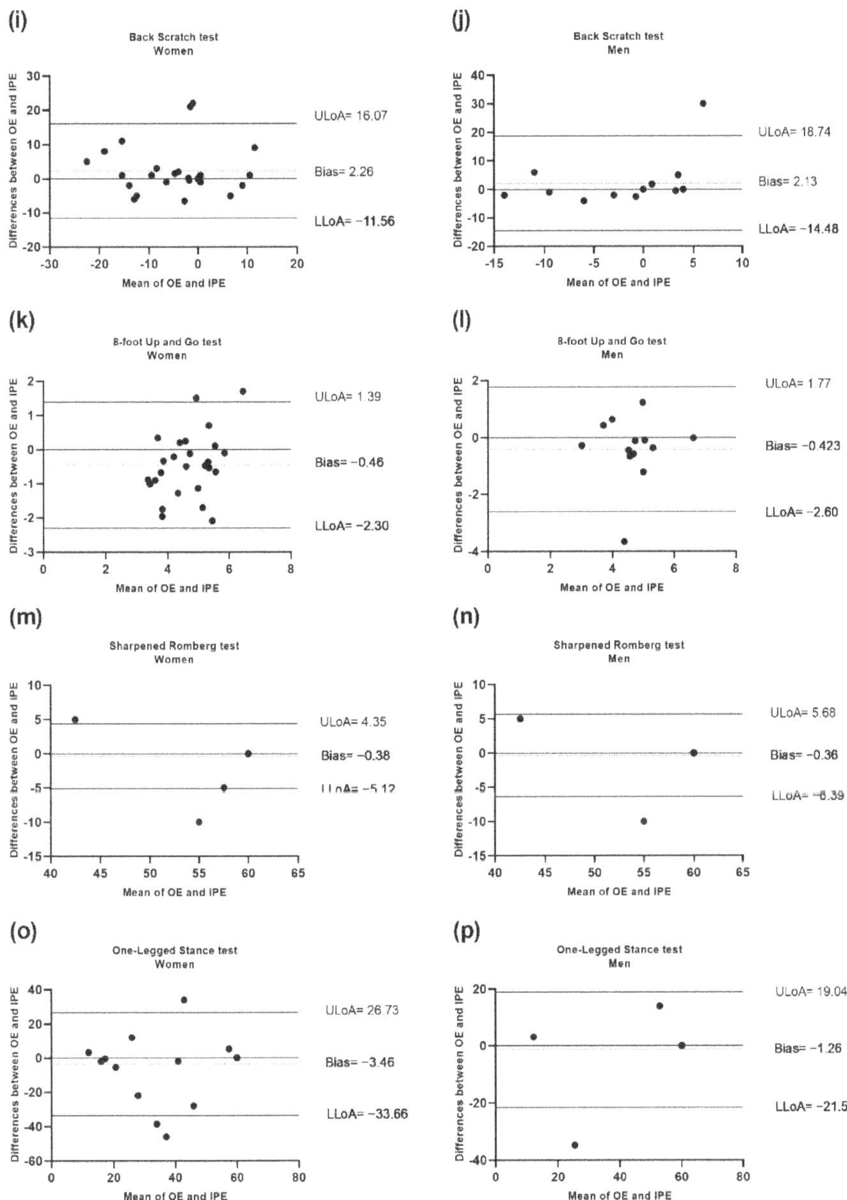

Figure 2. Bland-Altman plots for the difference between online evaluation and face-to-face evaluation of the SFT and the two balance test for participants divided by sex. (**i**) Back Scratch test for women; (**j**) Back Scratch test for men; (**k**) 8-foot Up and Go test for women; (**l**) 8-foot Up and Go test for men; (**m**) Sharpened Romberg test for women; (**n**) Sharpened Romberg test for men; (**o**) One-Legged Stance test for women; and (**p**) One-Legged Stance test for men. OE: online evaluation; IPE: face-to-face evaluation; ULoA: upper limit of agreement; LLoA: lower limit of agreement. The shaded areas represent the confidence interval limits for mean and agreement limits.

4. Discussion

The Senior Fitness test battery [13] consists of six physical fitness tests that assess strength, flexibility, cardiorespiratory fitness, and agility (30 s chair stand-up, arm curl, 2 min step, chair-sit and reach, back scratch test and 8-foot up-and-go). This battery is a useful tool to evaluate people with LBP since the ACSM recommends the evaluation of these physical capacities, concretely cardiorespiratory fitness, strength, and flexibility [12]. In addition, previous studies have shown balance impairment in people with LBP [14,15]. Thus, the validity of an OE in contrast to IPE through the level of agreement in different functionality tests was investigated. The results achieved with the present investigation showed reasonable levels of agreement regarding the ICC values [38], the Bland-Altman plots between the variables of the different evaluation methods [39], and the strong correlations obtained in all tests except for the arm curl test and the 8-foot up-and-go test, where moderate correlations were found. Nevertheless, all variables showed significant relationships between OE and IPE. Therefore, the initial hypothesis proposed should be accepted. These results may prove to be of great relevance in the clinical field in order to have objective tools to face current practical challenges.

With the COVID-19 pandemic [19], the growth of telemedicine, and the emergence of new technologies and Internet access, physical fitness assessments have considered new ways to monitor and control the functionality and physical condition of participants. The validation of online functional tests in comparison to the gold standard face-to-face test, as performed in the present study, can be of great help in clinical rehabilitation units, especially when hospitals are overcrowded or when, for health reasons, patients cannot be assessed face-to-face [42]. In this way, rehabilitation staff can remotely assess the physical function of patients without direct contact, and it would be possible to assess patients more objectively and quickly prior to their incorporation into rehabilitation or training programs. This clinical implication, that the incorporation of online testing in patients with LBP could foster the autonomy and self-determination of those suffering from LBP, is necessary in the effectiveness of their care [43].

Physical exercise home-based interventions have emerged as an effective way to maintain physical and mental health [44,45]. However, these home-based physical exercise interventions should monitor and control the participants' physical fitness. In this regard, previous studies have reviewed [21] and investigated the reliability of different physical fitness tests using remote evaluation [24]. Ogawa, Harris, Dufour, Morey, and Bean [24] showed high inter-rater reliability in three tests used in the present study, the 30 s arm curl, 30 s chair stand test, and 2 min step test, among community-dwelling older adults. However, remote physical fitness assessment was not compared and validated with a face-to-face evaluation. In the same line, Lin et al. [46], showed moderate-to-good reliability of the fitness assessments using a health app. Again, this remote evaluation was not compared with a face-to-face condition.

Nevertheless, some studies compared face-to-face evaluations with remote evaluations in populations different from people with LBP. Holland, et al. [47] examined the usability of the 6 min walking test at home. Results showed that this test underestimated the exercise capacity when conducted at home in people with chronic obstructive pulmonary disease. Similarly, Cox et al. [48] showed that exercise capacity assessment, using the 3 min step test, is feasible and accurate via remote videoconferencing in adults with cystic fibrosis. However, the OE conducted in these studies has not been compared to the IPE evaluation. In this regard, Winters-Stone, Lipps, Guidarelli, and Herrera-Fuentes [25] reported the validity of the chair stand test, showing a good correlation between an EO and IPE in older adults with cancer (r = 0.81). The results achieved with the present study showed a lower correlation value (correlation coefficient = 0.66) in people with LBP. Hoenemeyer, Cole, Oster, Pekmezi, Pye, and Demark-Wahnefried [26] reported the reliability and validity of the SFT (including the sit and reach test, the 30 s chair stand test, the back scratch test, 8-foot up-and-go test and go test, the timed 8 min walk test, and the 2 min step test) in cancer survivors and supportive partners. In this previous study, the authors reported the validity through the ICC, showing the highest value in the back scratch test (ICC = 0.95), classified as excellent. The other tests (30-s chair stand-up, 2 min step, chair-sit and reach, 8 min walk test, and 8-foot up-and-go) also showed high validity values. The results presented showed a lower ICC value (ICC = 0.79) in the back scratch, although validity can be considered good. In addition, they showed similar ICC values in the chair-sit and reach (ICC = 0.87) and the 2 min step (ICC = 0.93) compared to those in the Hoenemeyer, Cole, Oster, Pekmezi, Pye, and Demark-Wahnefried [26] study (ICC = 0.89 and ICC = 0.84 for the sit and reach and 2 min step test, respectively).

The greatest differences in validity were observed in the 8-foot up-and-go test with an ICC = 0.64 compared to an ICC = 0.80 observed in a previous study [26]. This could be due to the difference in the test performance, being higher in the present study than in the study focused on cancer survivors [26]. Furthermore, another possible explanation for this result can be extracted. In this regard, the lowest ICC values were obtained by the arm-curl test (ICC = 0.55) and the 8-foot up-and-go test (ICC = 0.64). In the same line, these two tests are those that yielded the lowest correlation values (arm-curl test correlation coefficient = 0.38 and the 8-foot up-and-go test correlation coefficient = 0.48). Regarding the arm-curl test, participants in the OE condition were asked to start the test when their stopwatch was at second 10 and finished it at second 40. Since participants would have paid attention to the stopwatch, the participants would be less accurate in measuring their performance on this test. Similarly, in the 8-foot up-and-go test, the stopwatch was controlled by the participant themself in the OE condition, whereas in the IPE condition, the stopwatch was controlled by the research staff. This would make the time the participant took to hit play and stop longer. Future studies should take this issue into account.

This study has some limitations that should be acknowledged. First, the investigation conducted was observational and monocentric with a sample composed of people with LBP, so results cannot be generalized to other populations. In addition, the sample of men was relatively small. Second, a test-retest analysis was not conducted. Thus, future studies should analyze the reliability of the physical fitness battery for OE and IPE. Third, it would have been interesting to specify the camera angle and the distance at which the device was located to facilitate the replication of the conditions of this study, although this is a difficult task since the conditions of each device vary according to the manufacturer. Fourth, it is possible that in the arm curl test, there were slight variations in the total weight incorporated into the sturdy cloth or plastic bag. In the same vein, it is also possible that in the back scratch test, the chair sit and reach test, and the 8-foot up-and-go test, there were small alterations due to the difficulty in performing the test or the time invested in pressing the stopwatch, respectively. Considering these limitations, results can be taken with caution. Although this study has some limitations, strengths should also be highlighted. In this regard, to the best of our knowledge, this is the first study that analyzed the validity between OE and IPE in an LBP population. Furthermore, unlike previous studies, the present investigation included a wide range of physical fitness tests: strength, flexibility, agility, cardiorespiratory fitness, and balance. In addition, the present results provide relevant information to clinical and physical specialists who want to implement online or home-based physical exercise interventions. In this regard, the physical fitness tests have been modified so that patients can be self-assessed. This would allow healthcare professionals and patients to economize time and aid resources while maintaining validity.

5. Conclusions

The results achieved in the current investigation showed reasonable levels of agreement that support the validity between OE and IPE in all tests regarding the ICC values, the Bland-Altman plots between the variables of the different evaluation methods, and the strong correlations obtained, except for the arm curl test and the 8-foot up-and-go test, where moderate correlations and lower ICC values were found. However, it is important to take into account the range of fluctuation of the ICC and the significant values obtained through correlations.

Supplementary Materials: The following supporting information can be downloaded at: https://www.mdpi.com/article/10.3390/healthcare11071019/s1, Supplementary Table S1. Agreement under online and face-to-face evaluations divided by women ($n = 26$) and men ($n = 14$). Supplementary Figure S1. Bland-Altman plots of the SFT and the two balance tests for participants divided by sex. OE: online evaluation; IPE: face-to-face evaluation; ULoA: upper limit of agreement; LLoA: lower limit of agreement. The shaded areas represent the confidence interval limits for the mean and agreement limits. Supplementary Figure S2. Bland-Altman plots of the SFT and the two balance tests for participants divided by sex. OE: online evaluation; IPE: face-to-face evaluation; ULoA: upper limit of agreement; LLoA: lower limit of agreement. The shaded areas represent the confidence interval limits for the mean and agreement limits.

Author Contributions: Conceptualization, F.J.S.C., D.C.-M. and P.G.C.; Data curation, A.M.L.-P. and F.J.S.C.; Formal analysis, A.M.L.-P., P.G.C. and S.V.; Investigation, A.M.L.-P., J.L.L.-L., P.G.C. and S.V.; Methodology, J.L.L.-L. and F.J.S.C.; Project administration, F.J.S.C., R.L.d.l.H. and P.G.C.; Resources, D.C.-M., R.L.d.l.H. and P.G.C.; Software, J.L.L.-L.; Supervision, D.C.-M. and S.V.; Validation, D.C.-M.; Visualization, R.L.d.l.H.; Writing—original draft, A.M.L.-P. and S.V.; Writing—review & editing, D.C.-M., J.L.L.-L. and S.V. All authors have read and agreed to the published version of the manuscript.

Funding: The author S.V. was supported by a grant from the Universities Ministry of Spain and the European Union (NextGen-erationUE) "Ayuda del Programa de Recualificación del Sistema Universitario Español, Modalidad de ayudas Margarita Salas para la formación de jóvenes doctores" (MS-03). The author J.L.L.L. was supported by a grant from the Spanish Ministry of Education, Culture and Sport (FPU18/05655). The author A.M.L.-P. is supported by the Industrial Doctorate Spanish National grant program, part of the Strategic Plan on Science and Innovation Support of the Spanish Ministry of Science, Innovation, and Universities. The predoctoral industry grant identification number is DIN2018-010129.

Institutional Review Board Statement: The study was conducted in accordance with the Declaration of Helsinki and approved by the Research Ethics Committee of the University of Extremadura (approval reference: 41//2021).

Informed Consent Statement: Informed consent was obtained from all subjects involved in the study.

Data Availability Statement: Data will be available upon reasonable request to the corresponding author.

Acknowledgments: The study has been carried out thanks to the Sport Medicine Service of Alcobendas, Spain.

Conflicts of Interest: The authors declare no conflict of interest. The funders had no role in the design of the study; in the collection, analyses, or interpretation of data; in the writing of the manuscript; or in the decision to publish the results.

References

1. Cieza, A.; Causey, K.; Kamenov, K.; Hanson, S.W.; Chatterji, S.; Vos, T. Global estimates of the need for rehabilitation based on the Global Burden of Disease study 2019: A systematic analysis for the Global Burden of Disease Study 2019. *Lancet* **2020**, *396*, 2006–2017. [CrossRef] [PubMed]
2. Hoy, D.; Bain, C.; Williams, G.; March, L.; Brooks, P.; Blyth, F.; Woolf, A.; Vos, T.; Buchbinder, R. A systematic review of the global prevalence of low back pain. *Arthritis Rheum.* **2012**, *64*, 2028–2037. [CrossRef] [PubMed]
3. Wu, A.; March, L.; Zheng, X.; Huang, J.; Wang, X.; Zhao, J.; Blyth, F.M.; Smith, E.; Buchbinder, R.; Hoy, D. Global low back pain prevalence and years lived with disability from 1990 to 2017: Estimates from the Global Burden of Disease Study 2017. *Ann. Transl. Med.* **2020**, *8*, 299. [CrossRef] [PubMed]
4. Grabovac, I.; Dorner, T.E. Association between low back pain and various everyday performances. *Wien. Klin. Wochenschr.* **2019**, *131*, 541–549. [CrossRef] [PubMed]
5. Macías-Toronjo, I.; Rojas-Ocaña, M.J.; Sánchez-Ramos, J.L.; García-Navarro, E.B. Pain catastrophizing, kinesiophobia and fear-avoidance in non-specific work-related low-back pain as predictors of sickness absence. *PLoS ONE* **2020**, *15*, e0242994. [CrossRef]
6. Alsufiany, M.B.; Lohman, E.B.; Daher, N.S.; Gang, G.R.; Shallan, A.I.; Jaber, H.M. Non-specific chronic low back pain and physical activity: A comparison of postural control and hip muscle isometric strength: A cross-sectional study. *Medicine* **2020**, *99*, e18544. [CrossRef]
7. Alonso-García, M.; Sarría-Santamera, A. The economic and social burden of low back pain in Spain: A national assessment of the economic and social impact of low back pain in Spain. *Spine* **2020**, *45*, E1026–E1032. [CrossRef]

8. Wun, A.; Kollias, P.; Jeong, H.; Rizzo, R.R.N.; Cashin, A.G.; Bagg, M.K.; McAuley, J.H.; Jones, M.D. Why is exercise prescribed for people with chronic low back pain? A review of the mechanisms of benefit proposed by clinical trialists. *Musculoskelet. Sci. Pract.* **2021**, *51*, 102307. [CrossRef]
9. Buckinx, F.; Croisier, J.L.; Reginster, J.Y.; Dardenne, N.; Beaudart, C.; Slomian, J.; Leonard, S.; Bruyère, O. Reliability of muscle strength measures obtained with a hand-held dynamometer in an elderly population. *Clin. Physiol. Funct. Imaging* **2017**, *37*, 332–340. [CrossRef]
10. Picelli, A.; Buzzi, M.G.; Cisari, C.; Gandolfi, M.; Porru, D.; Bonadiman, S.; Brugnera, A.; Carone, R.; Cerbo, R.; Del Carro, U. Headache, low back pain, other nociceptive and mixed pain conditions in neurorehabilitation. Evidence and recommendations from the Italian Consensus Conference on Pain in Neurorehabilitation. *Eur. J. Phys. Rehabil. Med.* **2016**, *52*, 867–880.
11. Kim, J.G.; Park, S.-M.; Kim, H.-J.; Yeom, J.S. Development and Validation of a Risk-Prediction Nomogram for Chronic Low Back Pain Using a National Health Examination Survey: A Cross-Sectional Study. *Healthcare* **2023**, *11*, 468. [CrossRef]
12. Pescatello, L.S.; Riebe, D.; Thompson, P.D. *ACSM's Guidelines for Exercise Testing and Prescription*; Lippincott Williams & Wilkins: Philadelphia, PA, USA, 2014.
13. Rikli, R.E.; Jones, C.J. *Senior Fitness Test Manual*; Human Kinetics: Champaign, IL, USA, 2013.
14. Mientjes, M.I.V.; Frank, J.S. Balance in chronic low back pain patients compared to healthy people under various conditions in upright standing. *Clin. Biomech.* **1999**, *14*, 710–716. [CrossRef]
15. Caffaro, R.R.; França, F.J.R.; Burke, T.N.; Magalhães, M.O.; Ramos, L.A.V.; Marques, A.P. Postural control in individuals with and without non-specific chronic low back pain: A preliminary case–control study. *Eur. Spine J.* **2014**, *23*, 807–813. [CrossRef] [PubMed]
16. Luoto, S.; Aalto, H.; Taimela, S.; Hurri, H.; Pyykkö, I.; Alaranta, H. One-footed and externally disturbed two-footed postural control in patients with chronic low back pain and healthy control subjects: A controlled study with follow-up. *Spine* **1998**, *23*, 2081–2089. [CrossRef]
17. Da Silva, R.A.; Vieira, E.R.; Carvalho, C.E.; Oliveira, M.R.; Amorim, C.F.; Neto, E.N. Age-related differences on low back pain and postural control during one-leg stance: A case–control study. *Eur. Spine J.* **2016**, *25*, 1251–1257. [CrossRef]
18. da Silva, R.A.; Vieira, E.R.; Fernandes, K.B.P.; Andraus, R.A.; Oliveira, M.R.; Sturion, L.A.; Calderon, M.G. People with chronic low back pain have poorer balance than controls in challenging tasks. *Disabil. Rehabil.* **2018**, *40*, 1294–1300. [CrossRef]
19. Narici, M.; Vito, G.D.; Franchi, M.; Paoli, A.; Moro, T.; Marcolin, G.; Grassi, B.; Baldassarre, G.; Zuccarelli, L.; Biolo, G. Impact of sedentarism due to the COVID-19 home confinement on neuromuscular, cardiovascular and metabolic health: Physiological and pathophysiological implications and recommendations for physical and nutritional countermeasures. *Eur. J. Sport Sci.* **2021**, *21*, 614–635. [CrossRef]
20. Lai, B.; Chiu, C.-Y.; Pounds, E.; Tracy, T.; Mehta, T.; Young, H.-J.; Riser, E.; Rimmer, J. COVID-19 modifications for remote teleassessment and teletraining of a complementary alternative medicine intervention for people with multiple sclerosis: Protocol for a randomized controlled trial. *JMIR Res. Protoc.* **2020**, *9*, e18415. [CrossRef]
21. Holland, A.E.; Malaguti, C.; Hoffman, M.; Lahham, A.; Burge, A.T.; Dowman, L.; May, A.K.; Bondarenko, J.; Graco, M.; Tikellis, G. Home-based or remote exercise testing in chronic respiratory disease, during the COVID-19 pandemic and beyond: A rapid review. *Chronic Respir. Dis.* **2020**, *17*, 1479973120952418. [CrossRef]
22. Hwang, R.; Fan, T.; Bowe, R.; Louis, M.; Bertram, M.; Morris, N.R.; Adsett, J. Home-based and remote functional exercise testing in cardiac conditions, during the COVID-19 pandemic and beyond: A systematic review. *Physiotherapy* **2022**, *115*, 27–35. [CrossRef]
23. Blair, C.K.; Harding, E.; Herman, C.; Boyce, T.; Demark-Wahnefried, W.; Davis, S.; Kinney, A.Y.; Pankratz, V.S. Remote Assessment of Functional Mobility and Strength in Older Cancer Survivors: Protocol for a Validity and Reliability Study. *JMIR Res. Protoc.* **2020**, *9*, e20834. [CrossRef] [PubMed]
24. Ogawa, E.F.; Harris, R.; Dufour, A.B.; Morey, M.C.; Bean, J. Reliability of Virtual Physical Performance Assessments in Veterans During the COVID-19 Pandemic. *Arch. Rehabil. Res. Clin. Transl.* **2021**, *3*, 100146. [CrossRef] [PubMed]
25. Winters-Stone, K.; Lipps, C.; Guidarelli, C.; Herrera-Fuentes, P. Converting physical function testing to the remote setting: Adapting our research protocol during COVID-19. *Innov. Aging* **2020**, *4*, 936–937. [CrossRef]
26. Hoenemeyer, T.W.; Cole, W.W.; Oster, R.A.; Pekmezi, D.W.; Pye, A.; Demark-Wahnefried, W. Test/Retest Reliability and Validity of Remote vs. In-Person Anthropometric and Physical Performance Assessments in Cancer Survivors and Supportive Partners. *Cancers* **2022**, *14*, 1075. [CrossRef] [PubMed]
27. Rozenberg, S. Chronic low back pain: Definition and treatment. *Rev Prat* **2008**, *58*, 265–272.
28. McGuigan, M. Principles of test selection and administration. In *Essentials of Strength Training and Conditioning*, 4th ed.; Haff, G.G., Triplett, N.T., Eds.; Human Kinetics: Champaign, IL, USA, 2016; pp. 249–258.
29. Franchignoni, F.; Tesio, L.; Martino, M.; Ricupero, C. Reliability of four simple, quantitative tests of balance and mobility in healthy elderly females. *Aging Clin. Exp. Res.* **1998**, *10*, 26–31. [CrossRef]
30. Izquierdo, M. Multicomponent physical exercise program: Vivifrail. *Nutr. Hosp.* **2019**, *36*, 50–56.
31. Lee, C. Sharpening the Sharpened Romberg. *SPUMS J.* **1998**, *28*, 125–132.
32. da Silva, R.A.; Vieira, E.R.; Léonard, G.; Beaulieu, L.-D.; Ngomo, S.; Nowotny, A.H.; Amorim, C.F. Age-and low back pain-related differences in trunk muscle activation during one-legged stance balance task. *Gait Posture* **2019**, *69*, 25–30. [CrossRef]
33. Balestroni, G.; Bertolotti, G. EuroQol-5D (EQ-5D): An instrument for measuring quality of life. *Monaldi Arch. Chest Dis.* **2012**, *78*, 155–159. [CrossRef]

34. Badia, X.; Roset, M.; Montserrat, S.; Herdman, M.; Segura, A. The Spanish version of EuroQol: A description and its applications. European Quality of Life scale. *Med. Clin.* **1999**, *112*, 79–85.
35. Thomas, J.; Silverman, S.; Nelson, J. *Research Methods in Physical Activity*, 7th ed.; Human Kinetics: Champaign, IL, USA, 2015.
36. Bellamy, N. Principles of Clinical Outcome Assessment. In *Rheumatology*; Mosby: Philadelphia, PA, USA, 2014.
37. Koo, T.K.; Li, M.Y. A guideline of selecting and reporting intraclass correlation coefficients for reliability research. *J. Chiropr. Med.* **2016**, *15*, 155–163. [CrossRef]
38. Portney, L.G.; Watkins, M.P. *Foundations of Clinical Research: Applications to Practice*; Pearson/Prentice Hall: Upper Saddle River, NJ, USA, 2009; Volume 892.
39. Bland, J.M.; Altman, D. Statistical methods for assessing agreement between two methods of clinical measurement. *Lancet* **1986**, *327*, 307–310. [CrossRef]
40. Giavarina, D. Understanding bland altman analysis. *Biochem. Med.* **2015**, *25*, 141–151. [CrossRef]
41. Cohen, J. *Statistical Power Analysis forthe Behavioural Sciences*; Baskı: Hillsdale, NJ, USA, 1988.
42. Plagg, B.; Piccoliori, G.; Oschmann, J.; Engl, A.; Eisendle, K. Primary health care and hospital management during COVID-19: Lessons from lombardy. *Risk Manag. Healthc. Policy* **2021**, *14*, 3987–3992. [CrossRef]
43. Podlog, L.W.; Brown, W.J. Self-determination theory: A framework for enhancing patient-centered care. *J. Nurse Pract.* **2016**, *12*, e359–e362. [CrossRef]
44. Borrega-Mouquinho, Y.; Sánchez-Gómez, J.; Fuentes-García, J.P.; Collado-Mateo, D.; Villafaina, S. Effects of High-Intensity Interval Training and Moderate-Intensity Training on Stress, Depression, Anxiety, and Resilience in Healthy Adults During Coronavirus Disease 2019 Confinement: A Randomized Controlled Trial. *Front. Psychol.* **2021**, *12*, 270. [CrossRef]
45. Schneider, S.; Brümmer, V.; Carnahan, H.; Kleinert, J.; Piacentini, M.F.; Meeusen, R.; Strüder, H.K. Exercise as a countermeasure to psycho-physiological deconditioning during long-term confinement. *Behav. Brain Res.* **2010**, *211*, 208–214. [CrossRef]
46. Lin, I.I.; Chen, Y.-L.; Chuang, L.-L. Test-Retest Reliability of Home-Based Fitness Assessments Using a Mobile App (R Plus Health) in Healthy Adults: Prospective Quantitative Study. *JMIR Form. Res.* **2021**, *5*, e28040. [CrossRef]
47. Holland, A.E.; Rasekaba, T.; Fiore, J.F., Jr.; Burge, A.T.; Lee, A.L. The 6-minute walk distance cannot be accurately assessed at home in people with COPD. *Disabil. Rehabil.* **2015**, *37*, 1102–1106. [CrossRef]
48. Cox, N.S.; Alison, J.A.; Button, B.M.; Wilson, J.W.; Holland, A.E. Assessing exercise capacity using telehealth: A feasibility study in adults with cystic fibrosis. *Respir. Care* **2013**, *58*, 286–290. [CrossRef] [PubMed]

Article

An Early Indicator in Evaluating Cardiac Dysfunction Related to Premature Ventricular Complexes: Cardiorespiratory Capacity

Xiaozhu Ma, Jiangtao Yan and Wanjun Liu *

Division of Cardiology, Department of Internal Medicine, Tongji Hospital, Tongji Medical College, Huazhong University of Science & Technology, Wuhan 430030, China; xzma2023@hust.edu.cn (X.M.); jtyan@tjh.tjmu.edu.cn (J.Y.)
* Correspondence: wjliu@tjh.tjmu.edu.cn; Tel.: +86-27-8366-3280

Abstract: Cardiac dysfunction induced by premature ventricular complexes (PVCs) is relatively controversial and challenging to detect in the early stage. In this observational study, we retrospectively analyzed the cardiopulmonary exercise test (CPET) data of 94 patients with frequent premature ventricular beats (47 males, 49.83 ± 13.63 years) and 98 participants (55 males, 50.84 ± 9.41 years) whose age and gender were matched with the patient with PVCs. The baseline information and routine echocardiography detection were recorded on admission. PVCs were diagnosed by 24 h Holter monitoring, and cardiorespiratory capacity was assessed using peak oxygen uptake (V'O$_2$peak), anaerobic threshold (AT), and other CPET parameters with an individualized bicycle ramp protocol according to the predicted workload and exercise situation of each participant. There were no statistically significant differences in most baseline characteristics between the two groups. Indicators that reflect cardiopulmonary capacity, such as V'O$_2$peak, AT, and ΔO2 pulse/Δwork rate(ΔV'O$_2$/ΔWR), were all significantly lower in the PVC group (p = 0.031, 0.021, and 0.013, respectively) despite normal and nondiscriminatory left ventricular ejection fractions between the two groups. However, there was no statistically significant difference among subgroups based on the frequency of PVCs, which was <10,000 beats/24 h, 10,000–20,000 beats/24 h, and >20,000 beats/24 h. The cardiorespiratory capacity was lower in patients with frequent PVCs, indicating that CPET could detect early signs of impaired cardiac function induced by PVCs.

Keywords: premature ventricular complexes; cardiac capacity; cardiopulmonary exercise test; oxygen uptake

Citation: Ma, X.; Yan, J.; Liu, W. An Early Indicator in Evaluating Cardiac Dysfunction Related to Premature Ventricular Complexes: Cardiorespiratory Capacity. *Healthcare* **2023**, *11*, 2940. https://doi.org/10.3390/healthcare11222940

Academic Editors: Wilhelm Mistiaen, Rafael Oliveira and Joào Paulo Brito

Received: 30 August 2023
Revised: 4 November 2023
Accepted: 8 November 2023
Published: 10 November 2023

1. Introduction

Premature ventricular complexes (PVCs) refer to early depolarization of the ventricles with or without mechanical contraction and are one of the most common types of arrhythmia [1]. With the widespread availability of diagnostic technologies, it is not surprising that an increasing number of patients with PVCs have been timely diagnosed in recent years. Initially, PVCs were considered benign without cardiac structural alterations. However, several research studies have revealed that an increasing burden of PVCs is associated with reduced left ventricular systolic function [2–4] and higher short-term mortality [5]. Apart from reduced cardiac function, a special kind of cardiomyopathy, namely PVC-induced cardiomyopathy, may also develop due to long-term burden of PVCs [6]. Unfortunately, it is challenging to detect these adverse effects on the cardiac function in the early stage of arrhythmia given that patients in the early stages of the disease usually do not exhibit obvious clinical symptoms.

Cardiopulmonary exercise testing (CPET) is a dynamic and noninvasive clinical test that provides an integrative and objective evaluation of multiple organs and systems, particularly pulmonary and cardiovascular function under physiological stress. It appears to be a promising method for assessing cardiac function before the detection of virtual structural and functional damages. CPET was initially proposed as a useful tool for grading the severity of heart failure in the 1980s [7] and has since been utilized in the management of various cardiovascular diseases. According to the latest 2021 European Society of Cardiology (ESC) Guidelines [8] for the diagnosis and treatment of acute and chronic heart failure, a cardiopulmonary exercise test is recommended as part of the evaluation and management of heart transplantation and/or mechanical circulatory support, as it helps to optimize the prescription of exercise training. However, the use of CPET in current clinical practice primarily focuses on heart failure and ischemic heart disease, while its application to evaluate the effects of PVCs on cardiorespiratory capacity is uncommon.

Therefore, the purpose of this study was to identify early cardiopulmonary impairment caused by PVCs and explore the advantage of CPET in evaluating cardiorespiratory performance in real-world patient care.

2. Materials and Methods

2.1. Population

This was a retrospective cohort study conducted at Tongji Hospital, affiliated with Tongji Medical College, Huazhong University of Science and Technology. Patients with symptoms such as palpitation, chest pain, chest tightness, or abnormal electrocardiography findings, who had undergone 24 h Holter monitoring and consented to CPET, were included in this study. Thereinto, patients with more than 1000 PVCs in the 24 h Holter monitoring and CPET performed within three days of the monitoring were classified as the PVC group. The exclusion criteria were as follows: (1) apparent wheezing and dyspnea; (2) severe arrhythmia; (3) acute coronary syndrome within the past three months; (4) chronic heart failure; (5) pregnancy; (6) uncontrolled severe obstructive lung disease; (7) severe hepatic or kidney dysfunction; and (8) moderate to severe aortic and mitral stenosis. Ninety-four patients with PVCs met these criteria and were included in the subsequent analysis. Simultaneously, patients who were age-, gender-, and body mass index (BMI)-matched with the PVC groups were enrolled as controls. Although controls had been referred to our hospital for similar reasons, their PVC frequency was less than 1000 times per 24 h.

The demographic and clinical information, including age, height, weight, BMI, cardiovascular risk factors, and medication use, presented in Table 1, were collected from the Tongji Hospital information system. Echocardiography was performed by two professional sonographers according to standard protocol. The routine parameters reflecting cardiac systolic function (left ventricular ejection fraction and fractional shortening) and diastolic function (E/E' and E/A value) were recorded. In addition, the morphology features of PVC beats, including the QRS duration, pseudo delta waves, the QRS axis in limb leads, and bundle branch-like morphology were evaluated by two professional cardiologists. This research study was approved by the ethics committee of the institutional review board of Tongji Hospital and written informed consent was obtained from all participants.

Table 1. Baseline characteristics of the study population.

Subject	Controls (n = 98)	PVCs (n = 94)	*p* Value
Age (yrs)	50.84 ± 9.41	49.83 ±13.63	0.551
Male (n,%)	55 (56.1)	47 (50.0)	0.395
Height (cm)	166.54 ± 7.58	165.97 ± 6.89	0.590
Weight (kg)	69.39 ± 13.38	65.92 ± 10.95	0.052
Body mass index (kg/m^2)	24.89 ± 3.65	23.87 ± 3.36	0.050
Smoking (n,%)	15 (15.3)	11 (11.8)	0.484
CAD (n,%)	18 (18.1)	16 (17.0)	0.156
ACS (n,%)	10 (10.2)	9 (9.6)	0.892
Number of lesional coronary artery			0.771
1-vessel CAD (n,%)	6 (6.1)	8 (8.5)	

Table 1. *Cont.*

Subject	Controls (n = 98)	PVCs (n = 94)	*p* Value
2-vessel CAD (n,%)	4 (4.1)	3 (3.1)	
3-vessel CAD (n,%)	8 (8.2)	5 (5.3)	
Gensini score	16.77 ± 11.45	19.20 ± 17.01	0.672
Hypertension (n,%)	33 (33.7)	47 (50.0)	0.071
Hyperlipidemia (n,%)	23 (23.5)	24 (25.5)	0.997
Diabetes (n,%)	12 (12.2)	14 (14.9)	0.758
Medication use			
Beta-blocker (n,%)	44 (44.9)	37 (37.8)	0.204
Calcium channel blocker (n,%)	19 (19.4)	28 (28.6)	0.178
Laboratories			
NT-proBNP (pg/mL)	35.5 (49.5)	62.1 (83.0)	0.044 [†]
cTnI (ng/mL)	5.32 ± 14.30	4.24 ± 6.81	0.509
eGFR (mL/min/1.73 m^2)	91.86 ± 12.83	89.34 ± 20.97	0.532

[†] Comparison between the two groups conducted using the Mann–Whitney U test. Abbreviations: PVC, premature ventricular complex; CAD, coronary artery disease; ACS, acute coronary syndrome; NT-proBNP, N-terminal pro-brain natriuretic peptide.

2.2. CPET Method

All subjects had been referred for CPET at our hospital settings to estimate their cardiopulmonary function. The participants were evaluated for indications by a professional physician and signed an informed consent form regarding relevant risks and benefits before CPET. The participants were instructed to abstain from smoking for at least 8 h, drinking coffee, and eating a full meal. Oxygen consumption, carbon dioxide production, ventilatory capacity, and hemodynamic indices were continuously measured throughout the test using professional equipment (CARDIOVIT CS-200 Excellence). After the preliminary assessment of static lung function, the participants were assigned to complete symptom-limited maximum extreme exercise testing on a cycle ergometer with a personalized ramp exercise protocol until exhaustion. Specifically, we recorded the initial resting state data for three minutes, and then the participants started warm-up cycling without load at a pedal speed of 60 r/min for three minutes. The increasing power rate was set to 10–30 W/min based on the patients' age, sex, height, usual exercise habits, and estimated functional state. Patients reached the maximum exercise limit with symptom restriction within 6–10 min and their recovery was recorded for more than five minutes. The prespecified criteria for interrupting the exercise test were as follows: (1) adverse clinical symptoms, such as fatigue, dyspnea, severe chest pain, dizziness, faintness, sudden pallor, and loss of coordination; (2) abnormal ECG presentation, such as pathological Q waves, severe arrhythmia, ≥ 2 mm ST depression or ≥ 1 mm ST elevation in at least two adjacent leads; (3) respiratory exchange ratio (RER) $\geq$ 1.15; and (4) upon request of the participant.

A large number of variables were typically measured. $V'O_2$ refers to the ability of the human body to inhale and utilize oxygen during physical activity. $V'O_2$peak, which describes the highest achieved $V'O_2$ value during an exercise test, is the most important parameter to reflect cardiopulmonary capacity. Here, we defined $V'O_2$peak as the average value of the last ten seconds of CPET. $V'CO_2$ is used to quantify the amount of carbon dioxide produced or exhaled by an individual, reflecting metabolic state during exercise. As an alternative to $V'O_2$peak in determining exercise tolerance, the anaerobic threshold (AT) was identified using a V-slope analysis of $V'O_2$ and $V'CO_2$ and was confirmed by specific trends of ventilatory equivalent for oxygen (VE/$V'O_2$), ventilatory equivalent for carbon dioxide (VE/$V'CO_2$), end-tidal pressure of oxygen, and end-tidal pressure of CO_2. Other routine variables, including $V'O_2$AT as a percentage of the predicted $V'O_2$, VE/$V'CO_2$ slope, peak O_2 pulse, the slope of change in $V'O_2$ to change in work rate, $V'CO_2$, VE, HR, blood pressure at peak, and AT were recorded synchronously. The Borg scale

(6–20 scale) which allowed individuals to subjectively rate their level of exertion during exercise, was performed after the test.

2.3. Statistical Analysis

All statistical analyses were conducted using IBM SPSS Statistics for Windows, version 29.0 (IBM Corp., Armonk, NY, USA) [9]. Numerical variables were expressed as mean ± standard deviation, while categorical variables were presented as counts (percentage). The normality of distribution for the data was assessed by the Kolmogorov–Smirnov test. Analysis on continuous variable data between the PVCs and control groups were evaluated by the unpaired t-test if the data was normally distributed. Otherwise, the comparison between the two groups would be conducted using the Mann–Whitney U test, such as NTpro-BNP. For the comparison of categorical variables, the Chi-square test was performed. Statistical significance was defined as $p < 0.05$. Effect size was expressed by the Cohen d whose absolute value greater than 0.2 representing the obvious difference between control and PVC groups. Three subgroup analyses were performed using univariate ANOVA followed by post hoc analysis using LSD Test. The correlation between PVC burden and related CPET parameters was assessed using Pearson's test. It was considered a strong correlation when the absolute value of r was greater than 0.6.

3. Results

3.1. Baseline Information

For the PVC group, 48 patients (51%) visited the hospital due to symptoms of palpitation or chest discomfort, while the remaining 46 cases (49%) were incidentally diagnosed as PVCs during a medical examination (Figure 1). Table 1 presents the specific baseline information of the PVCs and control groups. The basic demographic data, including age and gender ($p > 0.05$), showed no significant differences between the groups. There were also no statistically significant differences in the risk factors for cardiovascular disease, including hypertension, hyperlipidemia, diabetes, and diseases that may affect the cardiopulmonary capacity. A specific classification based on the severity of coronary heart disease indicated there were no significant differences in the degree of coronary artery stenosis between two groups. We also compared the medication history between the two groups, and our findings underlined that the use of beta blockers, which was highly related to heart rate and cardiac function and tended to be similar ($p = 0.204$).

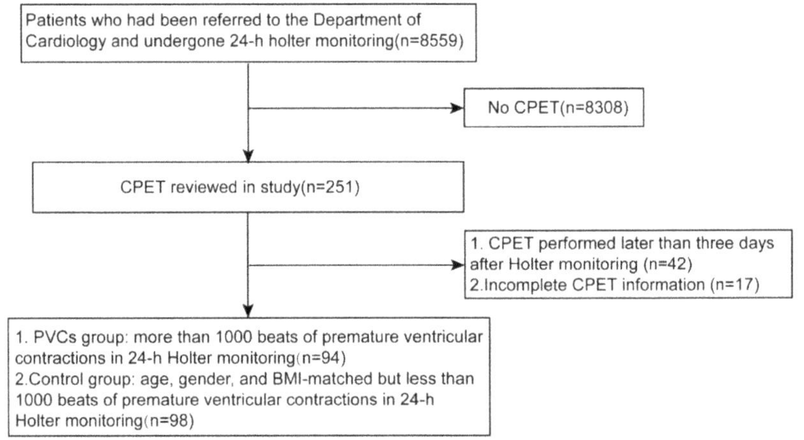

Figure 1. Flow diagram showing inclusion of participants in the study. CPET, cardiopulmonary exercise test; PVCs, premature ventricular complexes.

Table 2 showed the morphological features of arrhythmia. In addition, as shown in Table 3, there were no significant differences in ventricular wall thickness (interventricular septum, $p = 0.074$; left ventricular posterior wall, $p = 0.305$) and functional changes detected by echocardiography between the two groups.

Table 2. Morphological features of PVCs beats.

Characteristics	
QRS duration (ms)	140 ± 33
pseudo delta waves (n,%)	10 (10.6)
axis	
superior axis (n,%)	24 (25.5)
inferior axis (n,%)	70 (74.5)
bundle branch-like morphology	
left (n,%)	77 (81.9)
right (n,%)	17 (18.1)

Table 3. Echocardiographic indices of study patients.

Subject	Controls	PVCs	p Value
LV size (mm)	47.02 ± 4.68	47.93 ± 4.83	0.212
LA size (mm)	35.62 ± 4.99	35.88 ± 5.78	0.749
IVS(mm)	9.93 ± 1.80	9.44 ± 1.80	0.074
LVPW (mm)	9.46 ± 1.32	9.24 ± 1.52	0.305
LVEF (%)	63.11 ± 6.08	60.96 ± 8.1	0.055
FS (%)	34.47 ± 5.32	33.23 ± 6.19	0.157
E/A	1.17 ± 0.53	1.27 ± 0.58	0.224
E/E$'$	11.46 ± 3.61	12.6 ± 5.45	0.107

Abbreviations: LV, left ventricle; LA, left atrium; IVS, interventricular septum; LVPW, left ventricular posterior wall; LVEF, left ventricular ejection fraction; FS, fractional shortening.

3.2. CPET Data

Of the 94 PVCs patients and 98 controls enrolled based on the aforementioned inclusion and exclusion criteria, none of them terminated the exercise test due to severe clinical symptoms or abnormal ECG presentation. The Borg scale score of all participants was greater than 13. At the same time, suppression of frequent PVCs was noted in all patients. A comparison of CPET variables between groups was shown in Table 4 and Figure 2. Firstly, no significant differences were seen regarding the heart rate and blood pressure at rest. However, there were significant differences in V'O$_2$peak/kg (20.92 ± 4.99 in the control group vs. 19.32 ± 5.20 in the PVC group, $p = 0.031$) and V'O$_2$AT/kg (13.36 ± 2.70 in the control group vs. 12.40 ± 3.02 in the PVC group, $p = 0.021$). To our surprise, the important parameters representing impaired oxygen delivery (V'O$_2$peak (%pred), V'O$_2$AT(%pred), and ΔV'O$_2$/ΔWR) also decreased ($p = 0.009$, 0.005, and 0.013, respectively). No significant differences were found in heart rate at peak exercise between the two groups. We further divided the PVC patients into three groups based on the frequency of PVCs, namely < 10,000 beats/24 h, 10,000–20,000 beats/24 h, and >20,000 beats/24 h. Interestingly, significant difference was noted in V'O$_2$/KG peak but not in other main parameters among the three subgroups, which was shown in Table 5. Additionally, there was no significant correlation between the burden of PVCs and V'O$_2$peak (Figure 3). In our cohort, the origin of PVCs in most patients was the right ventricular outflow tract (80.7%).

Table 4. CPET parameters in PVCs and controls.

Subject	Controls (n = 98)	PVCs (n = 94)	p Value	Reference Value	Effect Size
AT (l/min)	0.91 ± 0.22	0.80 ± 0.24	0.002	-	-0.478
V′O_2 peak (l/min)	1.42 ± 0.39	1.25 ± 0.40	0.004	-	-0.430
V′E peak (l/min)	48.26 ± 12.16	42.37 ± 12.29	0.001	-	-0.482
SBP rest (mmHg)	129.74 ± 26.15	126.61 ± 21.74	0.370	-	-0.130
DBP rest (mmHg)	78.14 ± 15.64	78.52 ± 13.70	0.859	-	0.026
HR rest (bpm)	87.40 ± 13.81	87.96 ± 14.63	0.785	-	0.039
HR AT (bpm)	111.39 ± 14.97	108.18 ± 18.50	0.188	-	-0.191
HR peak (bpm)	142.66 ± 17.88	138.88 ± 26.58	0.247	-	-0.168
Load AT (W)	59.34 ± 21.06	52.49 ± 23.51	0.035	-	-0.307
Load peak (W)	101.97 ± 37.97	92.53 ± 37.35	0.084	-	-0.251
V′O_2/HR peak	0.010 ± 0.003	0.009 ± 0.003	0.016	-	-0.333
V′O_2 peak (%pred)	0.74 ± 0.14	0.68 ± 0.15	0.009	$\geq$85% pred	-0.414
V′O_2 AT (%pred)	0.48 ± 0.11	0.44 ± 0.10	0.005	40–60%	-0.380
V′E/V′CO_2 slope	26.44 ± 5.18	26.76 ± 7.58	0.750	<35	0.049
V′O_2/KG AT (ml/kg/min)	13.36 ± 2.70	12.40 ± 3.02	0.021	-	-0.336
V′O_2/KG peak (ml/kg/min)	20.92 ± 4.99	19.32 ± 5.20	0.031	-	-0.314
V′E/KG peak (ml/kg/min)	713.62 ± 155.25	657.38 ± 168.60	0.017	-	-0.347
ΔV′O_2/ΔWR ((ml/min)/W)	9.69 ± 1.92	8.84 ± 2.66	0.013	>8.6	-0.368

Abbreviations: PVCs, premature ventricular complexes; V′O_2, oxygen consumption; AT, anaerobic threshold; SBP, systolic blood pressure; DBP, diastolic blood pressure; V′CO_2, carbon dioxide output; V′E, ventilation; HR, heart rate; WR, work rate.

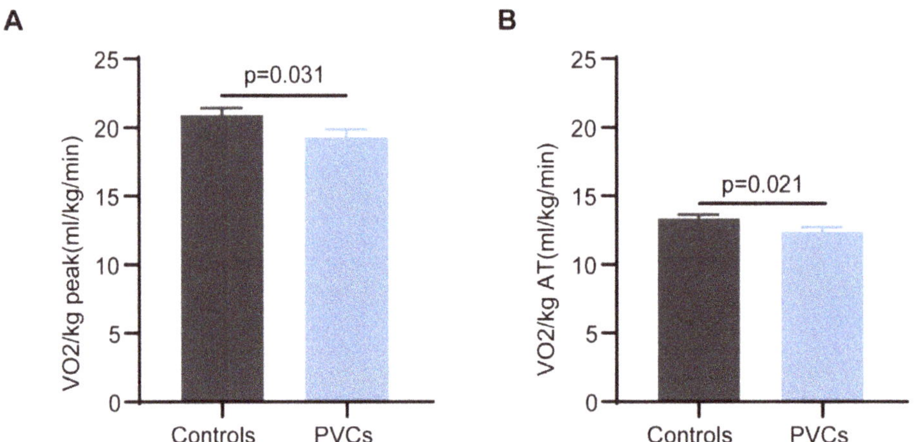

Figure 2. Weight-adjusted peak oxygen consumption between control group and PVC group. The comparison about VO_2/kg at peak (**A**) and anaerobic threshold (**B**) between the two groups. Abbreviations: V′O_2, oxygen consumption; AT, anaerobic threshold.

Table 5. Comparison of main parameters among different subgroups *.

Subject	<10,000/24 h (n = 55)	10,000–20,000/24 h (n = 23)	>20,000/24 h (n = 16)	p value	Post Hoc <10,000/24 h vs. >20,000/24 h	10,000–20,000/24 h vs. >20,000/24 h	<10,000/24 h vs. 10,000–20,000/24 h
AT (l/min)	0.78 ± 0.25	0.88 ± 0.25	0.79 ± 0.21	0.270			
V′O_2 peak (l/min)	1.18 ± 0.41	1.36 ± 0.36	1.35 ± 0.36	0.110			
V′O_2 AT (%pred)	0.45 ± 0.11	0.43 ± 0.1	0.40 ± 0.08	0.289			

Table 5. *Cont.*

					Post Hoc		
Subject	<10,000/24 h (n = 55)	10,000–20,000/24 h (n = 23)	>20,000/24 h (n = 16)	*p* value	<10,000/24 h vs. >20,000/24 h	10,000–20,000/24 h vs. >20,000/24 h	<10,000/24 h vs. 10,000–20,000/24 h
V′O_2 peak (%pred)	0.67 ± 0.16	0.66 ± 0.12	0.70 ± 0.15	0.726			
V′O_2/KG AT (ml/kg/min)	12.01 ± 3.03	13.23 ± 3.61	12.7 ± 2.38	0.282			
V′O_2/KG peak (ml/kg/min)	18.18 ± 5.29	20.46 ± 5.31	21.67 ± 4.34	0.040	0.023	0.490	0.090

* Groups compared with univariate ANOVA (LSD post hoc test). Abbreviations: V′O_2, oxygen consumption; AT, anaerobic threshold.

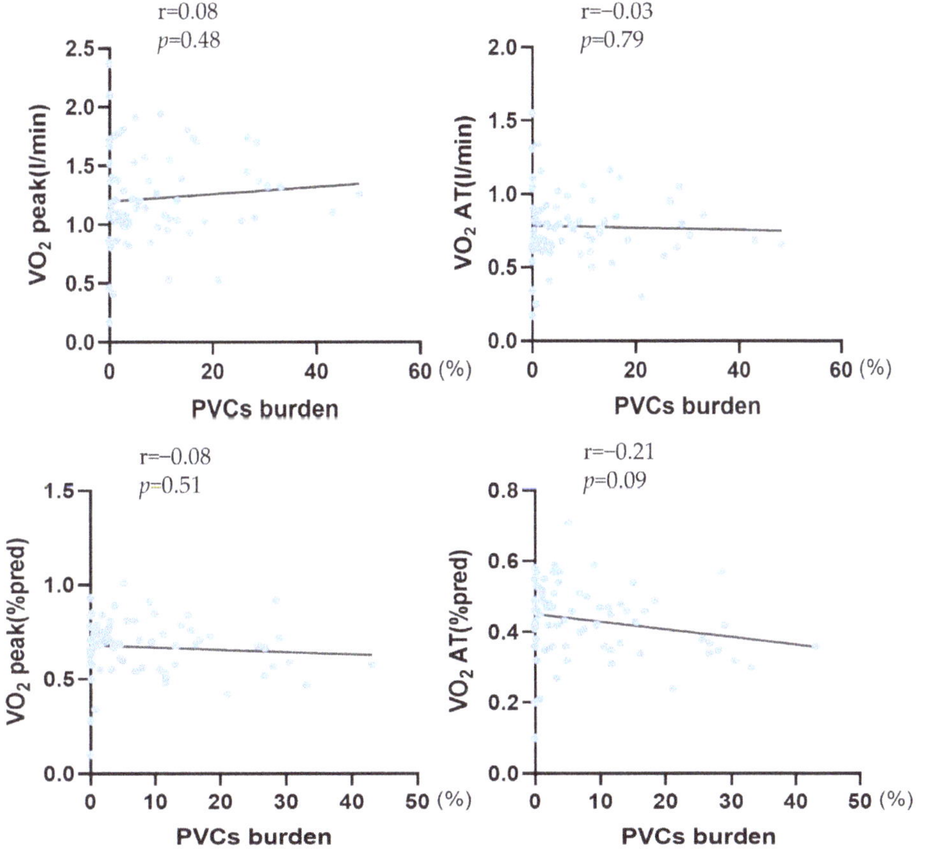

Figure 3. Correlation between PVCs burden and main parameters of CPET. Abbreviations: V′O_2, oxygen consumption; AT, anaerobic threshold. Statistical significance was defined as $p < 0.05$.

4. Discussion

In the present study, we identified that patients with frequent PVCs had lower V′O_2 peak compared to the controls, indicating a reduction in cardiopulmonary reserve function in patients with frequent PVCs and normal cardiac structure.

Premature ventricular complexes are common type of arrhythmia in clinical practice and frequent PVCs are considered risk factors for left ventricular dysfunction through long-term follow up observations. A previous cohort study on over 1100 participants without congestive heart failure (CHF) revealed that the population-level risk for incident heart failure attributed to PVCs was 8.1% [4]. Meanwhile, a portion of PVCs patients developed PVC-induced cardiomyopathy, which refers to a reversible reduction in left ventricular systolic function. Several research groups have indicated that LV dysfunction in patients with frequent PVCs could be improved to some extent after catheter ablation therapy [10,11]. However, the effects on cardiac function are not comparable in the early stages of frequent PVCs. In a previous study, most patients did not exhibit any clinical symptoms of reduced heart function, such as shortness of breath after exertion, and more than half of all patients failed to demonstrate any evident decrease in left ventricular ejection fraction (LVEF) [12]. Arrhythmogenic cardiomyopathy (ACM) is a myocardial disorder that can lead to sudden cardiac death. Ventricular arrhythmias may be an early phenotypic expression in ACM patients, characterized by PVCs which are inconsistent with the degree of cardiac dysfunction [13]. In general, patients with fewer symptoms are thought to have a benign prognosis. However, researchers have found that many asymptomatic patients were usually overlooked. As a result, they were more susceptible to long-term PVC burden and progression to PVC-induced cardiomyopathy [14]. Despite the significant decrease in LVEF among patients with highly frequent PVCs, the prognosis was found to be worse in patients with exercise-induced PVCs but without cardiac structural changes after long-term follow-up [15]. In our study, we were surprised to identify reduced cardiac capacity induced by PVCs prior to observing a decrease in routine echocardiography parameters. In our study, there were some differences in NT-proBNP levels between the two groups. However, its value was influenced by several factors, such as age, gender, renal function, and coronary artery disease. Of note, two cases in our cohort had significantly increased NT-proBNP values due to poor renal function. Consequently, NT-proBNP level could not reflect differences in cardiac function between the groups. Furthermore, the value of E/E′ in the PVC group was higher than in the controls group, but still within the normal range, which seemed to mean that there was impaired left ventricular diastolic function in the early stage. These further confirmed the special potential of CPET in identifying early cardiac dysfunction which is difficult to detect with routine echocardiography tests. Hence, this will enable the development of early prevention strategies and aggressive therapies of the underlying disease.

The mechanism of LV dysfunction observed in patients with frequent PVCs is not completely elucidated. Controversy still exists between findings in laboratory animals and human studies. Tachycardia-induced cardiac functional decline might be one of the important factors attributed to the progression of LV dysfunction [16] although the effect is probably nonessential. Nevertheless, bradycardia is also a potential confounder because cardiac output decreases with low heart rates. The concept of "PVCs-induced cardiomyopathy" has recently gained a lot of attention, and changes in cardiac rhythm do not appear to provide a complete explanation for this phenomenon. Left ventricular dyssynchrony is another plausible contributing factor. Long-term or high frequency PVCs alter the normal hemodynamic status, leading to increases in LV filling pressure and LA overload. As a result, the volume of blood pumped into peripheral tissues is insufficient to meet the metabolic demands of tissues. A study performed by Tomos [17] revealed the association between severe left ventricular dyssynchrony and PVCs-induced cardiomyopathy in an animal model. In addition, it is likely that some occult structural heart disease associated with the etiology of PVCs may have existed and was not detected by our conventional monitoring methods. Modifications in cardiac anatomy resulting from PVCs of the left ventricular outflow tract origin were confirmed. In our study, PVCs mainly originated on the right ventricle, but there was no statistically significant difference in V′O$_2$peak when dividing patients according to the site of PVCs origin. However, the classification of sites could have been refined further, allowing the identification of the precise origin and differences in cardiopulmonary function caused by ventricular premature complexes of different origins.

Genetic differences should be considered as a factor in the development of arrhythmia as it is challenging to explain the different responses noted among patients using the same treatment parameters. In a meta-analysis [18], genome-wide association studies were carried out and researchers identified that mutations in a certain gene locus contributed to the differences in the risk of morbidity due to supraventricular and ventricular ectopy. ACM is perceived to be highly correlated with genetic background. Consequently, this comes to underline that we need to be very careful when identifying potential cardiomyopathy through cardiac magnetic resonance and genetic testing and be hypervigilant against patients whose cardiac capacity has been impaired as detected by CPET.

The difference among the three subgroups classified by PVCs frequency indicates that the effect of varying frequency may not be apparent in the early stage of the disease. While there was significant difference in V'O_2/KG peak in patients with less than 10,000 PVC beats per 24 h as compared to patients with more than 20,000 PVC beats. However, it should be noted that PVCs-induced cardiac impairment is not solely attributable to its magnitude. Multiple factors, such as PVCs location, morphology, and QRS duration [19], may exert significant impacts. We believe that the correlation between the burden of PVCs and important indexes, which showed a negative trend but no statistical significance, could be ascribed to non-persistent arrhythmia and the relatively small sample size included in this study.

Moreover, identical findings were noted when the frequency of PVCs was suppressed by exercise during CPET in all patients, which was generally interpreted as an adrenergic response. A previous study demonstrated that PVCs were suppressed either during the recovery phase of exercise testing or after exercise testing during a mean follow up of 7.2 years in children [20], and this suppression was associated with a benign prognosis. In contrast, exercise-induced PVCs have been linked to death and the occurrence of cardiovascular events in adults [21]. Changes in electrocardiograms during CPET can be a useful method to assess the presence or absence of underlying ischemic heart disease or arrhythmia. In our study, PVCs were suppressed during exercise in all participants. Further investigation into the specific mechanism and different prognostic outcomes needs to be conducted in large sample studies.

5. Conclusions

In conclusion, we discovered patients with frequent PVCs have worse cardiorespiratory capacity than the controls, and further investigation on the association between PVCs burden and cardiorespiratory capacity is necessary to excavate the important role of CPET in disease diagnosis.

6. Limitations

The current study has several limitations. Firstly, this is an observational study with a relatively small sample size. Large long-term follow-up clinical trials are required to confirm and generalize our results in the future. Secondly, we mainly focused on cardiopulmonary capacity between patients with PVCs and those without PVCs. However, the large frequency variations within PVC group could not be avoided. Considering the research type, our conclusion has limited guidance for clinical decision-making in real clinical practice. Thirdly, global longitudinal strain is an appropriate parameter to evaluate subtle LV function changes. Unfortunately, it is not currently routine parameter of echocardiogram in our hospitals and is therefore not available in this observational study. Furthermore, inter-individual variation in exercise capacity should not be neglected, as individual differences in lifestyle may exert an impact on the CPET results, particularly in the case of a small sample size. In addition, the ECG recording times of some participants were not accurate for 24 h. Therefore, the comparison among patients with different frequencies of PVCs needs further verification.

Author Contributions: Writing—original draft preparation, X.M.; writing—review and editing, W.L.; methodology and software, W.L. and X.M.; validation, X.M., J.Y. and W.L.; formal analysis, J.Y. and X.M.; investigation, X.M.; resources, W.L.; supervision, J.Y.; funding acquisition, W.L. All authors have read and agreed to the published version of the manuscript.

Funding: This research received no external funding.

Institutional Review Board Statement: This project was approved by the ethics committee of the institutional review board of Tongji Hospital with No. TJ-IRB20220314.

Informed Consent Statement: Informed consent was obtained from all subjects involved in the study.

Data Availability Statement: Data are contained within the article.

Conflicts of Interest: The authors declare no conflict of interest.

References

1. Cronin, E.M.; Bogun, F.M.; Maury, P.; Peichl, P.; Chen, M.; Namboodiri, N.; Aguinaga, L.; Leite, L.R.; Al-Khatib, S.M.; Anter, E.; et al. 2019 HRS/EHRA/APHRS/LAHRS expert consensus statement on catheter ablation of ventricular arrhythmias. *Heart Rhythm* **2020**, *17*, e2–e154. [CrossRef] [PubMed]
2. Baman, T.S.; Lange, D.C.; Ilg, K.J.; Gupta, S.K.; Liu, T.-Y.; Alguire, C.; Armstrong, W.; Good, E.; Chugh, A.; Jongnarangsin, K.; et al. Relationship between burden of premature ventricular complexes and left ventricular function. *Heart Rhythm* **2010**, *7*, 865–869. [CrossRef] [PubMed]
3. Munoz, F.D.C.; Syed, F.F.; Noheria, A.; Cha, Y.-M.; Friedman, P.A.; Hammill, S.C.; Munger, T.M.; Venkatachalam, K.; Shen, W.-K.; Packer, D.L.; et al. Characteristics of premature ventricular complexes as correlates of reduced left ventricular systolic function: Study of the burden, duration, coupling interval, morphology and site of origin of PVCs. *J. Cardiovasc. Electrophysiol.* **2011**, *22*, 791–798. [CrossRef] [PubMed]
4. Dukes, J.W.; Dewland, T.A.; Vittinghoff, E.; Mandyam, M.C.; Heckbert, S.R.; Siscovick, D.S.; Stein, P.K.; Psaty, B.M.; Sotoodehnia, N.; Gottdiener, J.S.; et al. Ventricular Ectopy as a Predictor of Heart Failure and Death. *J. Am. Coll. Cardiol.* **2015**, *66*, 101–109. [CrossRef] [PubMed]
5. Lin, C.-Y.; Chang, S.-L.; Lin, Y.-J.; Lo, L.-W.; Chung, F.-P.; Chen, Y.-Y.; Chao, T.-F.; Hu, Y.-F.; Tuan, T.-C.; Liao, J.-N.; et al. Long-term outcome of multiform premature ventricular complexes in structurally normal heart. *Int. J. Cardiol.* **2015**, *180*, 80–85. [CrossRef] [PubMed]
6. Huizar, J.F.; Tan, A.Y.; Kaszala, K.; Ellenbogen, K.A. Clinical and translational insights on premature ventricular contractions and PVC-induced cardiomyopathy. *Prog. Cardiovasc. Dis.* **2021**, *66*, 17–27. [CrossRef] [PubMed]
7. Weber, K.T.; Kinasewitz, G.T.; Janicki, J.S.; Fishman, A.P. Oxygen utilization and ventilation during exercise in patients with chronic cardiac failure. *Circulation* **1982**, *65*, 1213–1223. [CrossRef] [PubMed]
8. McDonagh, T.A.; Metra, M.; Adamo, M.; Gardner, R.S.; Baumbach, A.; Böhm, M.; Burri, H.; Butler, J.; Čelutkienė, J.; Chioncel, O.; et al. 2021 ESC Guidelines for the diagnosis and treatment of acute and chronic heart failure. *Eur. Heart J.* **2021**, *42*, 3599–3726. [CrossRef] [PubMed]
9. IBM Corp. *IBM SPSS Statistics for Windows, Version 29.0*; IBM Corp.: Armonk, NY, USA, 2022.
10. Zalzstein, E.; Wagshal, A.; Zucker, N.; Levitas, A.; Ovsyshcher, I.E.; Katz, A. Ablation therapy of tachycardia-related cardiomyopathy. *Isr. Med. Assoc. J. IMAJ* **2003**, *5*, 64–65. [PubMed]
11. Sekiguchi, Y.; Aonuma, K.; Yamauchi, Y.; Obayashi, T.; Niwa, A.; Hachiya, H.; Takahashi, A.; Nitta, J.; Iesaka, Y.; Isobe, M. Chronic hemodynamic effects after radiofrequency catheter ablation of frequent monomorphic ventricular premature beats. *J. Cardiovasc. Electrophysiol.* **2005**, *16*, 1057–1063. [CrossRef]
12. Latchamsetty, R.; Bogun, F. Premature Ventricular Complex-Induced Cardiomyopathy. *JACC Clin. Electrophysiol.* **2019**, *5*, 537–550. [CrossRef]
13. Corrado, D.; Perazzolo Marra, M.; Zorzi, A.; Beffagna, G.; Cipriani, A.; Lazzari, M.; Migliore, F.; Pilichou, K.; Rampazzo, A.; Rigato, I.; et al. Diagnosis of arrhythmogenic cardiomyopathy: The Padua criteria. *Int. J. Cardiol.* **2020**, *319*, 106–114. [CrossRef]
14. Yokokawa, M.; Kim, H.M.; Good, E.; Chugh, A.; Pelosi, F.; Alguire, C.; Armstrong, W.; Crawford, T.; Jongnarangsin, K.; Oral, H.; et al. Relation of symptoms and symptom duration to premature ventricular complex-induced cardiomyopathy. *Heart Rhythm* **2012**, *9*, 92–95. [CrossRef]
15. Jouven, X.; Zureik, M.; Desnos, M.; Courbon, D.; Ducimetière, P. Long-term outcome in asymptomatic men with exercise-induced premature ventricular depolarizations. *N. Engl. J. Med.* **2000**, *343*, 826–833. [CrossRef]
16. Takemoto, M.; Yoshimura, H.; Ohba, Y.; Matsumoto, Y.; Yamamoto, U.; Mohri, M.; Yamamoto, H.; Origuchi, H. Radiofrequency catheter ablation of premature ventricular complexes from right ventricular outflow tract improves left ventricular dilation and clinical status in patients without structural heart disease. *J. Am. Coll. Cardiol.* **2005**, *45*, 1259–1265. [CrossRef]
17. Walters, T.E.; Rahmutula, D.; Szilagyi, J.; Alhede, C.; Sievers, R.; Fang, Q.; Olgin, J.; Gerstenfeld, E.P. Left Ventricular Dyssynchrony Predicts the Cardiomyopathy Associated With Premature Ventricular Contractions. *J. Am. Coll. Cardiol.* **2018**, *72*, 2870–2882. [CrossRef] [PubMed]

18. Napier, M.D.; Franceschini, N.; Gondalia, R.; Stewart, J.D.; Méndez-Giráldez, R.; Sitlani, C.M.; Seyerle, A.A.; Highland, H.M.; Li, Y.; Wilhelmsen, K.C.; et al. Genome-wide association study and meta-analysis identify loci associated with ventricular and supraventricular ectopy. *Sci. Rep.* **2018**, *8*, 5675. [CrossRef] [PubMed]
19. Park, K.-M.; Im, S.I.; Park, S.-J.; Kim, J.S.; On, Y.K. Risk factor algorithm used to predict frequent premature ventricular contraction-induced cardiomyopathy. *Int. J. Cardiol.* **2017**, *233*, 37–42. [CrossRef] [PubMed]
20. Cağdaş, D.; Celiker, A.; Ozer, S. Premature ventricular contractions in normal children. *Turk. J. Pediatr.* **2008**, *50*, 260–264. [PubMed]
21. Lee, V.; Hemingway, H.; Harb, R.; Crake, T.; Lambiase, P. The prognostic significance of premature ventricular complexes in adults without clinically apparent heart disease: A meta-analysis and systematic review. *Heart* **2012**, *98*, 1290–1298. [CrossRef] [PubMed]

Article

Maximum Heart Rate- and Lactate Threshold-Based Low-Volume High-Intensity Interval Training Prescriptions Provide Similar Health Benefits in Metabolic Syndrome Patients

Dejan Reljic [1,2,3,*], Fabienne Frenk [1,2], Hans Joachim Herrmann [1,2,3], Markus Friedrich Neurath [1,3] and Yurdagül Zopf [1,2,3]

1. Department of Medicine 1, University Hospital Erlangen, Friedrich-Alexander University Erlangen-Nürnberg, 91054 Erlangen, Germany
2. Hector-Center for Nutrition, Exercise and Sports, Department of Medicine 1, University Hospital Erlangen, Friedrich-Alexander University Erlangen-Nürnberg, 91054 Erlangen, Germany
3. German Center Immunotherapy (DZI), University Hospital Erlangen, Friedrich-Alexander University Erlangen-Nürnberg, 91054 Erlangen, Germany
* Correspondence: dejan.reljic@uk-erlangen.de; Tel.: +49-9131-85-45220

Citation: Reljic, D.; Frenk, F.; Herrmann, H.J.; Neurath, M.F.; Zopf, Y. Maximum Heart Rate- and Lactate Threshold-Based Low-Volume High-Intensity Interval Training Prescriptions Provide Similar Health Benefits in Metabolic Syndrome Patients. *Healthcare* **2023**, *11*, 711. https://doi.org/10.3390/healthcare11050711

Academic Editors: Rafael Oliveira and João Paulo Brito

Received: 13 January 2023
Revised: 23 February 2023
Accepted: 25 February 2023
Published: 28 February 2023

Abstract: Exercise is an integral part of metabolic syndrome (MetS) treatment. Recently, low-volume high-intensity interval training (LOW-HIIT) has emerged as a time-efficient approach to improving cardiometabolic health. Intensity prescriptions for LOW-HIIT are typically based on maximum heart rate (HR_{max}) percentages. However, HR_{max} determination requires maximal effort during exercise testing, which may not always be feasible/safe for MetS patients. This trial compared the effects of a 12-week LOW-HIIT program based on: (a) HR_{max} (HIIT-HR), or (b) submaximal lactate threshold (HIIT-LT), on cardiometabolic health and quality of life (QoL) in MetS patients. Seventy-five patients were randomized to HIIT-HR (5 × 1 min at 80–95% HR_{max}), HIIT-LT (5 × 1 min at 95–105% LT) groups, both performed twice weekly on cycle ergometers, or a control group (CON). All patients received nutritional weight loss consultation. All groups reduced their body weight (HIIT-HR: −3.9 kg, $p < 0.001$; HTT-LT: −5.6 kg, $p < 0.001$; CON: −2.6 kg, $p = 0.003$). The HIIT-HR and HIIT-LT groups similarly, improved their maximal oxygen uptake (+3.6 and +3.7 mL/kg/min, $p < 0.001$), glycohemoglobin (−0.2%, $p = 0.005$, and −0.3%, $p < 0.001$), homeostasis model assessment index (−1.3 units, $p = 0.005$, and −1.0 units, $p = 0.014$), MetS z-score (−1.9 and −2.5 units, $p < 0.001$) and QoL (+10 points, $p = 0.029$, and +11 points, $p = 0.002$), while the CON did not experience changes in these variables. We conclude that HIIT-LT is a viable alternative to HIIT-HR for patients who are not able/willing to undergo maximal exercise testing.

Keywords: obesity; cardiometabolic health; quality of life; interval training; exercise prescription; cardiorespiratory fitness; maximal oxygen uptake; lactate; heart rate; glycemic control

1. Introduction

The metabolic syndrome (MetS) is a pathology defined by the presence of several cardiometabolic disorders, including obesity (in particular excess abdominal fat), hypertension, dyslipidemia, hyperglycemia and insulin resistance [1]. The occurrence of MetS has significantly risen worldwide during the past decades [2], with the latest estimates suggesting that globally, ~13% to ~31% of adults are affected [3]. Recently, it has been reported that COVID-19 pandemic-related measures like quarantines, social distancing and lockdowns have further contributed to the spread of MetS [4,5]. This trend is alarming because MetS is associated with an increased risk of several serious secondary diseases, such as cardiovascular disease [6], different cancers [7], all-cause mortality [6] and diminished quality of life (QoL) [8]. Additionally, recent observations indicate that excess body weight and the existence of cardiometabolic risk factors constitute an increased risk of developing a critical or lethal disease progression following a COVID-19 infection [9,10]. Therefore, effective therapeutic measures for MetS treatment are probably more urgent now than ever before.

Dietary adaptations, particularly a reduction in caloric intake, and an increase in physical activity are cornerstones in obesity and MetS treatment [11]. While caloric intake restriction is of paramount importance to achieve weight loss [12], it has been demonstrated that physical activity independently lowers the risk of developing several chronic health conditions and premature death, regardless of body mass index (BMI) [13]. It has been suggested that the level of cardiorespiratory fitness (CRF), objectified by the determination of maximal oxygen uptake (VO_{2max}), is a major outcome for predicting cardiovascular and all-cause mortality, more significant than other well-established health risk factors like obesity, elevated blood pressure or nicotine abuse [14]. Despite a plethora of evidence on the wide range of health benefits associated with regular exercise, a large part of the global population [15], particularly obese individuals [16], do not meet the minimum physical activity guidelines of 75 min of vigorous-intensity or 150 min of moderate-intensity aerobic activity per week [17]. Over the last decade, surveys have consistently shown that time constraints are among the most frequently reported obstacles to regular exercise, both in the general population [18,19] as well as in clinical cohorts [20,21].

Thus, in the past few years, there has been an increasing scientific interest in designing and evaluating less time-consuming exercise approaches for preventing and treating chronic health conditions [22]. In this regard, low-volume high-intensity interval training (LOW-HIIT) has appeared as an innovative exercise modality to elicit comparable or even greater improvements in CRF and cardiometabolic outcomes in comparison to traditional continuous endurance training [23–26]. By definition, LOW-HIIT is a particular subtype of interval training involving a total duration of $\leq$10 min of intense interval bouts of $\geq$80% of maximum heart rate (HR_{max}), embedded in an overall exercise session of $\leq$30 min (when initial warm-up, recovery between intervals, and cool-down are added up) [23,24]. Recent research from our laboratory [27–30] and other researchers [31] has demonstrated that LOW-HIIT can effectively improve several cardiometabolic risk factors as well as subjective measures, such as QoL, in obese MetS patients.

Prescriptions for physical exercise are typically based on four main components: frequency, intensity, time, and type of exercise, also referred to as the FIIT principle [32]. Among these, intensity is considered the most important element of the physiological responses to exercise [32]. As with other cardiovascular training types, exercise intensity for LOW-HIIT is most commonly prescribed based on percentages of HR_{max}. For a rough estimation of exercise intensity, HR_{max} can be calculated using different formulas [33], most frequently via the "220—age" equation [34]. However, due to high interindividual variability in heart rate (HR) values [34,35], it is rather recommended to directly measure HR_{max} during an exhaustive exercise test in order to obtain more precise results. Although our own studies [27–30] and data from other research groups [31,36,37] indicate that guideline-based cardiopulmonary exercise testing (CPET) [38] is generally safe and tolerable in clinical settings, maximal exhaustion may be contraindicated in certain patient populations. Furthermore, in some individuals, true HR_{max} may not reached due to peripheral muscular fatigue or a lack of motivation.

Alternatively, exercise intensity can be prescribed using specific physiological thresholds based on ventilatory or blood lactate responses during incremental exercise. Ventilatory (VT) and lactate thresholds (LT) reflect specific submaximal metabolic inflection points of respiratory variables and blood lactate concentration [39]. Ventilatory thresholds and LT have been traditionally used to design training programs and predict performance in endurance and team sports athletes [40–42], but threshold-based exercise intensity prescription is also an interesting approach in clinical settings because it does not require maximal effort, making it potentially safer for high-risk patients. Additionally, it has been reported that threshold-based intensity prescriptions for traditional endurance training regimens were superior in improving VO_{2max} in sedentary healthy individuals [43] and cardiometabolic risk factors in MetS patients [44], when compared to intensity prescriptions based on percentages of HR_{max}. In this context, endurance training programs involving exercise intensities at or above the LT were found to be particularly effective in lowering blood pressure in type 2 diabetic patients [45] and visceral fat in obese women with MetS [46]. Furthermore, it has been suggested that threshold-based training seems to be related with a lower instance of non-responders to exercise compared to training programs based on maximum values [47]. However, to our knowledge, no research has yet been undertaken to compare the effects of a LOW-HIIT program using HR_{max}- versus LT-based exercise intensity prescriptions on cardiometabolic health status in a clinical setting.

Thus, the main objective of this investigation was to compare the effects of a 12-week LOW-HIIT intervention either prescribed based on: (a) percentages of HR_{max} (HIIT-HR) or, (b) LT (HIIT-LT), on various cardiometabolic health indices and QoL in a cohort of obese patients diagnosed with MetS. We hypothesized that both LOW-HIIT protocols would improve cardiometabolic health status and QoL compared with a physically inactive control group but, on the basis of previous research using traditional endurance training regimens [43,44,47], we expected HIIT-LT to provide superior improvements than HIIT-HR.

2. Materials and Methods

2.1. Study Design

The present investigation was a sub-group analysis of a larger clinical trial examining the impact of various exercise modalities on multiple health outcomes in MetS patients. Other parts of this research project have been previously published elsewhere [27–30]. The present sub-study presents previously unpublished data from the HIIT-LT group and a sub-sample of the HIIT-HR and a non-exercising control group (CON) that additionally received capillary blood sampling during the CPET for the measurement of lactate concentrations.

In the overall trial, patients were allocated at random to different interval training protocols that were performed 2 times per week for a duration of 12 weeks or to the CON. All patients were provided with standard care nutritional consultation to support their weight reduction. Analogous to the main trial, the key outcome of this investigation was VO_{2max}. Further outcomes of interest were cardiometabolic risk indices, body composition and QoL. The sample size determination and randomization process applied in the main trial were reported elsewhere [29]. Briefly, sample size calculation was based on the previous work of Reljic et al. [48], indicating a large effect of LOW-HIIT on VO_{2max} ($d = 0.97$), that resulted in an estimated number of 16 patients per group to yield a statistical power of 95%. To account for dropouts, the aim was to recruit at least 25 patients for randomization into each group. Randomization was preceded by stratification according to VO_{2max}, gender, age and BMI using the software MinimPy version 3.0 [49] to reduce the heterogeneity of patients' main characteristics between groups. The randomization was conducted by a co-worker not engaged in the acquisition and analysis of the data.

All patients were fully informed about the objectives and procedures of the study, which conformed to the Helsinki Declaration, and gave their written consent before being included in the study. The study protocol was approved by the Medical Ethical Committee of the Friedrich-Alexander University Erlangen-Nürnberg (approval number: 210_17B) and registered at ClinicalTrials.gov (ID-number: NCT03306069).

2.2. Patients

Patients were recruited via flyers that were posted in medical practices and newspaper advertisements. In a first step, all interested persons were screened for eligibility by phone call or personal visit. The inclusion criteria were: age ≥ 18 years, a self-reported mainly physically inactive lifestyle as defined previously [50] and clinical diagnosis of MetS as classified by the International Diabetes Federation [51,52]. Criteria for exclusion were: pregnancy, clinical diagnosis of heart disease, oncological diseases, substantial musculoskeletal disorders or other major health limitations that may constitute contraindications to safe participation in exercise. All patients agreed not to change their usual lifestyle habits, apart from the study intervention. Patients were required to attend at least 75% of the scheduled 24 LOW-HIIT sessions to be included in the final analysis.

2.3. Health Examinations

One week before starting the intervention, patients received the baseline examination, including several standardized assessments and measurements as described in detail below. The second examination was conducted during the week following the termination of the LOW-HIIT intervention with a minimum 3-day interval between the last training session.

Patients were instructed to appear overnight-fasted, to abstain from alcohol and to avoid strenuous physical activities for at least 24 h prior to each examination. If patients were required to take medication, care was taken to ensure that it was taken at the same time of day for both examinations. Prior to the second examination and regularly during the intervention period, patients were asked whether there had been any changes in the type of medication or dosage taken. The pre- and post-intervention examinations were scheduled at a similar daytime (08:00–08:30 a.m.) to minimize potential circadian bias and lasted approximately 2–3 h for each patient.

All measurements were carried out under stable and standardized laboratory conditions (temperature: 22–24 °C, and humidity: 30–50%) within the examination rooms of the Hector-Center for Nutrition, Exercise and Sports at the University Hospital Erlangen and in the standardized order described below. During the examinations, the patients were dressed in their casual clothes, with an exception for the anthropometric measurements, which were performed in underwear without shoes, and the CPET, for which the patients wore a sport dress or comparable clothing. All examinations were performed by a team of highly experienced personnel, consisting of two study nurses (>5 years of work experience) and an exercise physiologist and physician (>10 years of work experience) who were assisted by two medical students. All the staff involved in the data collection were blinded to the assignment of the patient's group.

2.3.1. Hydration Testing

After arriving at the research center, patients were first asked to provide a urine sample for a routine screening for urinary tract infections, kidney disorders and diabetes, and for measuring urine specific gravity (USG). Urine sample analyses were conducted within 30 min of collection using Multistix® 10 SG dipsticks (Siemens HealthCare, Erlangen, Germany).

2.3.2. Determination of Blood Pressure and Resting Heart Rate

Following urine collection, patients entered a quiet experimental room and after 5 min rest, resting HR (HR_{rest}) and blood pressure were recorded with an automatic upper arm blood pressure monitor (M5 professional, Omron, Mannheim, Germany) [53]. According to recent guidelines [54], systolic (SBP) and diastolic (DBP) blood pressure were measured twice at both upper arms at intervals of 60 s and the average value from the side with the higher blood pressure was recorded. In addition, mean arterial blood pressure (MAB) was estimated according to the following formula [55]:

$$MAB = DBP + (1/3 \, [SBP - DBP]).$$

2.3.3. Blood Collection

After blood pressure measurements, patients remained in the sitting position and venous blood samples were taken from the antecubital area. The blood collection tubes (Sarstedt, Nürmbrecht, Germany) were immediately further prepared and forwarded to the central laboratory of the University Hospital Erlangen for measurement of the serum concentrations of glucose, triglycerides, total cholesterol, low-density (LDL) and high-density lipoprotein cholesterol (HDL) using a photometrical determination method (Clinical Chemistry Analyzer AU700 or AU5800, Beckman Coulter, Brea, CA, USA), glycated hemoglobin

A_{1c} (HbA$_{1c}$) using turbidimetric immunoassays (COBAS Integra 400, Roche Diagnostics, Mannheim, Germany) and insulin using a chemiluminescence assay (Liaison XL, DiaSorin, Saluggia, Italy). The homeostasis model assessment index (HOMA-index) was calculated according to the following formula [56]:

$$\text{Homeostasis model assessment-index} = (\text{insulin} \times \text{glucose})/405.$$

2.3.4. Anthropometric Measurements

For standardization reasons, patients were asked again to empty their bladder, if necessary, before the measurements. Anthropometric evaluation included measurement of body weight and determination of body composition. More specifically, body weight, fat mass (FM), body fat percentage (FM%), fat free mass (FFM) and total body water (TBW) were determined using a multi-frequency segmental bioelectrical impedance analysis device (seca mBCA 515, Seca, Hamburg, Germany) with confirmed validity [57]. Patients' waist circumference was measured in the upright position with a flexible tape (Seca, Hamburg, Germany) to the nearest millimeter, at the approximate midpoint between the last touchable rib and the upper iliac crest, as previously described [50].

2.3.5. Determination of the Metabolic Syndrome Severity Score

Metabolic syndrome severity was assessed according to the MetS z-score. The score was calculated using sex-specific equations based on HDL, triglycerides, glucose, waist circumference and MAB, as previously suggested [58]:

$$\text{Males: } [(40 - \text{HDL})/9.0] + [(\text{triglycerides} - 150)/81.0] + [(\text{glucose} - 100)/11.3]$$
$$+ [(\text{waist circumference} - 102)/7.7] + [(\text{MAB} - 100)/9.1]$$

$$\text{Females: } [(50 - \text{HDL})/14.1] + [(\text{triglycerides} - 150)/81.0] + [(\text{glucose} - 100)/11.3]$$
$$+ [(\text{waist circumference} - 88)/9.0] + [(\text{MAB} - 100)/9.1]$$

2.3.6. Cardiopulmonary Exercise Testing

Cardiopulmonary exercise testing was performed on a stationary electronically braked cycle ergometer (Corival cpet, Lode, Groningen, The Netherlands) using two different standard exercise protocols [59,60]. Both protocols commenced with a brief familiarization period, followed by measurements from the resting 12-lead electrocardiogram (ECG, custo cardio 110, custo med, Ottobrunn, Germany), blood pressure and respiratory variables. Subsequently, the HIIT-HR and the CON performed a continuously incrementing ramp protocol, beginning at a workload of 50 W and then increasing by 1 W every 5 s (females) and 1 W every 4 s (males), respectively. Using this approach, maximal exertion was typically achieved within 8–12 min, as generally recommended for ramp protocols [59,60]. The HIIT-LT group performed a step incremental test, with a starting workload of 50 W, followed by a stepwise increase in the load by 25 W (females) and 30 W (males), respectively, every 3 min, as recommended to quantify the LT in untrained individuals [60,61]. With the step incremental test, maximum exertion was typically achieved within 10–14 min. Both protocols were performed with a constant cadence ranging between 60–80 rpm until volitional exhaustion.

During all CPET, exercise ECG was permanently monitored (custo cardio 110, custo med, Ottobrunn, Germany) and blood pressure was measured every 2 min with a standard cuff sphygmomanometer (ERKA, Bad Tölz, Germany). An open-circuit breath-by-breath spiroergometric system (Metalyzer 3B-R3, Cortex Biophysik, Leipzig, Germany) was used to continuously measure oxygen uptake (VO$_2$) and carbon dioxide output (VCO$_2$). At rest, immediately after termination of the exercise and at the 1st, 3rd and 5th min of recovery, 20 µL of capillary blood was sampled from the hyperemized earlobe to measure blood lactate concentrations. In the step incremental test (HIIT-LT group), capillary blood samples were additionally drawn within the last 20 s of each workload stage in order to determine the LT. Blood samples were immediately placed in collection tubes containing a hemolyzing solution, and subsequently measured in our laboratory using an enzymatic-amperometric method (LabTrend, BST Bio Sensor Technology, Berlin, Germany). Upon termination of the exercise, perceived exertion was requested from each patient using the 6–20 Borg scale [62]. Patients had to fulfill a minimum of two of the following criteria [63] in order to assume that maximum exertion had been achieved: a plateau in VO_2, reaching a peak respiratory exchange ratio (RER_{max}) of ≥ 1.1, a peak blood lactate level of ≥ 8.0 mmol/L, an age predicted HR_{max} of $\geq 90\%$ (according to the equation: 220—age) [34] and a perceived exertion value of ≥ 19 on the Borg scale [62].

2.3.7. Determination of Lactate and Ventilatory Thresholds

The lactate threshold was defined at the workload when blood lactate concentration had reached ≥ 4 mmol/L, as first established by Mader et al. [64] and later justified by Heck et al. [65]. Since then, the 4 mmol/L LT is also widely referred to as the "onset of lactate accumulation" (OBLA), and broadly used in exercise physiology and practice for performance diagnostics and training prescription [66]. Although it is clear that lactate accumulation does not occur suddenly at a sharp point but rather continuously in a transition zone [67], it is well accepted that the LT frequently corresponds with a blood lactate level of ~4 mmol/L in untrained individuals [68]. Moreover, the fixed 4 mmol/L LT was found to have high reproducibility and predictability in cycling endurance performance [69] and to be useful in prescribing exercise intensity for MetS patients [70]. Determination of HR (HR_{LT}) and workload (W_{LT}) at the 4 mmol/L LT was performed by applying the software Winlactat version 5.5.2.9 (Mesics, Münster, Germany). First (VT1) and second (VT2, also termed the respiratory compensation point, RCP) ventilatory thresholds were determined independently through visual inspection by two investigators from plots of VCO_2 and VO_2 (the V-slope method) [59]. In case of any discrepancy, a consensus was achieved by discussion. Heart rate and workload (W_{VT1} and W_{VT2}) at both VTs were identified using an automated software (MetaSoft Studio, Cortex Biophysik, Leipzig, Germany).

2.4. Assessment of Self-Reported Quality of Life

Self-reported QoL was measured with the validated EuroQol Group questionnaire (EQ-5D-5L) [71]. The questionnaire consists of the simple EQ visual analogue scale (VAS) ranging from 0–100 (higher ratings imply better QoL) and the EQ-5D index, composed of 5 sub-categories (mobility, self-care, usual activities, pain/discomfort, anxiety/depression, each categorized into 5 severity levels). The values of the 5 sub-categories are transformed into a single variable, with a score of 1.0 representing perfect subjective health and a score of 0 representing the poorest possible health status, respectively [71]. The questionnaires were completed by the patients in a separate waiting lounge at the Hector-Center for Nutrition, Exercise and Sports. Any questions or uncertainties about the questionnaire could be resolved immediately with the investigators.

2.5. Monitoring of Daily Nutrition and Nutritional Counseling

Before study enrolment and during the final intervention week, patients were instructed to track their daily food intake over a duration of 3 successive days before each of the two examinations, with the help of a standardized 24 h nutrition protocol (Freiburger Ernährungsprotokoll; Nutri-Science, Freiburg, Germany). After delivery, the protocols were evaluated by a registered dietitian using the software PRODI 6 expert (Nutri-Science, Freiburg, Germany). In addition, patients' resting metabolic rate (RMR) was estimated using the Harris–Benedict equation [72], as follows:

$$\text{Males: RMR (kcal/day)} = 66.5 + 13.8 \times \text{weight (kg)} + 5.0 \times \text{size (cm])} - 6.8 \times \text{age (years)}$$

$$\text{Females: RMR (kcal/day)} = 655 + 9.6 \times \text{weight (kg)} + 1.8 \times \text{size (cm)} - 4.7 \times \text{age (years)}$$

Based on the food record analysis, anthropometric data and the estimated RMR, patients received individual consultation during a personal conversation with a dietitian to support their weight loss. The dietary recommendations were made in accordance with the current obesity treatment guidelines, targeting a daily calorie reduction of 500 kcal [73]. Furthermore, patients were advised to consume at least 1.0 g/kg of protein per day to counteract a loss of muscle mass during caloric restriction, as previously recommended [74]. After the consultation, patients were provided with handouts, including recipes and nutrient lists, to increase adherence and to support them in the home-based implementation of the nutritional recommendations.

2.6. LOW-HIIT Protocols

Patients allocated to the exercise groups performed 2 supervised sessions per week of LOW-HIIT on electronically braked cycle ergometers (Corival cpet, Lode, Groningen, The Netherlands) in our exercise center with a minimum of 1 day recovery between sessions for a total of 12 weeks (24 sessions in total). In order to maximize adherence, patients had the option to schedule their sessions individually during the exercise center's opening hours. The structure of the LOW-HIIT intervention was in accordance with the protocol introduced by Reljic et al. [48].

In brief, the protocol commenced with a short low-intensity warm-up period of 2 min. Subsequently, patients performed 5 vigorous interval bouts of 1 min duration (by accelerating the cadence and/or increasing the ergometer watt load) divided by 1 min recovery phases. After the fifth interval bout, the protocol concluded with a cool-down of 3 min duration at low intensity, corresponding to an accumulated total duration of 14 min/session. In the HIIT-HR group, patients were instructed to reach a minimum exercise intensity of 80–85% HR_{max} during the intervals for the first 4 weeks. The target intensity during intervals was progressively increased as follows: week 5–8: 85–90% HR_{max} and week 9–12: 90–95% HR_{max}. In the HIIT-LT group, the initial minimum exercise intensity to be achieved during intervals was set at a HR corresponding to 95–100% of the LT for the first 4 weeks and then elevated to a HR corresponding to 100–105% of the LT (week 5–12). Patients were equipped with a chest strap HR monitor (Acentas, Hörgertshausen, Germany) in every exercise session, allowing them to follow their HR in real-time on a screen. The HR responses were recorded in every session and later analyzed using the software Heart Rate Monitoring Team System (Acentas, Hörgertshausen, Germany). Average power output and energy expenditure were recorded from the cycle ergometer's digital displays after each session. Certified sports- and physiotherapists monitored every single session to ensure that the imposed level of exercise intensity was reached.

2.7. Statistical Analysis

A priori sample size calculation was conducted using the software G*Power (Heinrich-Heine-University Düsseldorf, Düsseldorf, Germany). Data analyses were performed using the software package SPSS version 24.0 software (IBM Corp., Armonk, NY, USA). Initially, data normality was analyzed using the Shapiro–Wilk test. If the data were normally distributed, a 2×2 repeated measures ANOVA was conducted to examine the data for both the main effects (group and time) and interaction effects (group $\times$ time). In case of significant results, Holm–Sidak post hoc tests for multiple comparisons were performed. Significant main effects of time were followed by separate post hoc paired t-tests for each group. Levene's test was utilized to check and verify the homogeneity of variance. If no normal distribution of data was present, log or square root transformation was conducted and the respective statistical analyses were performed with the transformed data. If this procedure did not improve the data heterogeneity (i.e., HbA_{1c}, serum insulin concentration, HOMA-index and EQ-5D index), non-parametric tests were used for analysis, including the Friedman two-way analysis of variance by ranks, post hoc Dunn's Bonferroni tests for group comparisons and Wilcoxon's tests for within-group comparisons. Effect sizes were evaluated using partial eta-squared (ηp^2) for the ANOVAs and Kendall's coefficient of concordance (W) for the Friedman tests, respectively, and rated as small (0.01–0.05), medium (0.06–0.13) and large ($\geq$0.14) for ηp^2, and small ($\leq$0.10), medium ($\geq$0.30), and large ($\geq$0.50) for W [75]. For all analyses, the significance level was set at $p < 0.05$. Data are shown as means $\pm$ standard deviation (SD) and pre-/post-intervention changes of the outcome values are reported with 95% confidence intervals (95% CI).

3. Results

3.1. Study Flow

Fifty patients (25 each for the HIIT-HR and CON groups) were randomly selected from a larger cohort of the main trial [29] and agreed to additional blood draws during the CPET for the determination of lactate concentrations. Thirty-seven patients were additionally screened for eligibility to be included in the HIIT-LT group until 25 eligible patients were enrolled, resulting in a total sample of 75 patients (25 per group). Seventeen patients dropped out during the study (HIIT-HR = 5, HIIT-LT = 5 and CON = 7). The reasons for dropout are depictured in Figure 1. Consequently, the final analysis involved the data of 58 patients (HIIT-HR = 20, HIIT-LT = 20 and CON = 18). At the baseline, the three groups did not differ significantly in the primary outcome of VO_{2max} and the other main outcomes of interest. Moreover, we did not detect any significant gender effects and therefore, the data of females and males were jointly evaluated in all analyses. Compliance with the LOW-HIIT protocols (the number of scheduled vs. completed exercise sessions) was noticeably high in both exercise groups with 96 $\pm$ 6% in the HIIT-HR group and 94 $\pm$ 8% in the HIIT-LT group.

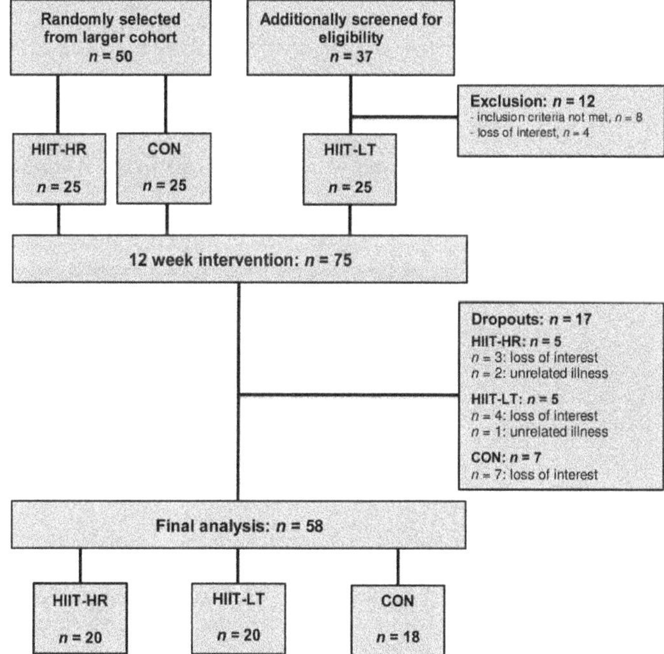

Figure 1. Study flow chart.

3.2. Training Data and Adverse Events

The HR_{LT} at the baseline examination corresponded to 94 $\pm$ 4% of the HR_{max} in the HIIT-LT group. The average peak HR recorded during each interval bout over the 12 weeks corresponded to 93 $\pm$ 7% of the HR_{max} in the HIIT-HR group and 96 $\pm$ 3% of the HR_{max} in the HIIT-LT group, respectively, verifying that the target exercise intensity was successfully reached in both groups. The mean session HR (including the warm-up and cool-down phase) corresponded to 79 $\pm$ 6% of the HR_{max} in the HIIT-HR group, and 82 $\pm$ 5% of the HR_{max} in the HIIT-LT group, respectively. The average peak HR reached during the single intervals and the average session HR were not significantly different between both groups (Table 1). Furthermore, there were no significant differences between both groups in the average power output and energy expenditure per session, with average values of 99.7 ± 24.6 W, 550 ± 149 kilojoules (kJ) and 8.5 ± 2.3 kJ/FFM in the HIIT-HR group, and 105.8 ± 33.2 W, 549 ± 164 kJ and 8.5 ± 2.3 kJ/FFM in the HIIT-LT group, respectively. There were no adverse events observed that were related to the LOW-HIIT.

Table 1. Average heart rate during intervals and during the whole exercise session [1].

Variable	HIIT-HR	HIIT-LT
Week 1–4		
Whole session (%HR_{max})	79.1 ± 5.7	81.8 ± 4.8
Intervals (%HR_{max})	92.6 ± 6.0	95.5 ± 3.4
Week 5–8		
Whole session (%HR_{max})	78.2 ± 7.7	81.6 ± 5.4
Intervals (%HR_{max})	92.3 ± 7.4	95.4 ± 4.1
Week 9–12		
Whole session (%HR_{max})	78.0 ± 7.3	82.0 ± 6.1
Intervals (%HR_{max})	92.7 ± 8.2	95.6 ± 3.7

Values are presented as mean $\pm$ SD. HR_{max} = maximum heart rate. [1] including warm up, intervals, recovery periods and cool-down.

3.3. Hydration Status and Anthropometric Data

During both examinations, the USG values were within the normal ranges for all patients, without significant group differences. There were main effects of time for body weight ($p < 0.001$, $\acute{\eta}^2 = 0.54$), BMI ($p < 0.001$, $\acute{\eta}^2 = 0.54$), FM ($p < 0.001$, $\acute{\eta}^2 = 0.42$), FM% ($p < 0.001$, $\acute{\eta}^2 = 0.29$), FFM ($p = 0.003$, $\acute{\eta}^2 = 0.15$), TBW ($p < 0.001$, $\acute{\eta}^2 = 0.21$) and waist circumference ($p < 0.001$, $\acute{\eta}^2 = 0.56$). Furthermore, there was a group-by-time interaction for waist circumference ($p < 0.001$, $\acute{\eta}^2 = 0.25$) and trend toward an interaction effect for FM ($p = 0.055$, $\acute{\eta}^2 = 0.10$) All groups significantly reduced their body weight (HIIT-HR: -3.9 kg, 95% CI: -5.5 to -2.3 kg, $p < 0.001$; HIIT-LT: -5.6 kg, 95% CI: -7.8 to -3.4 kg, $p < 0.001$; CON: -2.6 kg, 95% CI: -4.2 to -1.0 kg, $p = 0.003$). The quantity of weight loss was not significantly different between the three groups ($p = 0.064$), but compared to the exercise groups, the CON did not significantly reduce FM and waist circumference. Compared to the CON, the decrease in waist circumference was larger in the HIIT-HR (-7 cm, 95% CI: -9 to -1 cm, $p = 0.010$) and HIIT-LT (-8 cm, 95% CI: -11 to -3 cm, $p < 0.001$) groups. Table 2 displays all group specific pre-/post-intervention anthropometric variables.

Table 2. Anthropometric and hydration variables.

Variable	HIIT-HR (n = 20)		HIIT-LT (n = 20)		CON (n = 18)	
	Pre	Post	Pre	Post	Pre	Post
Weight (kg)	117.1 ± 30.3	113.2 ± 29.3 [c]	117.5 ± 24.5	111.9 ± 24.4 [c]	110.7 ± 21.8	108.1 ± 23.2 [b]
BMI (kg/m^2)	38.0 ± 7.7	36.8 ± 7.5 [c]	39.5 ± 8.2	37.5 ± 8.2 [c]	37.3 ± 5.1	36.4 ± 5.7 [b]
FM (kg)	50.0 ± 17.6	46.3 ± 16.4 [c]	52.5 ± 17.4	47.8 ± 18.2 [c]	49.8 ± 11.4	48.3 ± 13.7
FM (%)	42.3 ± 8.2	40.5 ± 8.5 [c]	44.0 ± 7.0	41.8 ± 7.9 [b]	45.0 ± 6.2	44.5 ± 7.1
FFM (kg)	67.1 ± 17.0	67.1 ± 18.0	65.1 ± 11.4	64.1 ± 11.1	60.9 ± 14.3	59.8 ± 13.8 [b]
TBW (L)	50.0 ± 13.0	49.8 ± 13.3	48.7 ± 8.4	47.8 ± 8.1 [a]	45.9 ± 10.3	44.9 ± 10.0 [c]
USG (mg/dL)	1025 ± 12	1022 ± 11	1026 ± 14	1023 ± 12	1024 ± 10	1023 ± 10
Waist (cm)	116.8 ± 21.3	110.2 ± 18.1 [c]	119.4 ± 13.6	111.0 ± 15.0 [c]	113.3 ± 13.4	111.8 ± 15.4

Values are presented as mean $\pm$ SD. BMI = body mass index, FM = fat mass, FFM = skeletal muscle mass, TBW = total body water, USG = urine specific gravity. [a] ($p < 0.05$), [b] ($p < 0.01$), [c] ($p < 0.001$): significantly different compared to pre-intervention.

3.4. Nutrition Data

There was a main effect of time for energy ($p = 0.024$, $\acute{\eta}^2 = 0.09$) and fat intake ($p = 0.005$, $\acute{\eta}^2 = 0.14$). Post hoc tests indicated that the reduction in energy and fat intake per day only reached statistical significance in the CON (-410 kcal, 95% CI: -747 to -72 kcal, $p = 0.020$, and -30 g, 95% CI: -46 to -8 g, $p = 0.007$, respectively), however, there were no significant differences in daily calorie reduction between the groups. Pre- and post-intervention, there were no significant group differences in nutritional intake (Table 3).

Table 3. Nutritional intake.

Variable	HIIT-HR ($n = 20$)		HIIT-LT ($n = 20$)		CON ($n = 18$)	
	Pre	Post	Pre	Post	Pre	Post
Energy (kcal/d)	2320 ± 669	2111 ± 875	2359 ± 1204	1966 ± 790	2206 ± 606	1796 ± 620 [a]
Protein (g/d)	96.8 ± 42.4	96.9 ± 37.9	107.3 ± 95.7	97.1 ± 60.9	94.0 ± 26.0	86.2 ± 30.5
Protein (g/kg/d)	0.9 ± 0.3	0.9 ± 0.3	1.0 ± 0.5	0.9 ± 0.6	0.9 ± 0.3	0.9 ± 0.3
Fat (g/d)	95.4 ± 39.4	83.3 ± 39.1	86.7 ± 41.3	76.7 ± 41.2	93.9 ± 32.1	66.6 ± 26.2 [b]
Fat (g/kg/d)	0.9 ± 0.3	0.8 ± 0.3	0.8 ± 0.3	0.7 ± 0.4	0.9 ± 0.3	0.7 ± 0.3 [a]
CHO (g/d)	211.3 ± 50.1	220.6 ± 106.5	233.7 ± 104.8	186.7 ± 73.8	212.2 ± 73.0	183.9 ± 83.6
CHO (g/kg/d)	1.9 ± 0.6	2.0 ± 0.7	2.1 ± 1.0	1.7 ± 0.7	2.0 ± 0.8	1.8 ± 0.9
Fiber (g/d)	24.3 ± 9.4	23.7 ± 12.0	21.1 ± 9.0	23.4 ± 11.4	22.8 ± 13.4	22.0 ± 8.8

Values are presented as mean $\pm$ SD. CHO = carbohydrates. [a] ($p < 0.05$), [b] ($p < 0.01$): significantly different compared to pre-intervention.

3.5. Cardiopulmonary Exercise Testing Data

During both CPET examinations, all patients fulfilled at least two maximal exertion criteria [63]. In all three groups, there were no significant differences in pre- and post-intervention resting lactate concentrations (HIIT-HR: 1.2 ± 0.3 and 1.1 ± 0.2 mmol/L, HIIT-LT: 1.1 ± 0.2 and 1.0 ± 0.3 mmol/L, CON: 1.2 ± 0.3 and 1.2 ± 0.2 mmol/L), maximal lactate levels (HIIT-HR: 7.5 ± 1.7 and 7.7 ± 2.1 mmol/L, HIIT-LT: 7.9 ± 1.5 and 7.9 ± 2.6 mmol/L, CON: 7.9 ± 1.8 and 7.5 ± 1.8 mmol/L) RER_{max} (HIIT-HR: 1.03 ± 0.1 and 1.04 ± 0.1, HIIT-LT: 1.02 ± 0.1 and 1.03 ± 0.1, CON: 1.01 ± 0.1 and 1.01 ± 0.1) and HR_{max} (HIIT-HR: 158 ± 17 and 160 ± 18 b/min, HIIT-LT: 156 ± 21 and 158 ± 19 b/min, CON: 158 ± 21 and 157 ± 21 b/min), indicating that maximal exhaustion levels were similar during both examinations. In the HIIT-LT group, there was a high agreement between W_{LT} and W_{VT2} ($p < 0.001$, $r = 0.78$).

A main effect of time and group-by-time interaction was found for relative ($p < 0.001$, $\acute{\eta}^2 = 0.50$, and $p < 0.001$, $\acute{\eta}^2 = 0.39$, respectively) and absolute VO_{2max} ($p < 0.001$, $\acute{\eta}^2 = 0.32$, and $p < 0.001$, $\acute{\eta}^2 = 0.42$, respectively), relative ($p < 0.001$, $\acute{\eta}^2 = 0.67$, and $p < 0.001$, $\acute{\eta}^2 = 0.48$, respectively) and absolute W_{max} ($p < 0.001$, $\acute{\eta}^2 = 0.66$, and $p < 0.001$, $\acute{\eta}^2 = 0.62$, respectively), W_{VT1} ($p < 0.001$, $\acute{\eta}^2 = 0.64$, and $p < 0.001$, $\acute{\eta}^2 = 0.58$, respectively) and W_{VT2} ($p = 0.026$, $\acute{\eta}^2 = 0.10$, and $p < 0.001$, $\acute{\eta}^2 = 0.39$, respectively). Additionally, there was a main effect of time for HR_{VT1} ($p = 0.002$, $\acute{\eta}^2 = 0.21$) and HR_{VT2} ($p < 0.001$, $\acute{\eta}^2 = 0.26$).

The HIIT-HR and HIIT-LT groups showed similar improvements in relative VO_{2max} (3.6 mL/kg/min, 95% CI: 2.5 to 4.7 mL/kg/min, $p < 0.001$, and 3.7 mL/kg/min, 95% CI: 2.3 to 5.0 mL/kg/min, $p < 0.001$), absolute VO_{2max} (301 mL/min, 95% CI: 194 to 409 mL/min, $p < 0.001$, and 257 mL/min, 95% CI: 154 to 360 mL/min, $p < 0.001$), relative W_{max} (0.3 W/kg, 95% CI: 0.2 to 0.4 W/kg, $p < 0.001$, and 0.3 W/kg, 95% CI: 0.2 to 0.4 W/kg, $p < 0.001$), absolute W_{max} (25 W, 95% CI: 20 to 30 W, $p < 0.001$, and 26 W, 95% CI: 20 to 31 W, $p < 0.001$), W_{VT1} (30 W, 95% CI: 24 to 35 W, $p < 0.001$, and 30 W, 95% CI: 22 to 38 W, $p < 0.001$) and W_{VT2} (17 W, 95% CI: 5 to 30 W, $p = 0.012$, and 22 W, 95% CI: 12 to 32 W, $p < 0.001$). In the HIIT-LT group, there was a significant increase in W_{LT} (12 W, 95% CI: 3 to 22 W, $p = 0.012$). None of these outcomes improved in the CON. By contrast, absolute VO_{2max} (-214 mL/min, 95% CI: -221 to -8 mL/min, $p = 0.037$) and W_{VT} (-17 W, 95% CI: -28 to -7 W, $p = 0.003$) decreased from pre- to post-intervention.

Compared to the CON, the HIIT-HR and HIIT-LT groups exhibited significantly greater increases in relative VO_{2max} (4.0 mL/kg/min, 95% CI: 2.0 to 5.9 mL/kg/min, $p < 0.001$, and 4.1 mL/kg/min, 95% CI: 2.1 to 5.9 mL/kg/min, $p < 0.001$), absolute VO_{2max} (415 mL/min, 95% CI: 237 to 594 mL/min, $p < 0.001$, and 371 mL/min, 95% CI: 196 to 547 mL/min, $p < 0.001$), relative W_{max} (0.3 W/kg, 95% CI: 0.2 to 0.4 W/kg, $p < 0.001$, and 0.3 W/kg, 95% CI: 0.2 to 0.4 W/kg, $p < 0.001$), absolute W_{max} (30 W, 95% CI: 21 to 38 W, $p < 0.001$, and 30 W, 95% CI: 21 to 39 W, $p < 0.001$), W_{VT1} (32 W, 95% CI: 12 to 53 W, $p < 0.001$, and 40 W, 95% CI: 19 to 60 W, $p < 0.001$) and W_{VT2} (35 W, 95% CI: 16 to 54 W, $p < 0.001$, and 40 W, 95% CI: 20 to 59 W, $p < 0.001$). Pre-/post-intervention CPET outcomes for each group are shown in Table 4.

Table 4. Cardiopulmonary exercise testing variables.

Variable	HIIT-HR ($n = 20$)		HIIT-LT ($n = 20$)		CON ($n = 18$)	
	Pre	Post	Pre	Post	Pre	Post
VO_{2max} (mL/kg/min)	21.6 ± 4.8	25.2 ± 5.3 [c]	22.0 ± 6.9	25.7 ± 7.4 [c]	21.6 ± 7.0	21.2 ± 7.4
VO_{2max} (L/min)	2.4 ± 0.6	2.7 ± 0.7 [c]	2.5 ± 0.6	2.8 ± 0.7 [c]	2.4 ± 0.8	2.2 ± 0.8 [a]
W_{max} (W/kg)	1.4 ± 0.4	1.7 ± 0.4 [c]	1.4 ± 0.5	1.7 ± 0.6 [c]	1.4 ± 0.5	1.4 ± 0.5
W_{max} (W)	156.6 ± 42.0	181.3 ± 38.8 [c]	156.7 ± 45.1	182.2 ± 49.6 [c]	153.3 ± 57.7	148.5 ± 53.7
W_{VT1} (W)	58.0 ± 25.3	87.3 ± 24.0 [c]	63.4 ± 26.2	93.1 ± 30.2 [c]	62.2 ± 33.8	57.1 ± 28.6
W_{VT2} (W)	125.2 ± 35.8	142.4 ± 33.4 [a]	132.6 ± 35.3	154.6 ± 38.7 [c]	136.0 ± 41.6	119.2 ± 41.0 [b]
W_{LT} (W) *	—	—	138.9 ± 38.8	151.3 ± 46.7 [a]	—	—

Values are presented as mean $\pm$ SD. VO_{2max} = maximal oxygen uptake, W_{max} = maximal power output, W_{VT1}, W_{VT2} and W_{LT} = power output at ventilatory threshold 1, power output at ventilatory threshold 2 and power output lactate threshold, respectively. [a] ($p < 0.05$), [b] ($p < 0.01$), [c] ($p < 0.001$): significantly different compared to pre-intervention. * only determined in the HIIT-LT group.

3.6. Cardiometabolic Data

A group-by-time interaction was found for SBP ($p < 0.001$, $\acute{\eta}^2 = 0.29$), DBP ($p < 0.001$, $\acute{\eta}^2 = 0.26$), MAB ($p < 0.001$, $\acute{\eta}^2 = 0.33$) and the MetS z-score ($p < 0.001$, $\acute{\eta}^2 = 0.30$). Additionally, a main effect of time was observed for HR_{rest} ($p < 0.001$, $\acute{\eta}^2 = 0.33$), SBP ($p < 0.001$, $\acute{\eta}^2 = 0.43$), DBP ($p < 0.001$, $\acute{\eta}^2 = 0.40$), MAB ($p < 0.001$, $\acute{\eta}^2 = 0.48$), HbA_{1c} levels ($p < 0.001$, $W = 0.23$), serum insulin concentration ($p < 0.001$, $W = 0.25$), HOMA-index ($p < 0.001$, $W = 0.28$) and the MetS z-score ($p < 0.001$ $\acute{\eta}^2 = 0.61$).

Post hoc tests showed that both in the HIIT-HR and HIIT-LT groups, there were significant reductions in HR_{rest} (-6 b/min, 95% CI: -9 to -2 b/min, $p = 0.006$, and -6 b/min, 95% CI: -7 to -3 b/min, $p < 0.001$), SBP (-11 mmHg, 95% CI: -15 to -7 mmHg, $p < 0.001$, and -13 mmHg, 95% CI: -16 to -9 mmHg, $p < 0.001$), DBP (-8 mmHg, 95% CI: -11 to -4 mmHg, $p < 0.001$, and -10 mmHg, 95% CI: -13 to -7 mmHg, $p < 0.001$), MAB (-9 mmHg, 95% CI: -12 to -6 mmHg, $p < 0.001$, and -11 mmHg, 95% CI: -14 to -8 mmHg, $p < 0.001$), HbA_{1c} levels (-0.2%, 95% CI: -0.4 to -0.1%, $p = 0.012$, and -0.3%, 95% CI: -0.4 to -0.2%, $p < 0.001$), serum insulin concentrations (-5 μU/mL, 95% CI: -8 to -1 μU/mL, $p = 0.007$, and -3 μU/mL, 95% CI: -8 to -2 μU/mL, $p = 0.019$), HOMA-index (-1.3 units, 95% CI: -2.4 to -0.2 units, $p = 0.005$, and -1.0 units, 95% CI: -2.1 to -0.2 units, $p = 0.014$) and the MetS z-score (-1.9 units, 95% CI: -2.6 to -1.8 units, $p < 0.001$, and -2.5 units, 95% CI: -3.0 to -2.0 units, $p < 0.001$). No significant changes occurred in the CON.

Compared to the CON, the HIIT-HR and HIIT-LT groups showed significantly greater pre-/post-intervention reductions in SBP (-12 mmHg, 95% CI: -19 to -4 mmHg, $p < 0.001$, and -13 mmHg, 95% CI: -20 to -5 mmHg, $p < 0.001$), DBP (-8 mmHg, 95% CI: -13 to -2 mmHg, $p = 0.007$, and -10 mmHg, 95% CI: -16 to -4 mmHg, $p < 0.001$), MAB (-9 mmHg, 95% CI: -15 to -4 mmHg, $p < 0.001$, and -11 mmHg, 95% CI: -17 to -6 mmHg, $p < 0.001$) and MetS z-score (-1.6 units, 95% CI: -2.6 to -0.5 units, $p < 0.001$, and -2.2 units, 95% CI: -3.2 to -1.1 units, $p < 0.001$). Pre-/post-intervention cardiometabolic variables for each group are shown in Table 5.

Table 5. Cardiometabolic variables.

Variable	HIIT-HR (n = 20)		HIIT-LT (n = 20)		CON (n = 18)	
	Pre	Post	Pre	Post	Pre	Post
HR_{rest} (b/min)	75.7 ± 9.8	70.3 ± 8.0 [b]	74.6 ± 12.4	69.4 ± 12.3 [c]	77.3 ± 8.8	74.7 ± 10.0
SBP (mmHg)	143.3 ± 13.1	132.0 ± 11.6 [c]	140.0 ± 13.6	127.4 ± 13.4 [c]	137.5 ± 10.9	137.8 ± 7.8
DBP (mmHg)	93.5 ± 7.8	86.0 ± 7.0 [c]	88.6 ± 9.7	78.4 ± 7.4 [c]	88.3 ± 8.9	88.4 ± 7.6
MAB (mmHg)	110.1 ± 8.3	101.2 ± 7.2 [c]	105.7 ± 10.2	94.8 ± 7.8 [c]	104.8 ± 8.8	105.0 ± 6.4
Glucose (mg/dL)	103.0 ± 18.6	101.0 ± 13.2	104.9 ± 14.1	99.8 ± 15.0	94.1 ± 17.4	92.6 ± 14.3
HbA_{1c} (%)	5.7 ± 0.5	5.5 ± 0.4 [b]	5.7 ± 0.4	5.4 ± 0.4 [c]	5.6 ± 0.9	5.6 ± 0.7
Triglycerides (mg/dL)	135.2 ± 59.2	130.0 ± 35.0	116.5 ± 45.1	119.4 ± 55.4	120.0 ± 90.0	119.1 ± 62.4
Cholesterol (mg/dL)	219.3 ± 35.1	215.3 ± 36.2	208.8 ± 36.4	211.5 ± 34.4	218.0 ± 37.2	217.0 ± 31.1
LDL (mg/dL)	147.8 ± 28.8	143.9 ± 26.8	140.7 ± 28.0	141.2 ± 29.3	148.7 ± 29.9	148.2 ± 22.6
HDL (mg/dL)	49.2 ± 10.0	49.4 ± 11.2	47.0 ± 9.4	49.8 ± 12.0	55.1 ± 12.3	53.0 ± 12.1
Insulin (μU/mL)	18.3 ± 11.7	12.7 ± 9.3 [b]	19.7 ± 10.8	17.3 ± 12.8 [a]	18.2 ± 12.6	15.9 ± 8.0
HOMA-index	4.8 ± 3.6	3.4 ± 2.1 [b]	5.2 ± 3.0	4.2 ± 3.0 [a]	4.4 ± 3.6	3.8 ± 2.4
MetS z-score	3.3 ± 4.0	1.4 ± 3.0 [c]	3.3 ± 2.6	0.8 ± 3.0 [c]	2.0 ± 2.8	1.5 ± 3.1

Values are presented as mean ± SD. HR_{rest} = resting heart rate, SBP = systolic blood pressure, DBP = diastolic blood pressure, MAB = mean arterial blood pressure, HbA_{1c} = glycated hemoglobin A_{1c}, LDL = low-density lipoprotein cholesterol, HDL = high-density lipoprotein cholesterol, HOMA = homeostasis model assessment, MetS = metabolic syndrome. [a] ($p < 0.05$), [b] ($p < 0.01$), [c] ($p < 0.001$): significantly different compared to pre-intervention.

3.7. Self-Reported Quality of Life Data

A main effect of time was detected for EQ-VAS ($p < 0.001$, η^2 = 0.24) and the EQ-5D index ($p = 0.031$, $W = 0.08$). Both the HIIT-HR and HIIT-LT groups experienced a pre-/post-intervention increase in EQ-VAS (10 points, 95% CI: 1 to 18 points, $p = 0.029$, and 11 points, 95% CI: 5 to 17 points, $p = 0.002$), whereas no significant changes were recorded in the CON (Table 6).

Table 6. Quality of life variables.

Variable	HIIT-HR (n = 20)		HIIT-LT (n = 20)		CON (n = 18)	
	Pre	Post	Pre	Post	Pre	Post
EQ-VAS	63.0 ± 15.3	72.6 ± 21.1 [a]	63.9 ± 15.7	74.7 ± 17.2 [b]	60.4 ± 25.1	65.7 ± 28.3
EQ-5D index	0.84 ± 0.16	0.86 ± 0.20	0.85 ± 0.12	0.88 ± 0.14	0.87 ± 0.14	0.86 ± 0.19

Values are presented as mean ± SD. BMI = body mass index, FM = fat mass, FFM = skeletal muscle mass, TBW = total body water. [a] ($p < 0.05$), [b] ($p < 0.01$): significantly different compared to pre-intervention.

4. Discussion

Exercise intensity is a crucial—if not the most pivotal—variable in exercise prescription [33]. Intensity prescriptions for (LOW-)HIIT programs are typically based on percentages of the HR_{max}, which, however, may be associated with several limitations in clinical populations. Given the rising popularity of LOW-HIIT in prevention programs and clinical exercise interventions, it is timely to investigate the viability of alternative approaches for exercise intensity prescriptions in individuals, where determination of the HR_{max} may not be feasible. To our knowledge, this investigation was the first to compare the effects of a LOW-HIIT intervention based on either the HR_{max} or the submaximal LT in obese patients with MetS. The major result was that the HIIT-HR and HIIT-LT produced similar improvements in key cardiometabolic outcomes and self-reported QoL after a period of 12 weeks.

The finding that the two LOW-HIIT protocols had similar beneficial effects on cardiometabolic health and QoL was in contrast to our hypothesis based on some previous research, reporting that threshold-based exercise intensity prescriptions are superior to relative percent concepts in improving various cardiometabolic outcomes [43,44,47]. When analyzing the training data (average HR and power output), however, it becomes evident that the physiological demands were comparable between both LOW-HIIT protocols. Furthermore, compliance with both protocols was similarly very high (HIIT-HR: 96 $\pm$ 6%, and HIIT-LT: 94 $\pm$ 8%) and thus, it is plausible that both protocols yielded similar benefits. In this context, it is noteworthy that 4 mmol/L LT data for obese MetS patients have rarely been described in the literature. We found that the HR_{LT} corresponded to 94 $\pm$ 4% of the HR_{max} in our patients, which is in accordance with the well-established 3-phase model introduced by Skinner et al. [76], illustrating that the HR at the 4 mmol/L LT typically exceeds 90% of the HR_{max}.

When comparing both LOW-HIIT protocols, it is notable that all patients in our study were able and willing to reach maximal exertion during the CPET. Thus, in general, if patients are physically able and no symptoms occur during exercise that would require premature termination, we recommend that CPET should be performed until exhaustion in order to acquire maximum performance data and to use the established criteria to verify that maximum exertion has been reached [63]. However, it is an important practical result of this investigation that exercise intensity prescription for the LOW-HIIT protocols can also be feasibly generated using a submaximal exercise test until the LT is reached, which may constitute a viable approach if maximal CPET is contraindicated or patients are not able/motivated to exercise until exertion.

Both LOW-HIIT protocols induced improvements in several health-related outcomes that can be considered clinically meaningful. First, patients involved in the LOW-HIIT improved VO_{2max} by ~3.7 mL/kg/min. The importance of CRF for health and longevity has been well-established in decades of research [13,14,77]. It has been reported, for example, that each VO_{2max} increase by 1 mL/kg/min is associated with a 9% risk decrease in overall mortality [78]. Recent large-scale research verified these findings, demonstrating that each 3.5 mL/kg/min improvement in CRF is related to a decreased risk of premature death due to cardiovascular disease and cancer each by 15% [79]. Second, the reduction in MetS z-score indicates an improvement in overall MetS severity, which was mainly related to reductions in blood pressure (-12 mmHg SBP/-9 mmHg DBP, on average) and waist circumference (-8 cm, on average). Large prospective cohort studies have indicated a reduced risk of coronary heart disease by 22% and stroke by 41%, respectively, per each -10 mmHg SBP/-5 mmHg DBP decrease [80] and an 8% reduction in all-cause mortality per -5 cm decrease in waist circumference [81]. Additionally, both LOW-HIIT protocols had beneficial effects on glucose metabolism as indicated by significant reductions in the HbA_{1c} levels, fasting insulin and HOMA-index. Improvements in these outcomes have been associated with improved cardiometabolic health [82] and a lower risk of colorectal cancer [83], for example. Third, self-reported QoL improved in response to both LOW-HIIT protocols. The mean pre-intervention EQ-VAS scores were markedly lower in our patient cohort than the values reported for the general population [71], which is in line with data from other researchers indicating a relationship between MetS and a diminished QoL [8]. The marked post-intervention improvement in EQ-VAS following LOW-HIIT underscores the well-established association between physical activity [84], CRF levels [85] and enhanced QoL.

Taken together, these findings highlight the pleiotropic effects of exercise on a broad range of important health markers and support the "exercise is medicine" message [86]. Although we clearly recommend that patients who are willing and capable of being more physically active should be encouraged to perform higher volumes of exercise in order maximize the health benefits, our observations provide further evidence [22–26] that even very small doses of targeted exercise can provide meaningful improvements in the physiological and psychological outcomes.

The three groups achieved an average weight loss of ~3.5% within the 12-week study period, which is in accordance with most lifestyle-intervention programs for obesity [87]. It is noteworthy that the relative weight loss amounts tended to be greater in the two exercise groups compared to the CON, but the total difference did not reach statistical significance (exercise groups vs. CON, $p = 0.066$). This finding is not surprising as the three groups did not significantly differ in the amount of caloric reduction, which is the key component to achieve a negative energy balance and to reduce body weight [88]. Although there is evidence that (LOW-)HIIT, compared to traditional continuous endurance training, may have different (more pronounced) effects on some physiological factors associated with weight loss, including higher excess post-exercise oxygen consumption [89], stronger post-exercise suppression of appetite perception [90] or greater changes in concentrations of distinctive gut hormones and leptin [91], our results suggest that the extremely low volume of exercise applied in the present study did not have a substantial impact on the daily overall energy balance. Thus, when it comes to pure weight loss, higher-volume exercise modalities with greater energy expenditure (e.g., longer-lasting endurance exercise or HIIT involving more and/or longer intervals) may be more effective compared to our very low-volume HIIT protocol. However, in agreement with previous reports, it is too short-sighted to define a successful obesity treatment solely in terms of pure weight loss because it is more important to improve the CRF and other cardiometabolic health outcomes than to strictly follow anthropometric measures to improve morbidity and longevity [13,14,77]. In this regard, we observed substantial differences between the patients allocated to the CON and those performing LOW-HIIT, with only the "exercisers" achieving significant improvements in cardiometabolic health and QoL, despite similar weight loss.

Finally, we note some potential limitations to this investigation. Firstly, we note that all patients received standard care nutritional counseling in addition to the LOW-HIIT, which may represent a confounding variable for the observed pre-/post intervention changes. However, we do not feel that the nutritional modification had any meaningful effect on the major research question of this study (HIIT-HR vs. HIIT-LT) because both groups received the same counseling and there were no significant differences in the nutritional intake between the HIIT-HR and HIIT-LT groups. Nevertheless, it cannot be completely ruled out that potential within- or between-group variations in nutrition might have affected the results to some extent.

Secondly, we are well aware that numerous LT as well as VT concepts exist [40,41,60,61,66,67,69] and one can argue why we used the fixed 4 mmol/L LT [64,65] to prescribe the exercise intensity to the HIIT-LT group. Specific reasons for selecting the 4 mmol/L LT are given in the methodology section, but we highlight that it was the major aim of this study to compare the effects of LOW-HIIT prescriptions based on maximal versus submaximal exercise parameters and not to investigate which threshold concept might be the best for obese MetS patients. Nevertheless, we do not rule out that another threshold concept/exercise prescription approach may have achieved even better results or might even have been superior to the HR_{max}-based prescription method. Future research may wish to explore this important question. Moreover, further research is necessary to investigate whether the findings obtained by this specific cohort of obese MetS patients may be transferred to other (clinical) populations. Lastly, it must be considered that all the examinations and the LOW-HIIT intervention were carried out in a well-controlled clinical environment. Thus, it remains to be elucidated to which degree our findings can be applied to non-clinical settings.

5. Conclusions

The HIIT-HR and HIIT-LT induced similar improvements in cardiometabolic health and QoL in obese MetS patients. Thus, the practical take-home message for clinicians and exercise physiologists who wish to implement LOW-HIIT in clinical populations, is that exercise intensity can feasibly and effectively be prescribed using a submaximal LT-based exercise test if patients are not willing or able to perform maximal CPET.

Author Contributions: Conceptualization, D.R.; methodology, D.R.; validation, D.R., H.J.H. and Y.Z.; formal analysis, D.R.; investigation, D.R. and F.F.; data curation, F.F.; writing—original draft preparation, D.R.; writing—review and editing, H.J.H., M.F.N. and Y.Z.; supervision, D.R.; project administration, D.R. and Y.Z.; funding acquisition, D.R., H.J.H. and Y.Z. All authors have read and agreed to the published version of the manuscript.

Funding: This study has been supported by the H.W. and J. Hector Foundation (funding number: MED1710), the Manfred Roth Foundation and Research Foundation for Medicine at the University Hospital Erlangen.

Institutional Review Board Statement: The study was conducted according to the guidelines of the Declaration of Helsinki, and approved by the Medical Ethical Committee of the Friedrich-Alexander University Erlangen-Nürnberg (approval number: 210_17B).

Informed Consent Statement: Informed consent was obtained from all subjects involved in the study.

Data Availability Statement: The datasets generated and analyzed during the current study are not publicly available but are available from the corresponding author on reasonable request.

Acknowledgments: We would like to thank Alisia Gerl, Melanie Klaußner, Maike Tobschall and Kerstin Weidlich for supervising and instructing the exercise sessions. We would also like to thank Kathinka Faustka and Julia Kratzer for their assistance in data collection. We are especially grateful to all patients for their willingness to participate in this study. Parts of this manuscript are used in the medical dissertation of co-author Fabienne Frenk. We acknowledge financial support by Deutsche Forschungsgemeinschaft and Friedrich-Alexander-Universität Erlangen-Nürnberg within the funding programme "Open Access Publication Funding".

Conflicts of Interest: The authors declare no conflict of interest.

References

1. Wu, S.H.; Liu, W.; Ho, S.C. Metabolic syndrome and all-cause mortality: A meta-analysis of prospective cohort studies. *Eur. J. Epidemiol.* **2010**, *25*, 375–384. [CrossRef]
2. Saklayen, M.G. The global epidemic of the metabolic syndrome. *Curr. Hypertens. Rep.* **2018**, *20*, 12. [CrossRef] [PubMed]
3. Noubiap, J.J.; Nansseu, J.R.; Lontchi-Yimagou, E.; Nkeck, J.R.; Nyaga, U.F.; Ngouo, A.T.; Tounouga, D.N.; Tianyi, F.L.; Foka, A.J.; Ndoadoumgue, A.L.; et al. Geographic distribution of metabolic syndrome and its components in the general adult population: A meta-analysis of global data from 28 million individuals. *Diabetes Res. Clin. Pract.* **2022**, *188*, 109924. [CrossRef]
4. Auriemma, R.S.; Pirchio, R.; Liccardi, A.; Scairati, R.; Del Vecchio, G.; Pivonello, R.; Colao, A. Metabolic syndrome in the era of COVID-19 outbreak: Impact of lockdown on cardiometabolic health. *J. Endocrinol. Investig.* **2021**, *44*, 2845–2847. [CrossRef]
5. Xu, W.; Li, Y.; Yan, Y.; Zhang, L.; Zhang, J.; Yang, C. Effects of coronavirus disease 2019 lockdown on metabolic syndrome and its components among Chinese employees: A retrospective cohort study. *Front. Public Health* **2022**, *10*, 885013. [CrossRef]
6. Mottillo, S.; Filion, K.B.; Genest, J.; Joseph, L.; Pilote, L.; Poirier, P.; Rinfret, S.; Schiffrin, E.L.; Eisenberg, M.J. The metabolic syndrome and cardiovascular risk a systematic review and meta-analysis. *J. Am. Coll. Cardiol.* **2010**, *56*, 1113–1132. [CrossRef] [PubMed]
7. Esposito, K.; Chiodini, P.; Colao, A.; Lenzi, A.; Giugliano, D. Metabolic syndrome and risk of cancer: A systematic review and meta-analysis. *Diabetes Care* **2012**, *35*, 2402–2411. [CrossRef]
8. Saboya, P.P.; Bodanese, L.C.; Zimmermann, P.R.; Gustavo, A.D.; Assumpção, C.M.; Londero, F. Metabolic syndrome and quality of life: A systematic review. *Rev. Lat. Am. Enferm.* **2016**, *24*, e2848. [CrossRef]
9. De Lorenzo, A.; Estato, V.; Castro-Faria-Neto, H.C.; Tibirica, E. Obesity-related inflammation and endothelial dysfunction inCOVID-19: Impact on disease severity. *J. Inflamm. Res.* **2021**, *14*, 2267–2276. [CrossRef]
10. Mauvais-Jarvis, F. Aging, male sex, obesity, and metabolic inflammation create the perfect storm for COVID-19. *Diabetes* **2020**, *69*, 1857–1863. [CrossRef]
11. Yang, M.; Liu, S.; Zhang, C. The related metabolic diseases and treatments of obesity. *Healthcare* **2022**, *10*, 1616. [CrossRef] [PubMed]
12. Liu, D.; Huang, Y.; Huang, C.; Yang, S.; Wei, X.; Zhang, P.; Guo, D.; Lin, J.; Xu, B.; Li, C.; et al. Calorie restriction with or without time-restricted eating in weight loss. *N. Engl. J. Med.* **2022**, *386*, 1495–1504. [CrossRef] [PubMed]
13. Barry, V.W.; Baruth, M.; Beets, M.W.; Durstine, J.L.; Liu, J.; Blair, S.N. Fitness vs. fatness on all-cause mortality: A meta-analysis. *Prog. Cardiovasc. Dis.* **2014**, *56*, 382–390. [CrossRef] [PubMed]
14. Myers, J.; McAuley, P.; Lavie, C.J.; Despres, J.P.; Arena, R.; Kokkinos, P. Physical activity and cardiorespiratory fitness as major markers of cardiovascular risk: Their independent and interwoven importance to health status. *Prog. Cardiovasc. Dis.* **2015**, *57*, 306–314. [CrossRef]
15. Guthold, R.; Stevens, G.A.; Riley, L.M.; Bull, F.C. Worldwide trends in insufficient physical activity from 2001 to 2016: A pooled analysis of 358 population-based surveys with 1·9 million participants. *Lancet Glob. Health* **2018**, *6*, e1077–e1086. [CrossRef]
16. Tudor-Locke, C.; Brashear, M.M.; Johnson, W.D.; Katzmarzyk, P.T. Accelerometer profiles of physical activity and inactivity in normal weight, overweight, and obese U.S. men and women. *Int. J. Behav. Nutr. Phys. Act.* **2010**, *7*, 60. [CrossRef]
17. World Health Organization. *Global Recommendations on Physical Activity for Health*; World Health Organization: Geneva, Switzerland, 2010.

18. Cavallini, M.F.; Callaghan, M.E.; Premo, C.B.; Scott, J.W.; Dyck, D.J. Lack of time is the consistent barrier to physical activity and exercise in 18 to 64 year-old males and females from both South Carolina and Southern Ontario. *J. Phys. Act. Res.* **2020**, *5*, 100–106. [CrossRef]

19. Hoare, E.; Stavreski, B.; Jennings, G.L.; Kingwell, B.A. Exploring motivation and barriers to physical activity among active and inactive Australian adults. *Sports* **2017**, *5*, 47. [CrossRef]

20. Sheill, G.; Guinan, E.; Brady, L.; Hevey, D.; Hussey, J. Exercise interventions for patients with advanced cancer: A systematic review of recruitment, attrition, and exercise adherence rates. *Palliat. Support. Care* **2019**, *17*, 686–696. [CrossRef]

21. Murphy, C.L.; Sheane, B.J.; Cunnane, G. Attitudes towards exercise in patients with chronic disease: The influence of comorbid factors on motivation and ability to exercise. *Postgrad. Med. J.* **2011**, *87*, 96–100. [CrossRef]

22. Gibala, M.J.; Little, J.P. Physiological basis of brief vigorous exercise to improve health. *J. Physiol.* **2020**, *598*, 61–69. [CrossRef]

23. Gibala, M.J.; Gillen, J.B.; Percival, M.E. Physiological and health-related adaptations to low-volume interval training: Influences of nutrition and sex. *Sport. Med.* **2014**, *44*, S1273–S1277. [CrossRef] [PubMed]

24. Gillen, J.B.; Gibala, M.J. Is high-intensity interval training a time-efficient exercise strategy to improve health and fitness? *Appl. Physiol. Nutr. Metab.* **2014**, *39*, 409–412. [CrossRef] [PubMed]

25. Skelly, L.E.; Bailleul, C.; Gillen, J.B. Physiological responses to low-volume interval training in women. *Sport. Med. Open* **2021**, *7*, 99. [CrossRef] [PubMed]

26. Sultana, R.N.; Sabag, A.; Keating, S.E.; Johnson, N.A. The effect of low-volume high-intensity interval training on body composition and cardiorespiratory fitness: A systematic review and meta-analysis. *Sport. Med.* **2019**, *49*, 1687–1721. [CrossRef] [PubMed]

27. Reljic, D.; Dieterich, W.; Herrmann, H.J.; Neurath, M.F.; Zopf, Y. "HIIT the Inflammation": Comparative effects of low-volume interval training and resistance exercises on inflammatory indices in obese metabolic syndrome patients undergoing caloric restriction. *Nutrients* **2022**, *14*, 1996. [CrossRef] [PubMed]

28. Reljic, D.; Frenk, F.; Herrmann, H.J.; Neurath, M.F.; Zopf, Y. Low-volume high-intensity interval training improves cardiometabolic health, work ability and well-being in severely obese individuals: A randomized-controlled trial sub-study. *J. Transl. Med.* **2020**, *18*, 419. [CrossRef] [PubMed]

29. Reljic, D.; Frenk, F.; Herrmann, H.J.; Neurath, M.F.; Zopf, Y. Effects of very low volume high intensity versus moderate intensity interval training in obese metabolic syndrome patients: A randomized controlled study. *Sci. Rep.* **2021**, *11*, 2836. [CrossRef]

30. Reljic, D.; Konturek, P.C.; Herrmann, H.J.; Siebler, J.; Neurath, M.F.; Zopf, Y. Very low-volume interval training improves nonalcoholic fatty liver disease fibrosis score and cardiometabolic health in adults with obesity and metabolic syndrome. *J. Physiol. Pharmacol.* **2021**, *72*, 927–938. [CrossRef]

31. Ramos, J.S.; Dalleck, L.C.; Borrani, F.; Beetham, K.S.; Wallen, M.P.; Mallard, A.R.; Clark, B.; Gomersall, S.; Keating, S.E.; Fassett, R.G.; et al. Low-volume high-intensity interval training is sufficient to ameliorate the severity of metabolic syndrome. *Metab. Syndr. Relat. Disord.* **2017**, *15*, 319–328. [CrossRef]

32. Garber, C.E.; Blissmer, B.; Deschenes, M.R.; Franklin, B.A.; Lamonte, M.J.; Lee, I.M.; Nieman, D.C.; Swain, D.P.; American College of Sports Medicine. American College of Sports Medicine position stand. Quantity and quality of exercise for developing and maintaining cardiorespiratory, musculoskeletal, and neuromotor fitness in apparently healthy adults: Guidance for prescribing exercise. *Med. Sci. Sport. Exerc.* **2011**, *43*, 1334–1359. [CrossRef] [PubMed]

33. Shookster, D.; Lindsey, B.; Cortes, N.; Martin, J.R. Accuracy of commonly used age-predicted maximal heart rate equations. *Int. J. Exerc. Sci.* **2020**, *13*, 1242–1250. [PubMed]

34. Robergs, R.; Landwehr, R. The surprising history of the "HRmax= 220-age" equation. *J. Exer. Physiol. Online* **2002**, *5*, 1–10.

35. Sarzynski, M.A.; Rankinen, T.; Earnest, C.P.; Leon, A.S.; Rao, D.C.; Skinner, J.; Bouchard, C. Measured maximal heart rates compared to commonly used age-based prediction equations in the Heritage Family Study. *Am. J. Hum. Biol.* **2013**, *25*, 695–701. [CrossRef]

36. Scheel, P.J., 3rd; Florido, R.; Hsu, S.; Murray, B.; Tichnell, C.; James, C.A.; Agafonova, J.; Tandri, H.; Judge, D.P.; Russell, S.D.; et al. Safety and utility of cardiopulmonary exercise testing in arrhythmogenic right ventricular cardiomyopathy/dysplasia. *J. Am. Heart. Assoc.* **2020**, *9*, e013695. [CrossRef]

37. Skalski, J.; Allison, T.G.; Miller, T.D. The safety of cardiopulmonary exercise testing in a population with high-risk cardiovascular diseases. *Circulation* **2012**, *126*, 2465–2472. [CrossRef]

38. American Thoracic Society; American College of Chest Physicians. ATS/ACCP Statement on cardiopulmonary exercise testing. *Am. J. Respir. Crit. Care Med.* **2003**, *167*, 211–277. [CrossRef]

39. Binder, R.K.; Wonisch, M.; Corra, U.; Cohen-Solal, A.; Vanhees, L.; Saner, H.; Schmid, J.P. Methodological approach to the first and second lactate threshold in incremental cardiopulmonary exercise testing. *Eur. J. Cardiovasc. Prev. Rehabil.* **2008**, *15*, 726–734. [CrossRef]

40. Amann, M.; Subudhi, A.W.; Foster, C. Predictive validity of ventilatory and lactate thresholds for cycling time trial performance. *Scand. J. Med. Sci. Sport.* **2006**, *16*, 27–34. [CrossRef]

41. Beneke, R.; Leithäuser, R.M.; Ochentel, O. Blood lactate diagnostics in exercise testing and training. *Int. J. Sport. Physiol. Perform.* **2011**, *6*, 8–24. [CrossRef]

42. Billat, L.V. Use of blood lactate measurements for prediction of exercise performance and for control of training. Recommendations for long-distance running. *Sport. Med.* **1996**, *22*, 157–175. [CrossRef] [PubMed]

43. Wolpern, A.E.; Burgos, D.J.; Janot, J.M.; Dalleck, L.C. Is a threshold-based model a superior method to the relative percent concept for establishing individual exercise intensity? a randomized controlled trial. *BMC Sport. Sci. Med. Rehabil.* **2015**, *7*, 16. [CrossRef] [PubMed]

44. Weatherwax, R.M.; Ramos, J.S.; Harris, N.K.; Kilding, A.E.; Dalleck, L.C. Changes in metabolic syndrome severity following individualized versus standardized exercise prescription: A feasibility study. *Int. J. Environ. Res. Public Health* **2018**, *15*, 2594. [CrossRef]

45. Asano, R.Y.; Sales, M.M.; Browne, R.A.; Moraes, J.F.; Coelho Júnior, H.J.; Moraes, M.R.; Simões, H.G. Acute effects of physical exercise in type 2 diabetes: A review. *World J. Diabetes* **2014**, *5*, 659–665. [CrossRef]

46. Irving, B.A.; Davis, C.K.; Brock, D.W.; Weltman, J.Y.; Swift, D.; Barrett, E.J.; Gaesser, G.A.; Weltman, A. Effect of exercise training intensity on abdominal visceral fat and body composition. *Med. Sci. Sport. Exerc.* **2008**, *40*, 1863–1872. [CrossRef] [PubMed]

47. Lehtonen, E.; Gagnon, D.; Eklund, D.; Kaseva, K.; Peltonen, J.E. Hierarchical framework to improve individualised exercise prescription in adults: A critical review. *BMJ Open Sport Exerc. Med.* **2022**, *8*, e001339. [CrossRef]

48. Reljic, D.; Wittmann, F.; Fischer, J.E. Effects of low-volume high-intensity interval training in a community setting: A pilot study. *Eur. J. Appl. Physiol.* **2018**, *118*, 1153–1167. [CrossRef]

49. Saghaei, M.; Saghaei, S. Implementation of an open-source customizable minimization program for allocation of patients to parallel groups in clinical trials. *J. Biomed. Sci. Eng.* **2011**, *4*, 734–739. [CrossRef]

50. American College of Sports Medicine. *ACSM's Guidelines for Exercise Testing and Prescription*, 8th ed.; Lippincott Williams & Wilkins: Philadelphia, PA, USA, 2010; pp. 26–27.

51. Alberti, K.G.; Zimmet, P.; Shaw, J. Metabolic syndrome–a new world-wide definition. A Consensus Statement from the International Diabetes Federation. *Diabet. Med.* **2006**, *23*, 469–480. [CrossRef]

52. Zhu, L.; Spence, C.; Yang, J.W.; Ma, G.X. The IDF definition is better suited for screening metabolic syndrome and estimating risks of diabetes in Asian American Adults: Evidence from NHANES 2011–2016. *J. Clin. Med.* **2020**, *9*, 3871. [CrossRef]

53. Tholl, U.; Lüders, S.; Bramlage, P.; Dechend, R.; Eckert, S.; Mengden, T.; Nürnberger, J.; Sanner, B.; Anlauf, M. The German Hypertension League (Deutsche Hochdruckliga) quality seal protocol for blood pressure-measuring devices: 15-year experience and results from 105 devices for home blood pressure control. *Blood Press. Monit.* **2016**, *21*, 197–205. [CrossRef] [PubMed]

54. Whelton, P.K.; Carey, R.M.; Aronow, W.S.; Casey, D.E., Jr.; Collins, K.J.; Dennison Himmelfarb, C.; DePalma, S.M.; Gidding, S.; Jamerson, K.A.; Jones, D.W.; et al. 2017 ACC/AHA/AAPA/ABC/ACPM/AGS/APhA/ASH/ASPC/NMA/PCNA Guideline for the prevention, detection, evaluation, and management of high blood pressure in adults: A report of the American College of Cardiology/American Heart Association Task Force on Clinical Practice Guidelines. *Hypertension* **2018**, *138*, e426–e483. [CrossRef]

55. DeMers, D.; Wachs, D. Physiology, Mean Arterial Pressure. In *StatPearls [Internet]*; StatPearls Publishing: Treasure Island, FL, USA, 2022.

56. Matthews, D.R.; Hosker, J.P.; Rudenski, A.S.; Naylor, B.A.; Treacher, D.F.; Turner, R.C. Homeostasis model assessment: Insulin resistance and beta-cell function from fasting plasma glucose and insulin concentrations in man. *Diabetologia* **1985**, *28*, 412–419. [CrossRef] [PubMed]

57. Bosy-Westphal, A.; Jensen, B.; Braun, W.; Pourhassan, M.; Gallagher, D.; Müller, M.J. Quantification of whole-body and segmental skeletal muscle mass using phase-sensitive 8-electrode medical bioelectrical impedance devices. *Eur. J. Clin. Nutr.* **2017**, *71*, 1061–1067. [CrossRef] [PubMed]

58. Johnson, J.L.; Slentz, C.A.; Houmard, J.A.; Samsa, G.P.; Duscha, B.D.; Aiken, L.B.; McCartney, J.S.; Tanner, C.J.; Kraus, W.E. Exercise training amount and intensity effects on metabolic syndrome (from Studies of a Targeted Risk Reduction Intervention through Defined Exercise). *Am. J. Cardiol.* **2007**, *100*, 1759–1766. [CrossRef] [PubMed]

59. Glaab, T.; Taube, C. Practical guide to cardiopulmonary exercise testing in adults. *Respir. Res.* **2022**, *23*, 9. [CrossRef] [PubMed]

60. Mezzani, A. Cardiopulmonary exercise testing: Basics of methodology and measurements. *Ann. Am. Thorac. Soc.* **2017**, *14*, S3–S11. [CrossRef]

61. Bentley, D.J.; Newell, J.; Bishop, D. Incremental exercise test design and analysis: Implications for performance diagnostics in endurance athletes. *Sport. Med.* **2007**, *37*, 575–586. [CrossRef]

62. Borg, G.A. Psychophysical bases of perceived exertion. *Med. Sci. Sport. Exerc.* **1982**, *14*, 377–381. [CrossRef]

63. Howley, E.T.; Bassett, D.R.; Welch, H.G. Criteria for maximal oxygen uptake: Review and commentary. *Med. Sci. Sport. Exerc.* **1995**, *27*, 1292–1301. [CrossRef]

64. Mader, A.; Liesen, H.; Heck, H.; Philippi, H.; Rost, R.; Schürch, P.; Hollmann, W. Zur Beurteilung der sportartspezifischen Ausdauerleistungsfähigkeit im Labor. *Sportarzt Sportmed.* **1976**, *27*, 109–112.

65. Heck, H.; Mader, A.; Hess, G.; Mücke, S.; Müller, R.; Hollmann, W. Justification of the 4-mmol/l lactate threshold. *Int. J. Sport. Med.* **1985**, *6*, 117–130. [CrossRef]

66. Wackerhage, H.; Gehlert, S.; Schulz, H.; Weber, S.; Ring-Dimitriou, S.; Heine, O. Lactate thresholds and the simulation of human energy metabolism: Contributions by the Cologne Sports Medicine Group in the 1970s and 1980s. *Front. Physiol.* **2022**, *13*, 899670. [CrossRef]

67. Faude, O.; Kindermann, W.; Meyer, T. Lactate threshold concepts: How valid are they? *Sport. Med.* **2009**, *39*, 469–490. [CrossRef]

68. Messonnier, L.A.; Emhoff, C.A.; Fattor, J.A.; Horning, M.A.; Carlson, T.J.; Brooks, G.A. Lactate kinetics at the lactate threshold in trained and untrained men. *J. Appl. Physiol.* **2013**, *114*, 1593–1602. [CrossRef]

69. Heuberger, J.A.; Gal, P.; Stuurman, F.E.; de Muinck Keizer, W.A.; Mejia Miranda, Y.; Cohen, A.F. Repeatability and predictive value of lactate threshold concepts in endurance sports. *PLoS ONE* **2018**, *13*, e0206846. [CrossRef]
70. Torres, G.; Crowther, N.J.; Rogers, G. Reproducibility and levels of blood lactate transition thresholds in persons with metabolic syndrome. *Metab. Syndr. Relat. Disord.* **2013**, *11*, 121–127. [CrossRef]
71. Grochtdreis, T.; Dams, J.; König, H.H.; Konnopka, A. Health-related quality of life measured with the EQ-5D-5L: Estimation of normative index values based on a representative German population sample and value set. *Eur. J. Health Econ.* **2019**, *20*, 933–944. [CrossRef]
72. Harris, J.A.; Benedict, F.G. A biometric study of basal metabolism in man. *Proc. Natl. Acad. Sci. USA* **1918**, *4*, 370–373. [CrossRef]
73. Carels, R.A.; Young, K.M.; Coit, C.; Clayton, A.M.; Spencer, A.; Hobbs, M. Can following the caloric restriction recommendations from the Dietary Guidelines for Americans help individuals lose weight? *Eat. Behav.* **2008**, *9*, 328–335. [CrossRef]
74. Al-Nimr, R.I. Optimal protein intake during weight loss interventions in older adults with obesity. *J. Nutr. Gerontol. Geriatr.* **2019**, *38*, 50–68. [CrossRef]
75. Cohen, J. *Statistical Power Analysis for the Behavioral Sciences*, 2nd ed.; Taylor and Francis: Routledge, UK, 1988.
76. Skinner, J.S.; McLellan, T.M. The transition from aerobic to anaerobic metabolism. *Res. Q. Exerc. Sport.* **1980**, *51*, 234–248. [CrossRef]
77. Gaesser, G.A.; Tucker, W.J.; Jarrett, C.L.; Angadi, S.S. Fitness versus fatness: Which influences health and mortality risk the most? *Curr. Sport. Med. Rep.* **2015**, *14*, 327–332. [CrossRef]
78. Kavanagh, T.; Mertens, D.J.; Hamm, L.F.; Beyene, J.; Kennedy, J.; Corey, P.; Shephard, R.J. Prediction of long-term prognosis in 12 169 men referred for cardiac rehabilitation. *Circulation* **2002**, *106*, 666–671. [CrossRef]
79. Vainshelboim, B.; Myers, J.; Matthews, C.E. Non-exercise estimated cardiorespiratory fitness and mortality from all-causes, cardiovascular disease, and cancer in the NIH-AARP diet and health study. *Eur. J. Prev. Cardiol.* **2022**, *29*, 599–607. [CrossRef]
80. Law, M.R.; Morris, J.K.; Wald, N.J. Use of blood pressure lowering drugs in the prevention of cardiovascular disease: Meta-analysis of 147 randomised trials in the context of expectations from prospective epidemiological studies. *BMJ* **2009**, *338*, b1665. [CrossRef]
81. Cerhan, J.R.; Moore, S.C.; Jacobs, E.J.; Kitahara, C.M.; Rosenberg, P.S.; Adami, H.O.; Ebbert, J.O.; English, D.R.; Gapstur, S.M.; Giles, G.G.; et al. A pooled analysis of waist circumference and mortality in 650,000 adults. *Mayo Clin. Proc.* **2014**, *89*, 335–345. [CrossRef]
82. González-González, J.G.; Violante-Cumpa, J.R.; Zambrano-Lucio, M.; Burciaga-Jimenez, E.; Castillo-Morales, P.L.; Garcia-Campa, M.; Solis, R.C.; González-Colmenero, A.D.; Rodríguez-Gutiérrez, R. HOMA-IR as a predictor of health outcomes in patients with metabolic risk factors: A systematic review and meta-analysis. *High Blood Press. Cardiovasc. Prev.* **2022**, *29*, 547–564. [CrossRef]
83. Xu, J.; Ye, Y.; Wu, H.; Duerksen-Hughes, P.; Zhang, H.; Li, P.; Huang, J.; Yang, J.; Wu, Y.; Xia, D. Association between markers of glucose metabolism and risk of colorectal cancer. *BMJ Open* **2016**, *6*, e011430. [CrossRef]
84. Marquez, D.X.; Aguinaga, S.; Vásquez, P.M.; Conroy, D.E.; Erickson, K.I.; Hillman, C.; Stillman, C.M.; Ballard, R.M.; Sheppard, B.B.; Petruzzello, S.J.; et al. A systematic review of physical activity and quality of life and well-being. *Transl. Behav. Med.* **2020**, *10*, 1098–1109. [CrossRef]
85. Sloan, R.A.; Sawada, S.S.; Martin, C.K.; Church, T.; Blair, S.N. Associations between cardiorespiratory fitness and health-related quality of life. *Health Qual. Life Outcomes* **2009**, *7*, 47. [CrossRef]
86. Pedersen, B.K.; Saltin, B. Exercise as medicine—Evidence for prescribing exercise as therapy in 26 different chronic diseases. *Scand. J. Med. Sci. Sport.* **2015**, *25*, 1–72. [CrossRef] [PubMed]
87. National Institute for Health and Care Excellence (NICE). *Weight Management: Lifestyle Services for Overweight or Obese Adults*; Public Health Guideline PH53; NICE: London, UK, 2014; Available online: www.nice.org.uk/guidance/ph53 (accessed on 7 December 2022).
88. Catenacci, V.A.; Wyatt, H.R. The role of physical activity in producing and maintaining weight loss. *Nat. Clin. Pract. Endocrinol. Metab.* **2007**, *3*, 518–529. [CrossRef] [PubMed]
89. Panissa, V.L.; Fukuda, D.H.; Staibano, V.; Marques, M.; Franchini, E. Magnitude and duration of excess of post-exercise oxygen consumption between high-intensity interval and moderate-intensity continuous exercise: A systematic review. *Obes. Rev.* **2021**, *22*, e13099. [CrossRef] [PubMed]
90. Hu, M.; Nie, J.; Lei, O.K.; Shi, Q.; Kong, Z. Acute effect of high-intensity interval training versus moderate-intensity continuous training on appetite perception: A systematic review and meta-analysis. *Appetite* **2022**, *182*, 106427. [CrossRef]
91. Larsen, P.; Marino, F.; Melehan, K.; Guelfi, K.J.; Duffield, R.; Skein, M. High-intensity interval exercise induces greater acute changes in sleep, appetite-related hormones, and free-living energy intake than does moderate-intensity continuous exercise. *Appl. Physiol. Nutr. Metab.* **2019**, *44*, 557–566. [CrossRef]

Systematic Review

Effects of Combined Training Programs in Individuals with Fibromyalgia: A Systematic Review

Mónica Sousa [1], Rafael Oliveira [1,2,3,*], João Paulo Brito [1,2,3,*], Alexandre Duarte Martins [1,2,4], João Moutão [1,2,3] and Susana Alves [1,2]

[1] Sports Science School of Rio Maior, Polytechnic Institute of Santarém, 2040-413 Rio Maior, Portugal; monicasousa@esdrm.ipsantarem.pt (M.S.); af_martins17@hotmail.com (A.D.M.); jmoutao@esdrm.ipsantarem.pt (J.M.); salves@esdrm.ipsantarem.pt (S.A.)

[2] Life Quality Research Centre, 2040-413 Rio Maior, Portugal

[3] Research Centre in Sport Sciences, Health Sciences and Human Development, 5001-801 Vila Real, Portugal

[4] Comprehensive Health Research Centre (CHRC), Departamento de Desporto e Saúde, Escola de Saúde e Desenvolvimento Humano, Universidade de Évora, Largo dos Colegiais, 7000-727 Évora, Portugal

*   Correspondence: rafaeloliveira@esdrm.ipsantarem.pt (R.O.); jbrito@esdrm.ipsantarem.pt (J.P.B.)

Abstract: Fibromyalgia is a rheumatic disease characterised by chronic widespread muscular pain and its treatment is carried out by pharmacological interventions. Physical exercise and a healthy lifestyle act as an important mechanism in reducing the symptoms of the disease. The aims of this study were to analyse and systematise the characteristics of combined training programs (i.e., type and duration of interventions, weekly frequency, duration and structure of training sessions and prescribed intensities) and to analyse their effects on people diagnosed with fibromyalgia. A systematic literature search was performed using the PRISMA method and then randomised controlled trial articles that met the eligibility criteria were selected. The Physiotherapy Evidence Database scale was used to assess the quality and risk of the studies. A total of 230 articles were selected, and in the end, 13 articles met the defined criteria. The results showed different exercise interventions such as: combined training, high-intensity interval training, Tai Chi, aerobic exercise, body balance and strength training. In general, the different interventions were beneficial for decreasing physical symptoms and improving physical fitness and functional capacity. In conclusion, a minimum duration of 14 weeks is recommended for better benefits. Moreover, combined training programs were the most effective for this population, in order to reduce the symptoms of the disease with a duration between 60 and 90 min, three times a week with a light to moderate intensity.

Keywords: fibromyalgia; exercise; multicomponent training; aerobic training; resistance training; strength training

Citation: Sousa, M.; Oliveira, R.; Brito, J.P.; Martins, A.D.; Moutão, J.; Alves, S. Effects of Combined Training Programs in Individuals with Fibromyalgia: A Systematic Review. *Healthcare* **2023**, *11*, 1708. https://doi.org/10.3390/healthcare11121708

Academic Editor: Gilbert Ramirez

Received: 4 May 2023
Revised: 2 June 2023
Accepted: 7 June 2023
Published: 11 June 2023

1. Introduction

Fibromyalgia (FM) is defined as a chronic rheumatic disease and is characterised by chronic widespread pain, muscle stiffness, sleep disturbances and cognitive problems [1–4]. In addition to these, the following symptoms are also observed: a feeling of fatigue and changes in the psychological state [5]. Moreover, FM can include muscle pain in the tender points, excessive fatigue, muscle strength loss and some psychological problems as mentioned before (i.e., sleep issues, anxiety, depression and reduced levels of satisfaction with life and self-esteem) [6,7]. Most of the time, the diagnosis is quite difficult to perform because there is no accurate (i.e., validated) diagnostic test to identify the disease. Thus, the diagnosis of this disease is carried out through palpation from tender points specific for FM [1].

Studies indicate that FM affects, on average, 2.1% of the world's population and 2.31% of the European population, implying a painful loss of quality of life for the people who suffer from it and high economic costs [8]. The literature also points out that FM is more prevalent in women with values between 2.4% and 6.8% and in urban areas between 0.7% and 11.4% [9]. In Portugal, the prevalence is estimated at 1.7% (1.1% to 2.1%) [10].

Scientifically, the exact cause of the origin of FM remains unknown, so all the treatments of this disease are directed towards the reduction in the signs and symptoms presented [11]. In addition, the clinical control of the patient is carried out mainly through pharmacological interventions [12]. However, this type of treatment is not effective in solving functional problems, namely the loss of mobility and muscle strength and power, which negatively interferes with the quality of life of patients [13,14]. In this sense, some studies have demonstrated the importance of including non-pharmacological treatments in this pathology, mainly the regular practice of physical exercise associated with a healthy lifestyle [5,15].

Physical exercise promotes several benefits on a physical and psychological level. A physical exercise program works as an important mechanism that positively influences this population, attenuating the main symptoms, such as: the feeling of fatigue, depression, anxiety, muscle stiffness and sleep disturbances [16]. In this way, physical exercise has been used as a form of non-pharmacological intervention [17].

The American College of Sports Medicine (ACSM) recommends performing strength exercises (2 to 3 days/week), aerobic exercises (2 to 4 days/week) and flexibility exercises (1 to 3 days/week) to attenuate or reduce the signs and symptoms of FM [18]. In this sense, a combined training program may adjust to the recommendations for this population [19]. due to the fact that it involves aerobic, strength and stretching exercises, simultaneously, inducing several important adaptations in order to cover a greater number of symptoms. Consequently, strength, power, and aerobic capacity and power improvements may occur [20]. This type of training can be performed in the same session or in different sessions [21]. In this sense, aerobic exercise induces adaptations in various functional capacities such as transport, capture and the use of oxygen [22]; strength training becomes essential for increasing muscle strength [23,24]; and stretching exercises are beneficial to reduce the loss of mobility due to its constant immobilisation associated with pain [25].

To better understand the benefits of different types of exercises and physical therapy in FM, a set of studies were reviewed to obtain a comprehensive guideline for the prescription of exercise in this population [26]. The results suggest that individuals with FM have different responses to different types of exercise programs (e.g., aerobic training or strength training), since these same individuals present a great diversity of signs and symptoms. Accordingly, preference should be given to more global exercise protocols that are able to provide positive effects to the greatest number symptoms possible. In this way, it is important to better understand the effects of combined training and recommendations regarding the prescription of physical exercise in individuals diagnosed with FM.

In this way, the objectives of this systematic review were to analyse and systematise the characteristics of combined training programs (i.e., type and duration of interventions, weekly frequency, duration and structure of training sessions and prescribed intensities) and to analyse their effects on people diagnosed with FM.

2. Materials and Methods

This systematic review was performed following Preferred Reporting Items for Systematic Reviews and Meta-Analyses (PRISMA) 2020 [27] and the guidelines for performing systematic reviews in sports sciences [28]. The systematic review protocol was a priori registered in the OSF platform with the associated project number osf.io/v37s4.

2.1. Eligibility Criteria

The studies included in the present systematic review had the following inclusion criteria: (i) participants ≥ 18 years old with FM and autonomy, without other diseases (e.g., diabetes, hypertension and/or cardiovascular diseases); (ii) studies with combined training programs (aerobic and strength) with duration ≥4 weeks; (iii) exercise training programs supervised by a multidisciplinary team including a fitness exercise professional; (iv) randomised clinical trials; (v) studies written in English because it is the universal language.

The following items were considered the exclusion criteria: (i) participants < 18 years old with other diseases (e.g., diabetes, hypertension and/or cardiovascular diseases); (ii) studies with durations lower than 4 weeks and/or without combined training; (iii) studies written in other languages than English; (iv) other studies than randomised clinical trials.

2.2. Information Sources and Search Strategy

A systematic search of three databases (Web of Science, PubMed and EBSCO) was performed until 14 September 2022. Additionally, a manual search on the references of the included articles was also performed.

The search strategy included the Boolean AND/OR and the following keywords: "fibromyalgia" AND "concurrent training" OR "combined training" OR "cross training". The search strategy and their specificities from each database are presented in Table 1.

Table 1. The complete search strategy for each database.

Database	Specificities of the Databases	Search Strategy	Number of Articles in Automatic Research
PubMed	Search for title and abstract also includes keywords	("fibromyalgia") AND ("concurrent training" OR "combined training" OR "cross training"	280
Web of Science	Search for title and abstract also includes keywords	("fibromyalgia") AND ("concurrent training" OR "combined training" OR "cross training"	160
EBSCO	Search for title and abstract also includes keywords	("fibromyalgia") AND ("concurrent training" OR "combined training" OR "cross training"	0

2.3. Selection and Data Collection Processes

All articles found by the aforementioned search strategy were evaluated by two authors (M.S. and A.D.M.) through titles and their abstracts in order to exclude duplicates and articles that did not meet the inclusion criteria. In addition, the abstracts that did not provide enough information were selected for a complete evaluation of the full article. In a second phase, the two authors evaluated all the articles in full to carry out a second selection according to the inclusion criteria. The lack of consensus between the two investigators was resolved in a meeting with the third investigator (R.O.). Then, M.S. and R.O. extracted the data while J.P.B reviewed the process.

2.4. Data Items

The following data were extracted from the selected articles: population characteristics such as sample size, sex, age, years of diagnosed FM, country, body mass index (BMI); intervention: characteristics of combined training programs (i.e., exercises and materials; weekly frequency and duration of training programs; intensity; sets and repetitions); outcomes: instruments/tools (type, manufacturer and questionnaires); aim and main results of the studies.

2.5. Study Risk-of-Bias Assessment

The Physiotherapy Evidence Database (PEDro) scale was applied to assess the risk of bias of the included studies. This PEDro scale was previously validated and its reliability confirmed [29]. The PEDro scale rates eleven criteria topics, in which 10 classify the overall score of the article, ranging from 0 (lowest quality) to 10 (highest quality). The classification of the scores was the following: "poor" (<4 points), "fair" (4–5 points), "good" (6–8 points) and "excellent" (9–10 points). Two authors (A.D.M. and R.O.) independently reviewed and rated the included articles, based on the PEDro scale. Then, the same authors shared the scores and discussed them on a point-by-point basis. When a consensus was not reached, a third author (J.P.B) was invited to its classification to make a final decision.

2.6. Certainty Assessment

Based on the physiotherapy evidence database scale, Tulder et al.'s [30] criteria were applied to assess the interventions' evidence. Thus, a study with a physiotherapy evidence database score of $\geq$6 is considered level 1 (high methodological quality) (6–8: good, 9–10: excellent) and a score of 5 or less is considered level 2 (low methodological quality) (4–5: moderate; <4: poor).

Due to the clinical and statistical heterogeneity of the results, a qualitative review was performed, conducting a best-evidence synthesis [31,32]. This classification indicates that if the number of studies displaying the same level of evidence for the same outcome measure or equivalent is lower than 50% of the total number of studies, no evidence can be concluded regarding any of the methods involved in the study.

3. Results

3.1. Study Identification and Selection

A total of 335 articles were found across the three databases. All studies were exported using reference management software (EndNoteTM 20.0.1, Clarivate Analytics, Philadelphia, PA, USA). A total of 105 duplicate articles were recorded and subsequently removed. The remaining 230 articles were analysed by their titles and abstracts, and when insufficient information was available, the article was read in full, resulting in the removal of 288 articles deemed not to be in the scope on this review. Finally, after a complete reading of all articles, 34 more articles were excluded for not meeting the eligibility criteria. Thus, 13 articles were included in this systematic review (Figure 1).

3.2. Study Characteristics

The sample of articles selected for this systematic review included 13 studies published between 2000 and 2020. The studies covered an adult population diagnosed with FM and females with ages ranging from 30 to 59 years. There were different exercise interventions such as combined training, high-intensity interval training, Tai Chi, aerobic exercise, body balance and strength training. Moreover, several instruments/tests were used to assess pain, sleep quality, health status and strength gains in the upper and lower limbs. The characteristics of the articles included in the systematic review are presented in Table 2.

3.3. Risk of Bias in Studies

Table 3 presents the assessment of risk of bias (PEDro scale). The criteria with lower scores were related to the blinding of all participants and blinding of all persons who administered the training protocols. Moreover, no study was classified with poor methodological quality.

3.4. Intervention Characteristics

The characteristics of the interventions exercise programs are presented in Table 4. Regarding the exercises included, only main phases of each training have been reported in the table.

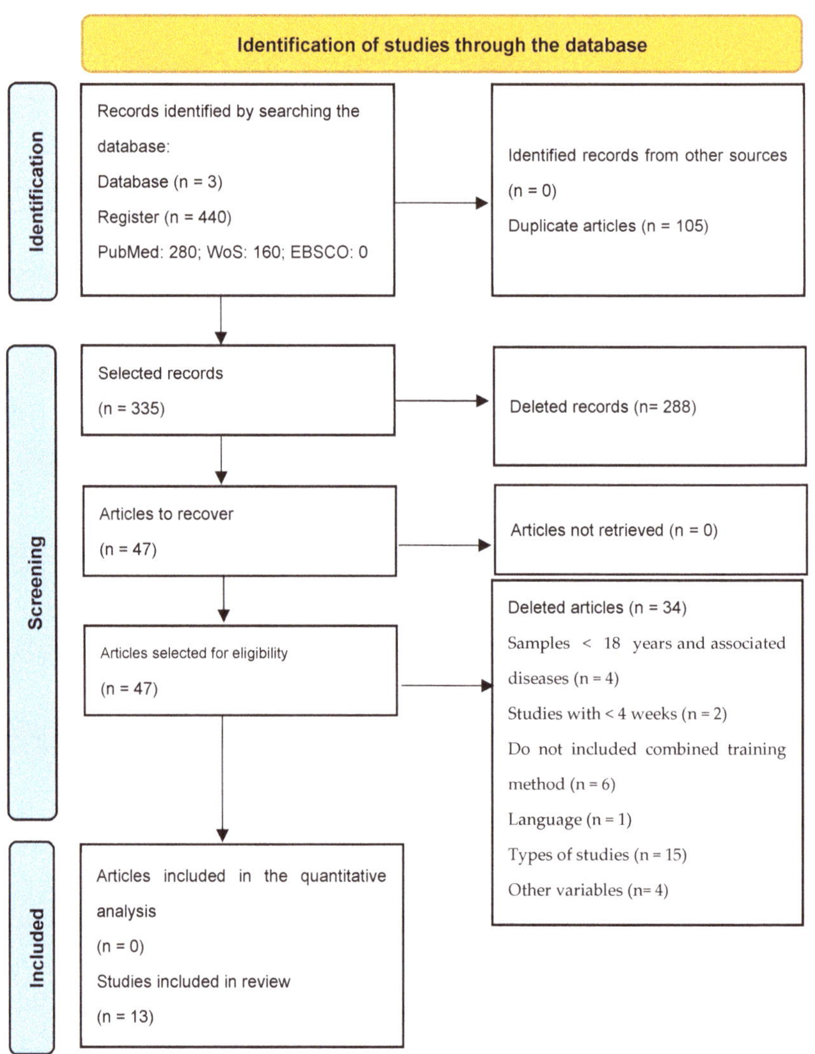

Figure 1. Flowchart of the Systematic Review.

Table 2. Characteristics of the articles included in the systematic review.

Author (Year)	Country	Objectives	Participants by Gender (N)	Age (M ± SD)	Years of Diagnosis	Instruments/Tests/Evaluation Tools and Variables
Gulsen et al., 2020 [33]	Turkey	To evaluate the effects of training combined with immersive virtual reality	N = 16; EG = 8 EG + Immersive Virtual Reality = 8	EG = 38.5 (29.5–50.0) EG + Immersive Virtual Reality = 46.5 (36.5–49.5)	EG = 4 (2–7.5) EG + Immersive Virtual Reality = 4 (2.5–8)	VAS: Pain Biodex Balance System (Shirley, NY, USA): Balance FIQ: Impact of FM Questionnaire IPAQ: Levels of PA Questionnaire 6-MWT: Aerobic Capacity SF-36: Health status of population Questionnaire
Atan and Karavelioglu et al., 2020 [1]	Turkey	To compare high-intensity interval training versus a combined training of continuous moderate intensity and strength plus stretching exercises	N = 45 HIIT Group = 19 MICT Group = 19 CG = 17	HIIT Group = 46.5 ± 9.4 MICT Group = 47.3 ± 8.0 CG = 52.7 ± 8.9	HIIT Group = 3.1 MICT Group = 2.0 CG = 2.3	FIQ VAS: Pain SF-36: Health status of population Questionnaire Maximal Cardiopulmonary Exercise Test InBody 720, Biospace: Body composition (Weight, waist circumference and BMI)
Wang et al., 2018 [17]	United States of America	To evaluate the effects of Tai Chi protocol versus aerobic exercise	N = 226 Tai Chi Group = 151 1 × 12 weeks = 39 2 × 12 weeks = 37 1 × 24 weeks = 39 2 × 24 weeks = 36 AEG 2 × 24 weeks = 75	Tai Chi Group: 1 × 12 weeks = 53 ± 12.6; 2 × 12 weeks = 52,1 ± 10.3; 1 × 24 weeks = 50.8 ± 11.8; 2 × 24 weeks = 52.1 ± 13.3 AEG 2 × 24 weeks = 50.9 ± 12.5	Tai Chi Group: 1 × 12 weeks = 11.1 ± 8.6; 2 × 12 weeks = 12.6 ± 12.1; 1 × 24 weeks = 12 ± 8.3; 2 × 24 weeks = 13.8 ± 10.4; AEG 2 × 24 weeks M = 11.3 ± 8.7	FIQ VAS: Pain Depression and Anxiety Questionnaire Pittsburgh Sleep Quality Questionnaire BDI-II: Depression and behavioural manifestations Questionnaire
Celenay et al., 2017 [15]	Germany	To compare the effectiveness of a 6-week combined exercise program with and without CMT on pain, fatigue, sleep problems, health status and quality of life	N = 20 EG N = 20 EG + connective issue massage = 20	EG = 39.9 ± 9.5 $\bar{x}$ = 42.5 ± 8.3	ND	IPAQ-7: Levels of PA Questionnaire VAS: Pain Sleep: Quality of sleep Questionnaire FIQ: Impact of FM Questionnaire SF-36: Health Status population Questionnaire

Table 2. *Cont.*

Author (Year)	Country	Objectives	Participants by Gender (N)	Age (M ± SD)	Years of Diagnosis	Instruments/Tests/Evaluation Tools and Variables
Sañudo et al., 2013 [34]	Spain	To determine the effect of body balance and dynamic strength of an exercise program complemented with WBV	N = 46 EG + WBV (WBVEX) = 15 EG = 15 CG = 16	EG + WBV = 57.1 ± 6.8 EG = 62.2 ± 9.8 CG = 55.5 ± 7.9	EG + WBV = 8.5 ± 7.4 EG = 9.2 ± 8.3 CG = 8.8 ± 8.2	Biodex F1C Stability System (BSS; Biodex, Inc. Shirley, NY, USA): Body Balance and Lower Limb Dynamic Strength
Sañudo et al., 2012 [35]	Spain	To analyse the effects of balance and strength through an exercise training program combined with WBV	N = 30 EG + WBV = 15 EG = 15 CG = 16	$\bar{x}$ = 59 ± 7.9	ND	Biodex Stability System (BSS, Biodex, Inc., Shirley, NY): Body Balance The Galileo Fitness Platform (Novotech, Germany): Evaluation of knee extensor muscle strength
Romero-Zurita et al., 2012 [36]	Spain	To analyse the effects of Tai Chi training in women	N = 23	$\bar{x}$ = 51.4 ± 6.8	ND	Body composition and anthropometric measurements: Weight, Waist Circumference, BMI FIQ SF-36: Health Status population Questionnaire Depression and Anxiety Questionnaire Vanderbiet Pain Management Inventory: Copping strategies Rosenbery Self-Esteem Scale: global self esteem General Self-Efficacy Scale: Beliefs in her/his own capabilities to attain aims
Sañudo et al., 2011 [37]	Spain	To analyse the effects on perceived health status, functional capacity and depression of a long-term exercise program versus usual care	N = 42 EG = 21 Usual Care CG = 21	EG = 55.4 ± 7.1 Usual Care CG = 56.1 ± 8.4	ND	FIQ SF-36: Health Status Population Questionnaire BDI: Attitudes and symptoms of stress Questionnaire

Table 2. *Cont.*

Author (Year)	Country	Objectives	Participants by Gender (N)	Age (M ± SD)	Years of Diagnosis	Instruments/Tests/Evaluation Tools and Variables
Sañudo et al., 2010 [38]	Spain	To determine the effects of supervised aerobic exercise and a supervised exercise program combined with aerobic exercise, strength and flexibility	N = 64 AEG2 CTG = 21 CG N = 21	AE group = 55.9 ± 1.6; CTG M = 55.9 ± 1.7; CG = 29.7 ± 1.1	ND	FIQ SF-36: Health Status Population Questionnaire BDI: Attitudes and symptoms of stress Questionnaire 6-MWT: Aerobic Capacity Hand-grip strength: Measure of muscular strength or the maximum force/tension by forearm muscles Flexion and extension (shoulders and hips): degrees
Carbonell-Balza et al., 2010 [39]	Spain	To analyse the effects on pain, body composition and physical fitness of a multidisciplinary intervention in women.	N = 75 EG = 41 CG = 34	EG = 50 ± 7.3 CG = 51.4 ± 7.3	ND	InBody 720; Biospace, Gateshead, UK: Body fat and muscle mass Functional Fitness Test Battery: lower and upper body strength and flexibility
Valkeineu et al., 2008 [23]	Finland	To determine the effects on muscle strength, aerobic and functional performance on postmenopausal symptoms of a combined strength and resistance training in women	N = 26 EG = 15 CG = 11	EG = 59 ± 3 CG = 58 3	ND	Health Assessment Questionnaire: Self-report functional status (disability) measures VO2 peak: Maximum oxygen carrying capacity with a bicycle ergometer test
King et al., 2002 [40]	Canada	To examine the effectiveness of a supervised aerobic exercise program, a self-management education program, and an exercise and education program for women	N = 152 EG = 46 Education group = 48 Exercise and Education Group = 37 CG = 39	EG = 45.2 ± 9.4 Education group = 44.9 ± 10.0 Exercise and Education Group = 47.4 ± 9.0 CG = 47.3 ± 7.3	EG= 7.8 ± 6.1 Education group= 10.9 ± 10.7 Exercise and Education Group = 8.9 ± 7.3 CG = 9.6 ± 7.9	FIQ 6-MWT: Aerobic Capacity

Table 2. *Cont.*

Author (Year)	Country	Objectives	Participants by Gender (N)	Age (M ± SD)	Years of Diagnosis	Instruments/Tests/Evaluation Tools and Variables
Mannerkorpi et al., 2000 [41]	Sweden	To determine the effects of a pool-based exercise training program combined with an education program	N = 69 EG = 37 CG = 32	EG = 45 ± 8.0 CG = 47 ± 11.6	EG = 8.9 ± 7.2 CG = 8.4 ± 6.0	FIQ 6 MW: Aerobic Capacity SF-36: Health Status population Questionnaire Arthritis Self Efficacy Scales: Pain and activities of daily living Questionnaire Arthritis Impact Measures Scales: Weight-bearing, posture and antigravity movement Questionnaire Quality of Life Questionnaire: Individual's physical, psychological and social well-being Questionnaire

$\bar{x}$: mean; RCT: Randomised Controlled Trial; VAS: Visual Analog Scale; FIQ: Fibromyalgia Impact Questionnaire; IPAQ: International Physical Activity Questionnaires; 6-MWT: 6-Minute Walk Test; SF-36: Short-Form 36; BDI-II: Beck Depression Inventory-II; BMI: body mass index; FM: Fibromyalgia; PA: Physical Activity; ND: Non-Described; EG, Exercise Group; AEG, Aerobic Group; CTG, Combined training Group; HIIT Group, High-Intensity Interval Training Group; MICT, moderate-intensity continuous training; CG, Control Group; WBV: Whole-Body Vibration; WBVEX: Whole-Body Vibration Exercise Group.

Table 3. Risk of bias assessment (PEDro scale)

Study	PEDro Scale											Total Score	Methodological Quality
	C1	C2	C3	C4	C5	C6	C7	C8	C9	C10	C11		
Gulsen et al. [33]	1	1	0	1	0	0	1	0	0	1	1	5	Moderate
Atan and Karavelioglu et al. [1]	1	1	1	1	0	0	1	1	0	1	1	7	Good
Wang et al. [17]	1	1	1	1	0	0	1	0	1	1	1	7	Good
Celenay et al. [15]	1	1	1	1	0	0	0	0	0	1	1	5	Moderate
Sañudo et al. [34]	1	1	1	1	0	0	0	1	1	1	1	7	Good
Sañudo et e al. [35]	1	1	1	1	0	0	0	1	0	1	1	6	Good
Romero-Zurita et al. [36]	1	1	1	1	0	0	0	1	1	1	1	8	Good
Sañudo et al. [37]	1	1	1	1	0	0	0	1	1	1	1	8	Good
Sañudo et al. [38]	1	1	1	0	0	0	0	1	1	1	1	6	Good
Carbonell-Balza et al. [39]	1	1	1	1	1	0	0	1	1	1	1	9	Excellent
Valkeineu et al. [23]	1	1	0	1	0	0	1	1	0	1	1	6	Good
King et al. [40]	1	1	0	1	0	0	1	0	1	1	1	6	Good
Mannerkorpi et al. [41]	1	1	0	1	0	0	1	0	0	1	1	5	moderate

C1: eligibility criteria were specified; C2: participants were randomly allocated to groups; C3: allocation was concealed; C4: the groups were similar at baseline regarding the most important prognostic indicators; C5: there was blinding of all participants; C6: there was blinding of all therapists who administered the therapy; C7: there was blinding of all assessors who measured at least one key outcome; C8: measures of at least one key outcome were obtained from more than 85% of the participants initially allocated to groups; C9: all participants for whom outcome measures were available received the treatment or control condition as allocated, or, where this was not the case, data for at least one key outcome were analysed according to "intention to treat"; C10: the results of between-group statistical comparisons are reported for at least one key outcome; C11: the study provides both point measures and measures of variability for at least one key outcome. Note: C1 values do not count for the total score.

Table 4. Characteristics of the Exercise Programs Present in the Systematic Review.

Author (Year)	Exercises	Frequency Program Length and Duration of Sessions	Intensity	Sets (N); Reps (N); Rest	Results
Gulsen et al., 2020 [33]	AEG: Treadmill; Pilates and IVR	Frequency: 2 × week Program length: 8 weeks Session duration: 80 min	AE = 60–80% HRmax	Non-Described	After the intervention, there were significant improvements in the exercise and Immersive Virtual Reality groups in pain, balance, impact of FM, fatigue, level of PA, functional exercise capacity and quality of life ($p < 0.05$). Exercise + IVR groups showed more improvements than exercise group in pain, fatigue, PA level, mental component and quality of life ($p < 0.05$).
Atan and Karavelioglu et al., 2020 [1]	HIIT—Cycle Ergometer MICT—Shoulder press with dumbbells or on machine; shoulder raises; bicep curl; squats; standing hip flexion and extension.	Frequency: 5 × week Program length: 6 weeks Time of the HIIT session—35 min; MICT—55 min Control Group—without exercise	HIIT: 4 min sets at 80–95% of peak HR interspersed with three 3 min of active recovery intervals at 70% of peak HR and 5 min of return to calm at 50% of peak HR; MICT: 45 min of ST performed for the main muscle groups.	MICT: 1 set, 8–10 reps; HIIT: 4 sets of 4 min with 3 sets of 3 min of active recovery intervals and 5 min cool down period cycling	Group-time interactions were significant for the FIQ between interventions and control ($p < 0.001$). There were significant group-time interactions for the pain, SF-36 and cardiopulmonary exercise test parameters between treatments and control (all, $p < 0.05$). Body weight, fat percentage, fat-mass and BMI improved significantly (all, $p < 0.05$) only in MICT group after treatment.
Wang et al., 2018 [15]	Tai Chi: choreographed AE	Frequency: 1 or 2× week (Tai Chi and EA) Program: 52 weeks Session duration: 60 min	AE: 20 min of AE 50–60% HRmax and 11–13 on the RPE. From the 10th session, AE progressed to 60–70% of HRmax	Non-Described	Tai Chi groups improved significantly more than AEG in FIQ, anxiety and self-efficacy at 24 weeks either training 1 or 2 times per week (all, $p < 0.05$). Tai Chi groups compared with AE administered with the same intensity and duration (24 weeks, 2 times per week group) had greater benefits in FIQ $p < 0.001$. The groups who received Tai Chi for 24 weeks showed greater improvements in FIQ than Tai Chi Group of 12 weeks ($p = 0.007$).

Table 4. *Cont.*

Author (Year)	Exercises	Frequency Program Length and Duration of Sessions	Intensity	Sets (N); Reps (N); Rest	Results
Celenay et al., 2017 [15]	AE: walk on the treadmill for 20 min; ST: Deep neck muscles; deltoid; latissimus dorsi; pectoralis; scapular retractor muscles; external rotators of the shoulder; erector spinae; abdominals; and gluteus muscles.	Frequency: 2 × week Program duration: 6 weeks Session duration: 60 min	AE: 65–70% HRmax and then 75–80% HRmax; ST: Light/medium band and progress to a strong band	1 set ST: 10 reps with a progression to 15 reps;	In the Exercise + connective tissue massage group, pain, fatigue and sleep problems decreased; health status and quality of life improved ($p < 0.05$). Exercises with connective tissue massage were superior in improving pain, fatigue, sleep problems and role limitations due to physical health compared to exercise alone. General health perceptions parameters related to quality of life improved in the Exercise group than in the connective tissue massage group ($p < 0.05$).
Sañudo et al., 2013 [34]	Exercise Group: 10–15 min of AE. This is followed by 15–20 min of ST. Exercise Group + WBV: stood on the platform on both legs, with both knees in isometric 120° flexion. Exercises such as unilateral static squats were used.		AE: 65–70% HRmax WBV: Vibration frequency of 30 Hz and at a peak-to-peak displacement of 4 mm ($71.1 \, \text{m/s}^{-2} \approx 7.2 \, \text{g}$).	ST: 1 set; 8–10 reps. WBV: 6 sets in the exercises in which the participants stand on the platform and 4 sets in the isometric 30 s exercises with 45 s of recovery.	There were no between-group differences in any outcome measures (all p-values > 0.05), except for MLMD with open eyes between both experimental groups ($p = 0.02$). The 8-week intervention of exercises and WBV resulted in a statistically significant improvement in MLS I, and significant differences for the WBVEX over the EX group ($p = 0.014$) and over the CG ($p = 0.029$).
Sañudo et e al., 2012 [35]	Exercise Group: combination of AE and ST. 10–15 min of AE; 15–20 min of ST Galileo Fitness_platform: stand up with both knees in isometric flexion plus unilateral static squats.	Frequency: 2 × week + 3 additional WBV sessions Program duration: 6 weeks Session duration: 45 min	Exercise Group: AE 65–70% HRmaxGalileo Fitness_platform: frequency of 20 Hz and variable amplitude of 2–3 mm.	ST: 1 set; 8 exercises; 8–10 reps; Galileo Fitness_platform: 3 sets of 45 s with a recovery of 120 s between sets and 4 sets of 15 s. The participants completed 15 s of the exercise on the right leg and then immediately completed 15 s on the left leg, and this was considered a set.	Exercise Groups of Medio-Lateral Stability Index improved balance when participants were assessed with eyes open and closed (all, $p < 0.05$).

Table 4. *Cont.*

Author (Year)	Exercises	Frequency Program Length and Duration of Sessions	Intensity	Sets (N); Reps (N); Rest	Results
Romero-Zurita et al., 2012 [36]	Yang style Tai-Chi forms,	Frequency: 3 × week Program: 28 weeks Session duration: 60 min	The average RPE value was 11 ± 1.	8 forms of Tai-Chi, Yang style	Patients showed improvements in pain threshold and total number of tender points (all, $p < 0.001$). Tai-Chi group improved the FIQ total score ($p < 0.001$) and six subscales: stiffness ($p = 0.005$), pain, fatigue, morning tiredness, anxiety and depression (all, $p < 0.001$). The intervention was also effective in six SF-36 subscales: bodily pain ($p = 0.003$), vitality ($p = 0.018$), physical functioning, physical role, general health and mental health (all, $p < 0.001$)
Sañudo et al., 2011 [37]	Exercise Group: muscle-strengthening exercises, consisting of a circuit of 8 exercise stations (shoulder press; shoulder raises; biceps curl; squats; hip flexion and extension; and standing abductors).	Frequency: 2 × week Program: 24 weeks Session duration: 60 min	AE: 65–70% HRmax	ST: 1 set, at each station 8–10 reps with dumbbells 1–3 kg;	Improvements in the Medio–Lateral Stability Index and Medio–Lateral Mean Deflection with open eyes were found in the whole-body vibration exercise group compared with the control group ($p = 0.02$).
Sañudo et al., 2010 [38]	AE: 15–20 min continuous walking with arm movements, aerobic dancing and jogging Combined exercise: 10–15 min AE; jogging; 15–20 min of muscle strengthening for 8 muscle groups: deltoids, biceps, neck, hips, back and chest.	Frequency: 2 × week Program: 24 weeks Session duration: 60 min	AEG: between 60% and 65% HRmax Combined Exercise: AE 65–70% HRmax	EA Group: 1 series for muscle strengthening and flexibility exercises. 6 exercises of 1.5 min of AE Combined Exercise: 1 series for muscle strengthening and flexibility exercises. Muscle strengthening consisting of 8–10 reps with a load between 1 and 3 kg and flexibility: 3 reps for 8–9 exercises, holding the static position for 30 s.	An improvement from baseline in total FIQ score was observed in the exercise groups ($p < 0.002$) and it was accompanied by decreases in BDI scores of 8.5 ($p < 0.001$) and 6.4 ($p < 0.001$) points in the AE and CE groups, respectively. Relative to non-exercising controls, CE evoked improvements in the SF-36 physical functioning ($p = 0.003$) and bodily pain ($p = 0.003$) domains and it was more effective than AE for evoking improvements in the vitality ($p = 0.002$) and mental health ($p = 0.04$) domains. Greater improvements were observed in shoulder/hip range of motion and handgrip strength in the CE group.

Table 4. *Cont.*

Author (Year)	Exercises	Frequency Program Length and Duration of Sessions	Intensity	Sets (N); Reps (N); Rest	Results
Carbonell-Balza et al., 2010 [39]	1st session, pool resistance exercises developed at a slow pace using water and aquatic materials as a means of resistance; 2nd session, pool-balance-oriented activities: position changes, walking backwards, coordination through aquatic exercises and dance exercises; 3rd session: land-based AE and coordination through an exercise circuit and 90 min of psychological follow-up.	Frequency: 3 × week Program: 12 weeks Session duration: 45 min of physical exercise classes; 90 min of psychological support	The average RPE was 12–13 AU.	Non-Described	A significant group×time effect for the left (L) and right (R) side of the anterior cervical ($p < 0.001$) and the lateral epicondyle R ($p = 0.001$) tender point. Pain threshold increased in the intervention group (positive) in the anterior cervical R ($p < 0.001$) and L ($p = 0.012$), and in the lateral epicondyle R ($p = 0.010$), whereas it decreased (negative) in the anterior cervical R ($p < 0.001$) and L ($p = 0.002$) in the usual care group.
Valkeineu et al., 2008 [23]	Concurrent training: in the 1st week, participants performed 2 strength training sessions and 1 endurance training session and in the 2nd week, 1 strength training and 2 endurance training sessions, and vice versa on alternate weeks. Strength training included isometric leg extension, concentric leg extension, elbow flexion and trunk flexion and extension.	Frequency: 3 × week Program: 21 weeks of combined training (strength and endurance training) Session duration: 60–90 min	Bicycle: 50 W with a load progression of 20 W until exhaustion Strength: 1 RM.	Bicycle: 3 min of heating with an intensity of 50 W and the load was increased to 20 W until exhaustion ST: 3 sets of exercises for legs, hips and knees and elbow flexors. Re-accelerate to maximum force for 3 to 5 s/1 min rest between exercises.	The concurrent training showed higher values in Wmax ($p = 0.001$), work time ($p = 0.001$), concentric leg extension force ($p = 0.043$), walking ($p = 0.001$), stair-climbing ($p < 0.001$) time and fatigue ($p = 0.038$) than strength training. The training led to an increase of 10% ($p = 0.004$) in Wmax and 13% ($p = 0.004$) than control group.
King et al., 2002 [40]	Education Group was based on self-management principles, where information was given about FM and individual goals and strategies for dealing with pain or other symptoms were established, guiding the participants toward a balanced and active life. Exercise Group included AE walking in deep or shallow water or low-impact AE such as walking outdoors. Education + EG was a combination of the previous protocols.	Frequency: 3 × week Program: 12 weeks Session duration: Exercise Group: 40 min Education Group: 60 min	AE: 60–75% HRmax	Non-Described	Only Exercise Group showed higher distance in the Six-Minute-Walk test ($p = 0.04$) when compared with Education Group.

Table 4. *Cont.*

Author (Year)	Exercises	Frequency Program Length and Duration of Sessions	Intensity	Sets (N); Reps (N); Rest	Results
Mannerkorpi et al., 2000 [41]	Exercise program in a heated pool and included resistance, flexibility, coordination and relaxation exercises.	Frequency: 1 × week Program: 24 weeks Session duration: 35 min	ND	Non-Described	Significant differences between the treatment group and the control group were found for the FIQ total score ($p = 0.017$) and the 6 min walk test ($p < 0.001$). Significant differences were also found for physical function, grip strength, pain severity, social functioning, psychological distress and quality of life for the treatment group.

VAS: Visual Analog Scale; FIQ: Fibromyalgia Impact Questionnaire; IPAQ: International Physical Activity Questionnaires; 6-MWT: 6-Minute-Walk Test; SF-36: Short-Form 36; CPET: Cardiopulmonary Exercise Test; BDI-II: Beck Depression Inventory-II; BMI: body mass index; PA: Physical Activity; EG, Exercise Group; AEG, Aerobic Group; CTG, Combined Training Group; HIIT Group, High-Intensity Interval Training Group; MICT, moderate-intensity continuous training; CG, Control Group; HRmax: heart rate maximal; RPE: rating of perceived exertion; AE: aerobic exercise; ST: strength training.

4. Discussion

This systematic review aimed to analyse and systematise the characteristics of combined training programs and their effects in individuals diagnosed with Fibromyalgia. In the studies that were analysed, significant values were found for at least one of the evaluated parameters in all studies: (i) physical fitness tests [1,23,33,36,37,39–41]; (ii) decreased symptoms and impact of FM on participants [1,5,15,17,23,33,36,37,41]; (iii) lower limb strength [1,15,33]. These results are in line with other authors who claim that physical exercise programs are important stimuli with positive influence, attenuating the symptoms of the disease, through changes in the hypothalamic–pituitary–adrenal axis (HPA)—resulting in the release of neurotransmitters due to exercise and controlling and/or reducing localised pain [16,42]. The major findings were the improvement of health-related life quality, pain intensity, stiffness, fatigue, physical function, withdrawals and absence of adverse events [17,43–45]. It is relevant to point out that the thirteen studies in the present review provided moderate- to good-quality evidence for the mental dimension of the SF-36, VAS, FIQ, IPAQ, improvement in %fat, FM, BMI, 6-MWT and decrease in BDI-II, pain, fatigue and sleep. For instance, Bidonde et al. [43], who used a 15% threshold for calculation of the clinically relevant differences between experimental and control groups, reported that eight trials provided low-quality evidence for pain intensity, fatigue, stiffness and physical function, and moderate-quality evidence for withdrawals and HRQL at completion of the intervention (6 to 24 weeks).

Regarding exercise programs, previous research has shown that combined exercise programs have more positive effects compared to single-type exercises in FM [5,46]. Both the aerobic, strength and combined training protocols showed positive effects on the participants as also reported by Bidonde et al. [43]. Results showed that strength training, aerobic training and combined exercise programs resulted in favourable effects on FM symptoms [1,5,15,23,33,35,37,40,41]. Among these, the aerobic and combined interventions presented the highest effects on reducing the FIQ. Flexibility interventions, however, were not significant for reducing it [47–49]. Exercise programs lasting between 13 and 24 weeks and training sessions lasting no more than 60 min seemed to be associated with greater improvements in pain relief. Regarding the durations of the exercise programs, it seems that the programs longer than 6 weeks have more positive effects in this population. The association of aerobic exercises and Tai Chi also revealed positive effects on the symptoms presented for this population [5,23]. Aerobic exercises seem to be fundamental stimuli in exercise programs as they induce adaptations in several systems, namely in the cardiovascular, energetic, neuromuscular and neuroendocrine systems [43]. The latter allows an increase in serotonin and norepinephrine concentrations, with a consequent improvement in mood and greater physical well-being [16,47]. The norepinephrine and serotonin are involved in the modulation of arousal and mood and have been related to a variety of affective functions as well as associated clinical dysfunctions [50,51]. The norepinephrine modulates drive and energy and exerts a fine regulation of specific processes including learning, memory, sleep, arousal and adaptation [52]. Further, exercise appears to reduce serotonin transporter expression, increase serotonin levels and increase opioid levels in central inhibitory pathways, suggesting that exercise can reduce pain by utilising our endogenous inhibitory systems [50–52].

Regarding studies that evaluated the effects of aerobic exercise combined with Tai Chi exercises, it was found that participants who performed Tai Chi exercises improved in all assessment parameters, such as: impact of Fibromyalgia (FIQ); pain threshold; anxiety and depression (Hospital Anxiety and Depression Scale and BDI-II); and finally, sleep quality (Pittsburgh sleep quality index) compared to participants who performed only aerobic exercise, who had less improvement. It should also be noted that the combined Tai Chi and aerobic exercise groups improved the impact of Fibromyalgia (FIQ) more in the 24-week Tai Chi groups than in the 12-week group [17,36]. In addition to these parameters, there were also enhancements in functional capacity tests, such as sit and reach, and in the 6-MWT, in terms of depression and anxiety [36]. Regarding the type of exercise programs in order to reduce the symptoms of the disease, it appears that combined training programs are the most effective for this population.

Regarding the effects of exercise programs on muscle strength, the studies analysed in this systematic review found that the combined exercise of aerobics and strength promotes higher improvements in strength of concentric extension in the legs compared to the groups that performed only aerobic exercise or strength training [23,34,35,40]. Exercise programs including strength training, endurance and aerobics are accepted as a standard treatment protocol in FM [33,46,53,54]. There is growing evidence and a strong recommendation of using aerobic and strength training together for FM treatment [55].

The studies present different types of exercise, such as aerobics, strength, flexibility and combined interventions, but some studies point to the beneficial effects of using aquatic exercises. The properties of water and the physical activity performed in warm water seem to positively affect FM symptoms [39,41]. Consequently, these physiological changes can help to relax the muscles of the body [56]. In view of the benefits, the practice of physical exercises in water has been indicated to reduce symptoms such as pain, anxiety and depression [1].

As evidenced in the literature, the main symptoms of the disease are pain and constant immobilisation due to this fact [57]. As such, participants, when immobilised in their daily lives, lose muscle strength, which can lead to their disability and reduced quality of life [58]. Therefore, muscle strengthening exercises are essential for gaining muscle mass to generate the strength needed for daily tasks in this population. Nevertheless, it was also verified that a training protocol lasting up to 6 weeks did not generate greater benefits in relation to the evaluated parameters, such as the pain threshold, the impact of FM (FIQ), the functional capacity and the muscle strength gains that were greater between 14 and 24 weeks [15,34,39]. This evidence can be justified by the fact that this type of program is carried out with light to moderate intensity, due to the type of pathology and, as such, the results are not immediately observable, rather in the long term [52]. In addition to these parameters, the distance covered (6-MWT) improved significantly for participants in the exercise groups [14,33,40]. It is relevant to highlight that according to the ACSM [18], light aerobic activity is considered as <40% and moderate is 40 to 59% of the HR reserve.

However, in the study by Atan and Karavelioglu [1], high-intensity interval training (HIIT) plus strengthening and stretching exercises and moderate-intensity continuous training (MICT) plus strengthening and stretching exercises interventions showed significant improvements for FM effect, pain degree, functional capacity and quality of life compared to the control group. HIIT was not superior to MICT, but body composition parameters improved significantly only for the MICT group. It is important to draw attention to the fact that this study was the only one to use an intensity higher than 59% of the HR reserve and, even so, some improvements were not noted.

Light to moderate intensity programs are recommended for this type of population, which could be a reason why the HIIT group has not achieved improvements in pain symptoms [59]. However, it should be noted that when muscle strengthening exercises are combined with flexibility exercises, improvements were observed in the pain threshold, as well as in the participants' quality of life [34,35]. These improvements are justified by the fact that flexibility exercises are used as a way to increase the range of motion of one or more joints, reinforcing once again that these participants lose their mobility due to their constant mobilisation associated with pain [60]. An important point to consider is related to the intensity of strength training, since the protocols were not revealed. One study used six sets [30], while the other used just one set [31]. Even so, the number of repetitions followed the ACSM guidelines [18,61].

Studies that evaluated the effects of aerobic exercise combined with a physical exercise program for muscle strengthening and flexibility and the group that performed only supervised aerobic exercise demonstrated improvements in depression and in the quality of life associated with health in both groups. However, beneficial improvements in quality of life (SF-36) were observed in both exercise groups, with the supervised aerobic exercise group demonstrating improvements in the dimensions of physical and social functioning, while the supervised aerobic exercise group combined with muscle strengthening and flexibility exercises demonstrated improvements in the dimensions of physical functioning, body pain, vitality and mental health [5].

This study reviews the effects of combined training programs in individuals with FM, summarising recent reviews and describing new advances in the research related to progressive exercise regimens (aerobic, strength and flexibility interventions), and other forms of physical activity applied to FM (e.g., Tai Chi, Yoga and Pilates). However, since the population with FM presents heterogeneity, including different years of diagnosis, further investigation with regard to the short- and long-term response patterns to exercise and physical activity prescription may lead to a better customisation program in order to improve exercise adherence and optimise the benefits of exercise and physical activity.

For instance, six studies did not present the years of diagnosis [15,23,35–39], while the remaining studies ranged from 1 to 13.8 years [1,17,33,34,40,41]. Such information should be considered by fitness professional and exercise physiologists when interpreting the positive effects.

Thus, future research examining the effects of exercise and physical activity for people with FM are needed to elucidate the best dose–response curve for exercise intensity, frequency and duration on symptoms. The assessment of the long-term effects on health in this population and how long the positive effects are sustained should also be studied. Another aspect to consider in future studies is the race description of participants. From the 13 studies included, only one described that information [15]. Finally, when conducting a future systematic review study on this population, a meta-analysis would be interesting to strengthen the results of the same intervention type, which was not possible in the present study.

5. Conclusions

According to the studies included in this systematic review, it was concluded that the practice of physical exercise is, in general, beneficial and essential for improving the symptoms of the disease, physical fitness and functional capacity and also in terms of anxiety and depression.

Through the analysed studies, the most appropriate training protocol for women diagnosed with FM, with the aim of mitigating the various symptoms of the disease, should contain the following parameters:

(i)   Minimum duration of 14 weeks and never less than 6 weeks. Programs lasting 6 weeks did not have such positive effects in terms of the symptoms of the disease.

(ii)   Combined training programs are the most effective for this population, in order to reduce the symptoms of the disease. Training programs composed of aerobic exercises, strength training and stretching are the most indicated.

(iii)   Duration of sessions between 60 and 90 min, with the objective of executing the program outlined according to the limitations of each participant.

(iv)   Carry out the exercise program at least up to 3 times a week.

(v)   Aerobic exercises should be performed at 60–65% HRmax.

(vi)   Perform one set of exercises for large muscle groups (associated with pain points), consisting of 8 exercises and performing 8–10 repetitions in an initial phase, progressing to 15 repetitions. Rest at least 1–2 min between exercises.

(vii)   Perform static stretching exercises lasting 30–60 s for pain points.

(viii) The intensity of the program should be light to moderate, following the ACSM guidelines for aerobic exercise.

The results of the present systematic review demonstrate that exercise programs that include aerobic exercises, strength training and stretching exercises are the most beneficial to reduce the symptoms of the disease. Aerobic exercises are essential for the transport and use of oxygen; strength training is essential for gaining muscle mass, generating strength for the patients' daily tasks; and stretching is essential for their mobility.

6. Future Lines of Research

For future studies, a follow-up of the program is recommended. It would be interesting to evaluate the participants over a year, where several assessments would be carried out at different times regarding their physical fitness and the impact of FM. In this way, it would be possible to collect larger sample data for a better comparison between moments.

Author Contributions: Conceptualization: M.S., R.O., A.D.M. and S.A.; Methodology: M.S., R.O., A.D.M. and S.A.; Software: M.S., R.O., A.D.M. and S.A.; Validation: M.S., R.O., A.D.M., J.M. and S.A.; Formal Analysis: M.S. and S.A.; Investigation: M.S., R.O., A.D.M., J.M. and S.A.; Resources: M.S., R.O., A.D.M., J.M. and S.A.; Data Curation: M.S., J.P.B. and S.A.; Writing—Original Draft: M.S., R.O., A.D.M., J.M., J.P.B. and S.A.; Writing—Review and Editing: M.S., R.O., A.D.M., J.M., J.P.B. and S.A.; Visualization: M.S., R.O., A.D.M., J.M. and S.A.; Supervision: R.O. and S.A.; Project Administration: M.S. and S.A.; Funding Acquisition: R.O. and S.A. All authors have read and agreed to the published version of the manuscript.

Funding: This research was funded by the Portuguese Foundation for Science and Technology, I.P., Grant/Award Number 2021.04598.BD and UIDP/04748/2020.

Institutional Review Board Statement: Not applicable.

Informed Consent Statement: Not applicable.

Data Availability Statement: Not applicable.

Conflicts of Interest: The authors declare no conflict of interest. The funders had no role in the design of the study; in the collection, analyses or interpretation of data; in the writing of the manuscript or in the decision to publish the results.

References

1. Atan, T.; Karavelioğlu, Y. Effectiveness of High-Intensity Interval Training vs Moderate-Intensity Continuous Training in Patients With Fibromyalgia: A Pilot Randomized Controlled Trial. *Arch. Phys. Med. Rehabil.* **2020**, *101*, 1865–1876. [CrossRef] [PubMed]
2. O'Dwyer, T.; Maguire, S.; Mockler, D.; Durcan, L.; Wilson, F. Behaviour Change Interventions Targeting Physical Activity in Adults with Fibromyalgia: A Systematic Review. *Rheumatol. Int.* **2019**, *39*, 805–817. [CrossRef]
3. Chica, A.; Gonzalez-Guirval, F.; Reigal, R.E.; Carranque, G.; Hernandez-Mendo, A. Efectos de un programa de danza española en mujeres con fibromialgia. *Cuad. Psicol. Deporte* **2019**, *1*, 52–69. [CrossRef]
4. Norouzi, E.; Hosseini, F.S.; Vaezmosavi, M.; Gerber, M.; Pühse, U.; Brand, S. Zumba Dancing and Aerobic Exercise Can Improve Working Memory, Motor Function, and Depressive Symptoms in Female Patients with Fibromyalgia. *Eur. J. Sport Sci.* **2020**, *20*, 981–991. [CrossRef]
5. Chafer, S.; Hamilton, F. Efficacy of Physical Exercise Therapy as a Treatment for Fibromyalgia Patients. *J. Phys. Ther. Health Promot.* **2015**, *3*, 1–10. [CrossRef]
6. Busch, A.J.; Webber, S.C.; Brachaniec, M.; Bidonde, J.; Bello-Haas, V.D.; Danyliw, A.D.; Overend, T.J.; Richards, R.S.; Sawant, A.; Schachter, C.L. Exercise Therapy for Fibromyalgia. *Curr. Pain Headache Rep.* **2011**, *15*, 358–367. [CrossRef]
7. Chang, K.V.; Hung, C.H.; Sun, W.Z.; Wu, W.T.; Lai, C.L.; Han, D.S.; Chen, C.C. Evaluating Soreness Symptoms of Fibromyalgia: Establishment and Validation of the Revised Fibromyalgia Impact Questionnaire with Integration of Soreness Assessment. *J. Formos. Med. Assoc.* **2020**, *119*, 1211–1218. [CrossRef]
8. Cabo-Meseguer, A.; Cerdá-Olmedo, G.; Trillo-Mata, J.L. Fibromyalgia: Prevalence, Epidemiologic Profiles and Economic Costs. *Med. Clin.* **2017**, *149*, 441–448. [CrossRef]
9. Marques, A.P.; Santo, A.; de Sousa do Espírito Santo, A.; Berssaneti, A.A.; Matsutani, L.A.; Yuan, S.L.K. Prevalence of Fibromyalgia: Literature Review Update. *Rev. Bras. Reumatol.* **2017**, *57*, 356–363. [CrossRef]
10. Branco, J.C.; Rodrigues, A.M.; Gouveia, N.; Eusébio, M.; Ramiro, S.; Machado, P.M.; Pereira Da Costa, L.; Mourão, A.F.; Silva, I.; Laires, P.; et al. Prevalence of rheumatic and musculoskeletal diseases and their impact on health-related quality of life, physical function and mental health in Portugal: Results from EpiReumaPt—A national health survey. *RMD Open* **2016**, *2*, e000166. [CrossRef]
11. Wolfe, F.; Allen, M.; Bennett, R.M.; Bombardier, C.; Broadhurst, N.; Cameron, R.S.; Carette, S.; Chalmers, A.; Cohen, M.; Crook, J.; et al. The Fibromyalgia Syndrome: A Consensus Report on Fibromyalgia and Disability. *J. Rheumatol.* **1996**, *23*, 534–539.
12. Sarac, A.J.; Gur, A. Complementary and Alternative Medical Therapies in Fibromyalgia. *Curr. Pharm. Des.* **2006**, *12*, 47–57. [CrossRef] [PubMed]
13. Santos, J.M.; Mendonça, V.A.; Ribeiro, V.G.C.; Tossige-Gomes, R.; Fonseca, S.F.; Prates, A.C.N.; Flor, J.; Oliveira, A.C.C.; Martins, J.B.; Garcia, B.C.C.; et al. Does Whole Body Vibration Exercise Improve Oxidative Stress Markers in Women with Fibromyalgia? *Braz. J. Med. Biol. Res.* **2019**, *52*. [CrossRef]

14. Mannerkorpi, K.; Nordeman, L.; Cider, Å.; Jonsson, G. Does moderate-to-high intensity Nordic walking improve functional capacity and pain in fibromyalgia? A prospective randomized controlled trial. *Arthritis Res. Ther.* **2010**, *12*, 1–10. [CrossRef] [PubMed]

15. Celenay, S.; Anaforoglu Kulunkoglu, B.; Yasa, M.E.; Sahbaz Pirincci, C.; Un Yildirim, N.; Kucuksahin, O.; Ugurlu, F.G.; Akkus, S. A Comparison of the Effects of Exercises plus Connective Tissue Massage to Exercises Alone in Women with Fibromyalgia Syndrome: A Randomized Controlled Trial. *Rheumatol. Int.* **2017**, *37*, 1799–1806. [CrossRef] [PubMed]

16. Bonnabesse, A.; Cabon, M.; L'Heveder, G.; Kermarrec, A.; Quinio, B.; Woda, A.; Marchand, S.; Dubois, A.; Giroux-Metges, M.A.; Rannou, F.; et al. Impact of a Specific Training Programme on the Neuromodulation of Pain in Female Patient with Fibromyalgia (DouFiSport): A 24-Month, Controlled, Randomised, Double-Blind Protocol. *BMJ Open* **2019**, *9*, e023742. [CrossRef]

17. Wang, C.; Schmid, C.H.; Fielding, R.A.; Harvey, W.F.; Reid, K.F.; Price, L.L.; Driban, J.B.; Kalish, R.; Rones, R.; McAlindon, T. Effect of Tai Chi versus Aerobic Exercise for Fibromyalgia: Comparative Effectiveness Randomized Controlled Trial. *BMJ* **2018**, *360*, k851. [CrossRef]

18. Pescatello, L.S.; Riebe, D.; Arena, R. *American College of Sports Medicine. ACSM's Guidelines for Exercise Testing and Prescription*, 9th ed.; Lippincott Williams & Wilkins: Baltimore, MD, USA, 2014.

19. Gómez-Hernández, M.; Gallego-Izquierdo, T.; Martínez-Merinero, P.; Pecos-Martín, D.; Ferragut-Garcías, A.; Hita-Contreras, F.; Martínez-Amat, A.; Montañez-Aguilera, F.J.; Achalandabaso Ochoa, A. Benefits of Adding Stretching to a Moderate-Intensity Aerobic Exercise Programme in Women with Fibromyalgia: A Randomized Controlled Trial. *Clin. Rehabil.* **2020**, *34*, 242–251. [CrossRef]

20. De Souza, E.O.; Tricoli, V.; Roschel, H.; Brum, P.C.; Bacurau, A.V.N.; Ferreira, J.C.B.; Aoki, M.S.; Neves-Jr, M.; Aihara, A.Y.; Da Rocha Correa Fernandes, A.; et al. Molecular Adaptations to Concurrent Training. *Int. J. Sports Med.* **2013**, *34*, 207–213. [CrossRef]

21. Murlasits, Z.; Kneffel, Z.; Thalib, L. The Physiological Effects of Concurrent Strength and Endurance Training Sequence: A Systematic Review and Meta-Analysis. *J. Sports Sci.* **2018**, *36*, 1212–1219. [CrossRef]

22. Worrel, L.M.; Krahn, L.E.; Sletten, C.D.; Pond, G.R. Treating Fibromyalgia with a Brief Interdisciplinary Program: Initial Outcomes and Predictors of Response. *Mayo Clin. Proc.* **2001**, *76*, 384–390. [CrossRef] [PubMed]

23. Valkeinen, H.; Alén, M.; Häkkinen, A.; Hannonen, P.; Kukkonen-Harjula, K.; Häkkinen, K. Effects of Concurrent Strength and Endurance Training on Physical Fitness and Symptoms in Postmenopausal Women With Fibromyalgia: A Randomized Controlled Trial. *Arch. Phys. Med. Rehabil.* **2008**, *89*, 1660–1666. [CrossRef] [PubMed]

24. Larsson, A.; Palstam, A.; Löfgren, M.; Ernberg, M.; Bjersing, J.; Bileviciute-Ljungar, I.; Gerdle, B.; Kosek, E.; Mannerkorpi, K. Resistance Exercise Improves Muscle Strength, Health Status and Pain Intensity in Fibromyalgia-a Randomized Controlled Trial. *Arthritis Res. Ther.* **2015**, *17*, 161. [CrossRef] [PubMed]

25. Triñanes, Y.; González-Villar, A.; Gómez-Perretta, C.; Carrillo-de-la-Peña, M.T. Profiles in Fibromyalgia: Algometry, Auditory Evoked Potentials and Clinical Characterization of Different Subtypes. *Rheumatol. Int.* **2014**, *34*, 1571–1580. [CrossRef] [PubMed]

26. Albuquerque, M.; Alvarez, M.; Monteiro, D.; Esteves, D.; Neiva, H. *Exercise: Physical, Physiological and Psychological Benefits*; Nova Science Publishers: New York, NY, USA, 2021.

27. Page, M.J.; McKenzie, J.E.; Bossuyt, P.M.; Boutron, I.; Hoffmann, T.C.; Mulrow, C.D.; Shamseer, L.; Tetzlaff, J.M.; Akl, E.A.; Brennan, S.E.; et al. The PRISMA 2020 Statement: An Updated Guideline for Reporting Systematic Reviews. *Int. J. Surg.* **2021**, *88*. [CrossRef] [PubMed]

28. Rico-González, M.; Pino-Ortega, J.; Clemente, F.M.; Arcos, A.L. Guidelines for Performing Systematic Reviews in Sports Science. *Biol. Sport* **2022**, *39*, 463–471. [CrossRef]

29. Maher, C.G.; Sherrington, C.; Herbert, R.D.; Moseley, A.M.; Elkins, M. Reliability of the PEDro Scale for Rating Quality of Randomized Controlled Trials. *Phys. Ther.* **2003**, *83*, 713–721. [CrossRef]

30. van Tulder, M.; Furlan, A.; Bombardier, C.; Bouter, L. Updated Method Guidelines for Systematic Reviews in the Cochrane Collaboration Back Review Group. *Spine* **2003**, *28*, 1290–1299. [CrossRef]

31. Kollen, B.J.; Lennon, S.; Lyons, B.; Wheatley-Smith, L.; Scheper, M.; Buurke, J.H.; Halfens, J.; Geurts, A.C.H.; Kwakkel, G. The Effectiveness of the Bobath Concept in Stroke Rehabilitation What Is the Evidence? *Stroke* **2009**, *40*, e89–e97. [CrossRef]

32. Vaughan-Graham, J.; Cott, C.; Wright, F.V. The Bobath (NDT) Concept in Adult Neurological Rehabilitation: What Is the State of the Knowledge? A Scoping Review. Part II: Intervention Studies Perspectives. *Disabil. Rehabil.* **2015**, *37*, 1909–1928. [CrossRef]

33. Gulsen, C.; Soke, F.; Eldemir, K.; Apaydin, Y.; Ozkul, C.; Guclu-Gunduz, A.; Akcali, D.T. Effect of Fully Immersive Virtual Reality Treatment Combined with Exercise in Fibromyalgia Patients: A Randomized Controlled Trial. *Assist. Technol.* **2020**, *34*, 256–263. [CrossRef]

34. Sañudo, B.; Carrasco, L.; de Hoyo, M.; Oliva-Pascual-vaca, Á.; Rodríguez-Blanco, C. Changes in Body Balance and Functional Performance Following Whole-Body Vibration Training in Patients with Fibromyalgia Syndrome: A Randomized Controlled Trial. *J. Rehabil. Med.* **2013**, *45*, 678–684. [CrossRef]

35. Sañudo, B.; de Hoyo, M.; Carrasco, L.; Rodríguez-Blanco, C.; Oliva-Pascual-Vaca, Á.; McVeigh, J.G. Effect of Whole-Body Vibration Exercise on Balance in Women with Fibromyalgia Syndrome: A Randomized Controlled Trial. *J. Altern. Complement. Med.* **2012**, *18*, 158–164. [CrossRef]

36. Romero-Zurita, A.; Carbonell-Baeza, A.; Aparicio, V.A.; Ruiz, J.R.; Tercedor, P.; Delgado-Fernández, M. Effectiveness of a Tai-Chi Training and Detraining on Functional Capacity, Symptomatology and Psychological Outcomes in Women with Fibromyalgia. *Evid.-Based Complement. Altern. Med.* **2012**, *2012*. [CrossRef]

37. Corrales, B.S.; Galiano, D.; Carrasco, L.; de Hoyo, M.; McVeigh, J.G. Effects of a Prolonged Exercise Programe on Key Health Outcomes in Women with Fibromyalgia: A Randomized Controlled Trial. *J. Rehabil. Med.* **2011**, *43*, 521–526. [CrossRef]
38. Sañudo, B.; Galiano, D.; Carrasco, L.; Blagojevic, M.; de Hoyo, M.; Saxton, J. Aerobic Exercise versus Combined Exercise Therapy in Women with Fibromyalgia Syndrome: A Randomized Controlled Trial. *Arch. Phys. Med. Rehabil.* **2010**, *91*, 1838–1843. [CrossRef]
39. Carbonell-Baeza, A.; Aparicio, V.A.; Ortega, F.B.; Cuevas, A.M.; Alvarez, I.C.; Ruiz, J.R.; Delgado-Fernandez, M. Does a 3-Month Multidisciplinary Intervention Improve Pain, Body Composition and Physical Fi Tness in Women with Fibromyalgia? *Br. J. Sports Med.* **2011**, *45*, 1189–1195. [CrossRef] [PubMed]
40. King, S.J.; Wessel, J.; Bhambhani, Y.; Sholter, D.; Maksymowych, W. The effects of exercise and education, individually or combined, in women with fibromyalgia. *J. Rheumatol.* **2002**, *29*, 2620–2627. [PubMed]
41. Mannerkorpi, K. Pool Exercise Combined with an Education Program for Patients with Fibromyalgia Syndrome. A Prospective, Randomized Study The al-Andalus Project View Project. *J. Rheumatol.* **2000**, *27*, 2473–2481.
42. Doerr, J.M.; Fischer, S.; Nater, U.M.; Strahler, J. Influence of Stress Systems and Physical Activity on Different Dimensions of Fatigue in Female Fibromyalgia Patients. *J. Psychosom. Res.* **2016**, *93*, 55–61. [CrossRef]
43. Bidonde, J.; Busch, A.J.; Schachter, C.L.; Overend, T.J.; Kim, S.Y.; Góes, S.M.; Boden, C.; Foulds, H.J.A. Aerobic Exercise Training for Adults with Fibromyalgia. *Cochrane Datab. Syst. Rev.* **2017**, *6*. [CrossRef]
44. Häuser, W.; Klose, P.; Langhorst, J.; Moradi, B.; Steinbach, M.; Schiltenwolf, M.; Busch, A. Efficacy of Different Types of Aerobic Exercise in Fibromyalgia Syndrome: A Systematic Review and Meta-Analysis of Randomised Controlled Trials. *Arthritis Res. Ther.* **2010**, *12*, 1–14. [CrossRef] [PubMed]
45. Catalá, P.; Peñacoba, C.; López-Roig, S.; Pastor-Mira, M.A. Effects of Walking as Physical Exercise on Functional Limitation through Pain in Patients with Fibromyalgia—How Does Catastrophic Thinking Contribute? *Int. J. Environ. Res. Public Health* **2023**, *20*, 190. [CrossRef] [PubMed]
46. Okifuji, A.; Gao, J.; Bokat, C.; Hare, B.D. Management of Fibromyalgia Syndrome in 2016. *Pain Manag.* **2016**, *6*, 383–400. [CrossRef]
47. Busch, A.J.; Barber, K.; Schachter, C.L.; Overend, T.J.; Peloso, P.M. Exercise for Fibromyalgia: A Systematic Review. *J. Rheumatol.* **2008**, *35*, 1130–1144. [PubMed]
48. Dupree Jones, K. A Randomized Controlled Trial of Muscle Strengthening versus Flexibility Training in Fibromyalgia [Microform]. *J. Rheumatol.* **2002**, *29*, 1041–1048.
49. Matsutani, L.A.; Marques, A.P.; Ferreira, E.A.G.; Assumpção, A.; Lage, L.V.; Casarotto, R.A.; Pereira, C.A.B.; Matsutani, L.A.; Marques, A.P.; Ferreira, E.A.G.; et al. Effectiveness of Muscle Stretching Exercises with and without Laser Therapy at Tender Points for Patients with Fibromyalgia Stretching Exercises and Laser Therapy for FM Patients/L.A. Matsutani et al. *Clin. Exp. Rheumatol.* **2007**, *25*, 410–415.
50. Singleton, O.; Hölzel, B.K.; Vangel, M.; Brach, N.; Carmody, J.; Lazar, S.W. Change in Brainstem Gray Matter Concentration Following a Mindfulness-Based Intervention Is Correlated with Improvement in Psychological Well-Being. *Front. Hum. Neurosci.* **2014**, *8*, 33. [CrossRef]
51. Nutt, D.J. The Neuropharmacology of Serotonin and Noradrenaline in Depression. *Int. Clin. Psychopharmacol.* **2002**, *17* (Suppl. S1), S1–S12. [CrossRef]
52. Montgomery, S. Reboxetine: Additional Benefits to the Depressed Patient. *J. Psychopharmacol.* **1997**, *11* (Suppl. S4), S9–S15.
53. Alarab, A.; Taqatqa, N. Resistance Exercises for Musculoskeletal Disorders. In *Resistance Training*; InTech: London, UK, 2023. [CrossRef]
54. Winkelmann, A.; Bork, H.; Brückle, W.; Dexl, C.; Heldmann, P.; Henningsen, P.; Krumbein, L.; Pullwitt, V.; Schiltenwolf, M.; Häuser, W. Physiotherapy, Occupational Therapy and Physical Therapy in Fibromyalgia Syndrome: Updated Guidelines 2017 and Overview of Systematic Review Articles. *Schmerz* **2017**, *31*, 255–265. [CrossRef]
55. Macfarlane, G.J.; Kronisch, C.; Dean, L.E.; Atzeni, F.; Häuser, W.; Flub, E.; Choy, E.; Kosek, E.; Amris, K.; Branco, J.; et al. EULAR Revised Recommendations for the Management of Fibromyalgia. *Ann. Rheum. Dis.* **2017**, *76*, 318–328. [CrossRef]
56. Zamunér, A.R.; Andrade, C.P.; Arca, E.A.; Avila, M.A. Impact of Water Therapy on Pain Management in Patients with Fibromyalgia: Current Perspectives. *J. Pain Res.* **2019**, *12*, 1971–2007. [CrossRef] [PubMed]
57. Rodríguez-Mansilla, J.; Mejías-Gil, A.; Garrido-Ardila, E.M.; Jiménez-Palomares, M.; Montanero-Fernández, J.; González-López-arza, M.V. Effects of Non-Pharmacological Treatment on Pain, Flexibility, Balance and Quality of Life in Women with Fibromyalgia: A Randomised Clinical Trial. *J. Clin. Med.* **2021**, *10*, 3826. [CrossRef] [PubMed]
58. Leon-Llamas, J.L.; Villafaina, S.; Murillo-Garcia, A.; Collado-Mateo, D.; Domínguez-Muñoz, F.J.; Sánchez-Gómez, J.; Gusi, N. Strength Assessment under Dual Task Conditions in Women with Fibromyalgia: A Test–Retest Reliability Study. *Int. J. Environ. Res. Public Health* **2019**, *16*, 4971. [CrossRef]
59. Paolucci, T.; Baldari, C.; Di Franco, M.; Didona, D.; Reis, V.; Vetrano, M.; Iosa, M.; Trifoglio, D.; Zangrando, F.; Spadini, E.; et al. A New Rehabilitation Tool in Fibromyalgia: The Effects of Perceptive Rehabilitation on Pain and Function in a Clinical Randomized Controlled Trial. *Evid.-Based Complement. Altern. Med.* **2016**, *2016*. [CrossRef] [PubMed]

60. Busch, A.; Webber, S.; Richards, R.; Bidonde, J.; Schacter, C.; Schafer, L.; Danylwin, A.; Sawant, A.; Dal Bello-Haas, V.; Rander, T.; et al. Resistance exercise training for fibromyalgia. *Cochrane Datab. Syst. Rev.* **2013**, *12*, CD010884. [CrossRef] [PubMed]
61. Liguori, G.; Feito, Y.; Fountaine, C.F.; Roy, B.A. *ACSM'S Guidelines for Exercise Testing and Prescription*, 11th ed.; Wolters Kluwer: Philadelphia, PA, USA, 2021.

Article

Effects of 24 Weeks of a Supervised Walk Training on Knee Muscle Strength and Quality of Life in Older Female Total Knee Arthroplasty: A Retrospective Cohort Study

Wei-Hsiu Hsu [1,2,3,†], Wei-Bin Hsu [1,†], Zin-Rong Lin [4], Shr-Hsin Chang [1], Chun-Hao Fan [1], Liang-Tseng Kuo [1,2,3] and Wen-Wei Robert Hsu [1,2,*]

1 Sports Medicine Center, Chang Gung Memorial Hospital, No. 6, West Section, Chia-Pu Road, Chia-Yi County, Pu-Tz City 61363, Taiwan

2 Department of Orthopaedic Surgery, Chang Gung Memorial Hospital, No. 6, West Section, Chia-Pu Road, Chia-Yi County, Pu-Tz City 61363, Taiwan

3 School of Medicine, College of Medicine, Chang Gung University, No. 259, Wenhua 1st Road, Guishan District, Taoyuan City 33302, Taiwan

4 Department of Athletic Sports, National Chung Cheng University, No. 168, University Road, Minhsiung Township, Chia-Yi County 62102, Taiwan

* Correspondence: wwh@cgmh.org.tw; Tel.: +886-5-362-1000 (ext. 2000)

† These authors contributed equally to this work.

Abstract: Poor supervision, impaired exercise adherence, and low compliance with exercise regimens result in inconsistent effects regarding exercise interventions. A supervised-walk training regimen (9 km/week) may have a positive effect on functional recovery in female total knee arthroplasty (TKA). This study aimed to evaluate the effect of a supervised walking regimen on lower limb muscle strength, functional fitness, and patient-reported outcomes in female TKA. Twenty-eight female TKA were allocated into a control (CON) ($n = 14$) or walk training (WT) ($n = 14$) group. WT on treadmills was initiated 12 weeks after TKA. All patients were examined for lower muscle strength (including extension and flexion of hip and knee), physical function (including a 6-min walk test, 8-foot up-and-go test, and 30-s chair stand test), and Knee Injury and Osteoarthritis Outcome Score (KOOS) questionnaire. Knee flexor (WT: CON; 64.4 ± 4.1 nm/kg: 43.7±3.3 nm/kg; $p = 0.001$; effect size: 5.62) and extensor strengths (WT: CON; 73.1 ± 7.5 nm/kg: 48.2 ± 2.4 nm/kg; $p = 0.001$; effect size: 4.47) statistically increased in the WT group compared to the CON group. The 6-min walk test (from 341.3 ± 20.5 m to 405.5 ± 30.7 m; $p = 0.001$; effect size: 2.46) and 8-foot up-and-go test (from 9.5 ± 0.7 s to 8.3 ± 0.7 s; $p = 0.002$; effect size: 1.71) tests also showed significant improvements in the WT group in the follow-up compared to the baseline. An increase in quality of life score according to the KOOS questionnaire (WT: CON; 91.0 ± 2.8: 68.1 ± 5.8; $p = 0.001$; effect size: 5.02) was noted in the WT group compared to the CON group in the follow-up. WT facilitated improvements in knee muscle strength and functional outcomes in TKA patients.

Keywords: total knee arthroplasty; walk training; knee muscle strength; quality of life

Citation: Hsu, W.-H.; Hsu, W.-B.; Lin, Z.-R.; Chang, S.-H.; Fan, C.-H.; Kuo, L.-T.; Hsu, W.-W.R. Effects of 24 Weeks of a Supervised Walk Training on Knee Muscle Strength and Quality of Life in Older Female Total Knee Arthroplasty: A Retrospective Cohort Study. *Healthcare* **2023**, *11*, 356. https://doi.org/10.3390/healthcare11030356

Academic Editors: Rafael Oliveira and João Paulo Brito

Received: 5 December 2022
Revised: 21 January 2023
Accepted: 22 January 2023
Published: 26 January 2023

1. Introduction

Total knee arthroplasty (TKA) is a well-established knee reconstruction procedure for treating advanced osteoarthritis (OA) and a reliable surgical procedure for alleviating pain, accelerating functional recovery, and improving quality of life (QoL) with respect to patient-reported outcomes [1,2]. The number of patients undergoing TKA has steadily increased every year, and primary TKA patients aged <65 years are estimated to comprise more than half of all TKA patients in the United States [3]. Postoperative rehabilitation following TKA is an important issue. A decrease in the strength of both the quadriceps and hamstring occurs following TKA. This decline in the muscle strength of the lower limbs is associated with a decline in functional outcomes [4,5]. Postoperative rehabilitation aims to increase the range of motion of the knee, muscle strength, and proprioceptive functions after TKA. Various exercise interventions have been associated with different levels of positive effects in patients after TKA [6–9]; however, these effects are inconsistent because of poor supervision, impaired exercise adherence, and low compliance with exercise regimens [10–12].

Walking is the most common and essential physical activity of daily living (ADL). It is also a complex activity involving all levels of the neuromuscular and cardiorespiratory systems and has numerous health benefits [13]. Meanwhile, walking training has a positive effect on the health-related and perceived quality of Life in older persons [14,15]. Thus, we performed an easy training regimen, walk training (WT), on patients in TKA recovery for the purpose of reducing the bias of poor supervision. Because the number of female patients who had undergone TKA was approximately twice that of men between 2001 and 2010 [16], this study focused only on female patients who had undergone TKA. Additionally, patient-reported outcome measurements are routinely used in clinical research for assessing patients' health statuses [17]. The Knee Injury and Osteoarthritis Outcome Score (KOOS) questionnaire is the most reliable and relevant patient-reported outcome measurement to assess knee-specific and related problems, including patients' quality of life [18]. Therefore, we used the KOOS questionnaire to evaluate the health status of TKA. This study aimed to evaluate the effect of a supervised walking regimen on lower limb muscle strength, functional fitness, and patient-reported outcomes in female patients with TKA. We hypothesized that the WT would have positive effects on low limb muscle strength, performance on the functional fitness test, and KOOS in female TKA.

2. Materials and Methods

2.1. Participants

The retrospective cohort study was approved by the Ethics Committee and Institutional Review Board (IRB: 102-0979B, Approval Date: 26 April 2013). The participants were recruited in the orthopedic clinic of the hospital, and the orthopedic physician explained the protocol to the participants. All patients provided informed consent. The eligibility criteria were female patients aged 60~80 who had been diagnosed with Ahlbäck stage III IV knee OA [19] without infectious joint disease or rheumatic disease, and who had undergone TKA without additional surgeries such as ligament reconstruction or corrective osteotomy. Patients with diabetes, a neuromusculoskeletal disorder, history of fracture of the lower limb, or the presence of an artificial limb, as well as those who were unsuitable for exercise training, were excluded.

The participants were allocated into two groups by their willingness. Patients in the control (CON) and WT groups followed the routine postoperative rehabilitation protocol, including a self-administered straight leg raise (in a sitting position, static stretching exercises for hamstrings and gastrosoleus muscles), and a range of motion exercise consisting of terminal knee extension with weight; straight leg raises with weight in the supine and side-lying positions; and prone, hip, and knee flexion–extension with weight in supine as well as knee flexion–extension with weight in prone. Moreover, participants in the WT group practiced an extra walking intervention for 24 weeks.

WT was started 3 months after TKA, because patients could tolerate exercise training as the pain had notably decreased at this time [8]. All assessments in both the CON and WT groups were performed at the following time points (Supplementary Figure S1): preoperatively, 12 weeks postoperatively (baseline), 12 weeks after the start of training (mid-exercise), after training (post-exercise), and 12 weeks after completing the training (follow-up). All assessments in both groups were conducted on the same day by the same experienced investigator, who was blinded to the participant allocation, at the Sports Medicine Center.

From August 2015 to August 2017, 46 female patients with end-stage OA were identified from the author's clinics. Among these, nine patients were excluded because they did not meet the inclusion criteria. In total, 37 female participants were enrolled. After the initial allocation, there were 18 patients in the CON group and 19 patients in the WT group. Because four patients in the CON group declined the following assessments and five patients in the WT group discontinued the intervention, nine patients were excluded from the final analysis. Thus, 28 patients (14 each in the CON and WT groups) completed the treatment regimen and were included in the final analysis (Figure 1; the template of the flow chart was adopted from the reference [20]).

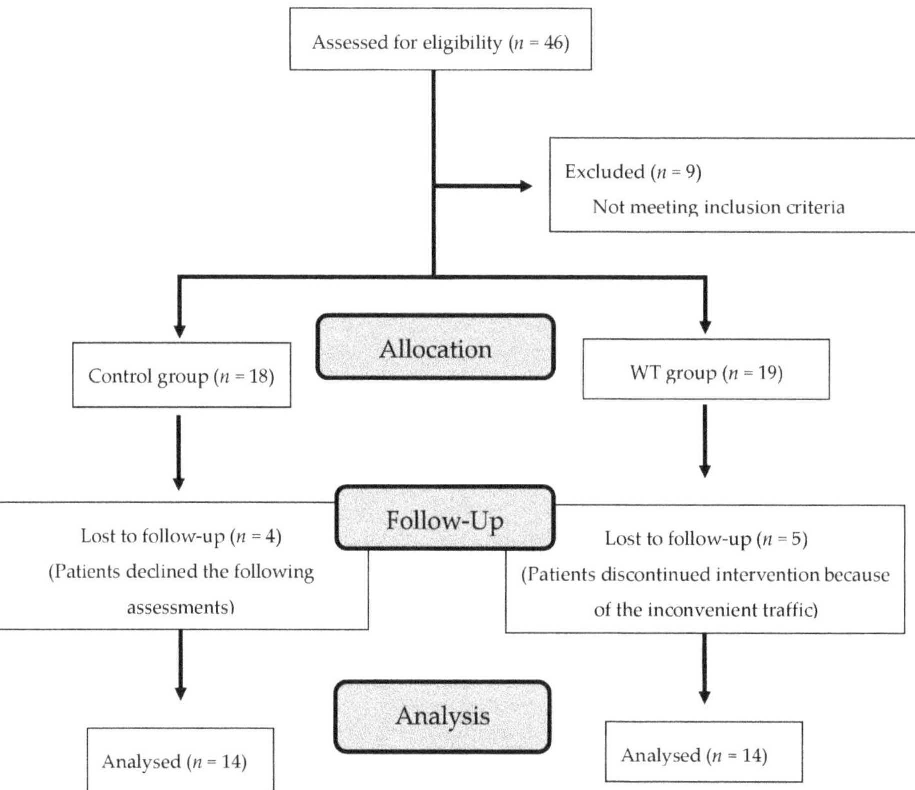

Figure 1. Flowchart of patient enrollment.

2.2. Sample Size

The sample size was determined using G*power software version 3.1.9.7. (Heinrich Heine University, Dusseldorf, Germany). The input parameters used for the t test were alpha = 0.05, effect size d = 0.8, and the power was calculated as 0.8. The total sample size was calculated to be 30 participants, with fifteen participants in each group. In previous patients' surveys, most patients were divided into different groups, with the number of people in each group ranging from 14 to 20 for statistical analysis [7,8,21]. The intraclass correlation coefficient (ICC), typical error (TE), and percentual coefficient of variation (CV) were calculated to access intertest correlation (Supplementary Table S1).

2.3. WT Course

Schimpl and his colleagues reported that the average walking speed of elders aged over 60 years is about 1.2 m/s (~4.32 km/h) [22]. Additionally, Serwe and his colleagues reported that the effective duration (30 minutes/day × 5 days) of walk training for increasing physical activity in women is a total of 2.5 hours a week [23]. Thus, a total distance of 10.8 km (4.32 km/h × 2.5 h = 10.8 km) per week might be an effective prescript of walk training for healthy women. Considering the patients' baseline 6-minute walk testing results (average walking speed about 3.4 km/h in the two groups), a total distance of 8.5 km (3.4 km/h × 2.5 h = 8.5 km) a week was determined to be a more effective prescript of walk training for female TKA patients. Therefore, we determined that the training target was to finish a total of 9 km a week. The participants in the WT group were asked to complete a total distance of 9 km a week on the treadmill with unlimited walking speed, at an intensity of 50–70% of the target heart rate, monitored by a POLAR FT40 monitor (Polar Electro Oy, Kempele, Finland). The walking speed was dependent on the individual, and the entire training process was recorded and supervised by a trained physical therapist at the Sports Medicine Center. Maximum heart rate (100% target heart rate) was calculated using the 220-age formula [24]. During the training period, the patients' motion and physical conditions were supervised by at least one trained physical therapist for the WT group. To determine the distance and reduce injury risk from the environment, such as uneven or sloping ground, WT was performed using the JOHNSON Matrix T50X treadmill (Johnson Health Tech, Taichung, Taiwan).

2.4. Muscle strength

The isokinetic dynamometer HUMAC NORM system (CSMi®, Stoughton, MA, USA) was used to assess the lower extremity muscle strength of the affected leg, including the extension and flexion of the hip and knee [7]. The muscle strength of the hip was evaluated when participants were in the supine position, and that of the knee was tested in the seated position. Angular velocity of 60° per second was set in the concentric contraction model. Isokinetic tests were performed five times for each examination, and the maximum force was recorded. There was a 3-min rest period between hip and knee trials. Body composition is associated with muscle strength, especially fat-free mass. To avoid the influence of body composition, the value of muscle strength was normalized by body weight. Normalized force is an effective and useful method to compare persons of different individual sizes [25].

2.5. Functional Fitness Test

The 6-minute walk test (6MWT), the 8-foot up-and-go test (8UG), and the 30-second chair stand test (30CST) were used to assess exercise capacity, motor agility, and lower-body muscle strength, respectively. The details of these tests, such as criterion-referenced standard, validity, and reliability, have been described in a previous study [26]. The procedure of 6MWT is to measure the distance that the subject can walk over a total of six minutes on a hard and flat surface. The protocol of the 8UG is to record the time that the subjects are asked to rise from a chair placed next to a wall, walk to a traffic cone at 8 feet in front of the chair, turn, walk back, and sit down. The 30CST is intended to record the number of times the subjects stand up with a straight back and hands on the opposite shoulder crossed at the wrists from a chair without armrests in 30 seconds.

2.6. Knee Injury and Osteoarthritis Outcome Score (KOOS) Questionnaire

Clinical knee scoring was used to assess outpatients' self-reported outcomes by using the Chinese version of the KOOS questionnaire [27]. KOOS contains forty-two items in the five subscales, named Symptoms (seven items), Pain (nine items), Activity of daily livings (ADL; seventeen items), Sport (five items), and knee-related quality of life (QoL; four items). All items have five possible answer options presented with a Likert scale. The scores range from 0 (Good) to 4 (Worse). Each subscale score is calculated using $100 - \frac{Mean\ score\ (All\ items\ in\ each\ subscale)\ \times\ 100}{4}$ formulas, and scores are transformed to a 0–100 scale. A zero score indicates an extreme problem.

2.7. Statistical Analysis

All data were analyzed using the Statistical Package for the Social Sciences, Windows version 17.0 (SPSS®, Chicago, IL, USA). The difference in demographic data between the two groups was tested by the unpaired Student's *t*-test. Generalized estimating equations (GEEs) [28] were used to resolve the differences of each assessment outcome between the

WT and CON groups, as well as within the groups. The Shapiro–Wilk results showed no abnormal distributions in either group in pre-operation (Supplementary Table S2). The effect size (ES) was used to measure the difference between group means. The relative ES was calculated using Cohen's d, which is defined as the difference between two means divided by a standard deviation for the data. The primary outcome of the study was the QoL subscale of KOOS. The effect size between the CON and WT groups at follow-up was 5.02, which indicates a huge effect size [29]. The ideal power was considered to be 80%, with a 0.05 type I error. All continuous data are presented as the mean ± standard deviation, and p-values of <0.05 were considered statistically significant. To reduce bias, we attempted to identify controls with matching ages and body mass indices.

3. Results

The patients' demographic data are shown in Table 1. The mean age was 68.5 ± 6.6 and 69.7 ± 2.8 years in the CON and WT groups, respectively. Height and weight were similar between the CON and WT groups. Lower extremity muscle strength, functional test results, and KOOS scores were similar between the CON and WT groups ($p > 0.05$).

Table 1. The demographic characters of patients at the pre-operative assessment.

Characters	CON (N = 14)	WT (N = 14)	p	ES
Age (years)	68.5 ± 6.6	69.7 ± 2.8	0.372	0.23
Height (cm)	152.0 ± 5.8	152.8 ± 4.9	0.877	0.14
Weight (kg)	65.9 ± 11.5	66.4 ± 9.8	0.682	0.04
BMI (kg/m^2)	28.4 ± 4.0	28.4 ± 3.4	0.567	0.01
	Pre-operation			
Muscle strength				
HE (Nm/kg)	72.0 ± 8.2	79.0 ± 9.2	0.571	0.80
HF (Nm/kg)	32.4 ± 3.5	31.3 ± 4.1	0.844	0.28
KE (Nm/kg)	42.1 ± 2.9	41.5 ± 5.3	0.941	0.14
KF (Nm/kg)	41.8 ± 3.0	45.2 ± 2.4	0.376	1.25
Functional test				
6MWT (m)	309.1 ± 31.4	326.1 ± 29.7	0.693	0.55
8UG (sec)	9.5 ± 1.6	10.7 ± 1.3	0.541	0.82
30s-CST (Times)	10.6 ± 1.1	10.4 ± 0.9	0.836	0.19
KOOS				
Sym	63.9 ± 5.1	63.7 ± 6.3	0.983	0.03
Pain	65.6 ± 4.8	63.8 ± 5.6	0.817	0.31
ADL	64.5 ± 4.9	68.4 ± 6.2	0.611	0.69
Sports	40.6 ± 7.1	35.5 ± 4.4	0.550	0.86
QoL	61.1 ± 5.4	51.5 ± 5.8	0.218	1.71

Data are presented as mean ± SD. Abbreviations: CON: control; WT: walk training; HE: hip extension; HF: hip flexion; KE: knee extension; KF: knee flexion; 6MWT: 6-min walk test; 8UG: 8-foot up-and-go; 30 s-CST: 30-s chair stand test; Sym: symptom; ADL: activity of daily living; QoL: quality of life; ES: effect size. There was no difference between the two groups.

In the lower extremity muscle strength assessment, the WT group exhibited a significant increase in hip extensor, knee extensor, and knee flexor muscle strengths compared to the CON group at the post-exercise assessment, and the differences in knee extensor and flexor muscle strengths lasted until the follow-up (Table 2). When participants were compared within each group in a temporal manner (Table 3), the CON group exhibited a significant increase only in knee extensor muscle strength at the final follow-up compared with the baseline measurement. However, the WT group exhibited a significant increase in hip extensor, knee extensor, and knee flexor measurements at the post-exercise assessment compared with the baseline. The knee extensors even exhibited a significant increase in muscle strength during the mid-exercise assessment.

Table 2. Comparison of muscle strength between two groups of TKA patients.

(Nm/kg)	Baseline				Mid-Exercise				Post-Exercise				Follow-Up			
	CON	WT	ES	p	CON	WT	ES	p	CON	WT	ES	p	CON	WT	ES	p
HE	71.5 ± 5.4	85.9 ± 9.9	1.58	0.199	74.6 ± 9.6	102.6 ± 12.5	2.51	0.074	77.5 ± 8.0	116.9 ± 11.3 §	4.02	0.004 *	87.6 ± 10.6	106.8 ± 9.4 §	1.91	0.177
HF	27. 6 ± 2.8	33.4 ± 3.6	1.79	0.190	26.4 ± 4.3	34.5 ± 4.2	1.90	0.172	27.4 ± 3.6	37.1 ± 4.7	2.32	0.094	29.6 ± 3.1	33.9 ± 5.4	0.97	0.452
KE	40.3 ± 5.1	45.2 ± 4.9	1.17	0.476	45.9 ± 4.1	57.6 ± 5.7 §	2.36	0.091	48.2 ± 2.4	73.1 ± 7.5 §	4.47	0.001 *	54.2 ± 3.2 §	71.4 ± 6.6 §	3.31	0.019 *
KF	43.6 ± 4.0	46.1 ± 3.0	0.70	0.615	49.8 ± 3.6	54.1 ± 4.0	1.13	0.406	43.7 ± 3.3	64.4 ± 4.1 §	5.62	0.001 *	43.1 ± 3.0	65.0 ± 2.6 §	1.27	0.001 *

Data are presented as mean ± SD. *: A significant difference between two groups. §: Significant difference compared to baseline within the individual group ($p < 0.05$). Abbreviations: CON: control; WT: walk training; HE: hip extension; HF: hip flexion; KE: knee extension; KF: knee flexion; ES: effect size. There were significant differences in knee extension and flexion between the two groups.

Table 3. The effect sizes and p-values of comparison of muscle strength between different time points and baselines within the WT and CON groups.

	Variable	ES			p		
		Mid-Exercise	Post-Exercise	Follow-Up	Mid-Exercise	Post-Exercise	Follow-Up
WT	HE	1.48	2.92	2.17	0.211	0.002	0.024
	HF	0.28	0.88	0.11	0.751	0.514	0.154
	KE	2.33	4.40	4.51	0.032	0.001	0.012
	KF	2.26	5.09	6.73	0.301	0.002	0.001
CON	HE	0.40	0.88	1.91	0.468	0.216	0.165
	HF	0.33	0.06	0.68	0.532	0.151	0.205
	KE	1.21	1.98	3.26	0.541	0.304	0.023
	KF	1.63	0.03	0.14	0.665	0.501	0.487

Abbreviations: CON: control; WT: walk training; HE: hip extension; HF: hip flexion; KE: knee extension; KF: knee flexion; ES: effect size.

The functional outcomes of the functional fitness tests (Table 4) revealed that the results of the 6MWT, 8UG, and 30CST did not differ significantly between the CON and WT groups at each assessment time point. However, in the temporal comparison (Table 5), the results of the 6MWT and 8UG showed significant improvements at the mid-exercise, post-exercise, and follow-up measurements compared with baseline measurements in the WT group, but not in the CON group.

Table 4. Comparison of physical function fitness between the two groups of TKA patients.

	Baseline				Mid-Exercise				Post-Exercise				Follow-Up			
	CON	WT	ES	p	CON	WT	ES	p	CON	WT	ES	p	CON	WT	ES	p
6MWT (m)	331.0 ± 16.5	341.3 ± 20.5	0.55	0.696	346.6 ± 14.8	375.2 ± 21.4 §	1.55	0.271	346.1 ± 16.8	400.1 ± 23.8 §	2.62	0.40	345.8 ± 16.5	405.5 ± 30.7 §	2.42	0.093
8UG (sec)	10.4 ± 0.7	9.5 ± 0.7	1.28	0.386	9.1 ± 0.5	8.5 ± 0.5 §	1.20	0.427	9.1 ± 0.8	8.6 ± 0.8 §	0.62	0.692	9.2 ± 0.6	8.3 ± 0.7 §	1.38	0.302
30s-CST (Times)	11.9 ± 0.9	12.2 ± 0.7	0.37	0.764	12.6 ± 0.6	14.0 ± 1.0	1.69	0.238	12.6 ± 0.8	14.21 ± 1.0	1.76	0.211	12.8 ± 0.7	14.0 ± 1.2	1.22	0.370

Data are presented as mean ± SD. §: Significant difference compared to baseline within the individual group ($p < 0.05$). Abbreviations: CON: control; WT: walk training; 6MWT: 6-minute walk test; 8UG: 8 feet up-and-go; 30s-CST: 30-second chair stand test; ES: effect size. 6MWT and 8UG tests were statistically improved after WT in female TKA patients at the post-exercise and follow-up evaluations.

The self-explanatory KOOS questionnaire revealed (Table 6) that the QoL subscale was significantly increased in the WT group compared to the CON group at follow-up. When a temporal comparison was performed within each group (Table 7), scores for the subscales of symptoms, pain, and ADL exhibited significant improvements in both the CON and WT groups at the follow-up compared with the baseline. The improvement in scores in the pain and ADL subscales in the WT group occurred at an earlier time point (pain scores decreased at the post-exercise evaluation; ADL scores increased at the mid-and post-exercise evaluations). Among these subcategories, scores for the QoL subscale in the WT group exhibited a significant increase in the post-exercise and follow-up assessments compared with the baseline; however, these improvements were not observed in the CON group.

Table 5. The effect sizes and *p*-values of comparison of physical function fitness between different time points and baselines within the WT and CON groups.

		ES			*p*		
	Variable	Mid-Exercise	Post-Exercise	Follow-Up	Mid-Exercise	Post-Exercise	Follow-Up
WT	6MWT	1.62	2.65	2.46	0.031	0.012	0.001
	8UG	1.64	1.20	1.71	0.029	0.031	0.002
	30s-CST	2.09	2.33	1.83	0.376	0.471	0.389
CON	6MWT	0.96	0.90	0.85	0.588	0.546	0.437
	8UG	2.14	1.73	1.84	0.632	0.652	0.594
	30s-CST	0.92	0.82	1.12	0.417	0.374	0.360

Abbreviations: CON: control; WT: walk training; 6MWT: 6-minute walk test; 8UG: 8 feet up-and-go; 30s-CST: 30-second chair stand test; ES: effect size.

Table 6. Comparison of Knee Injury and Osteoarthritis Outcome Score (KOOS) subscales between two groups of TKA patients.

	Baseline				Mid-Exercise				Post-Exercise				Follow-Up			
	CON	WT	ES	*p*	CON	WT	ES	*p*	CON	WT	ES	*p*	CON	WT	ES	*p*
Sym	62.1 ± 4.9	71.1 ± 5.4	1.74	0.200	71.2 ± 5.7 §	76.0 ± 4.7	0.91	0.487	70.9 ± 4.9	78.2 ± 3.9	1.64	0.253	83.0 ± 4.1 §	86.7 ± 3.8 §	0.93	0.514
Pain	78.2 ± 3.4	79.5 ± 4.4	0.33	0.794	84.8 ± 3.5	87.6 ± 4.0	0.74	0.584	84.3 ± 4.3	90.5 ± 2.0 §	1.84	0.185	91.1 ± 3.1 §	97.3 ± 1.5 §	2.54	0.072
ADL	76.3 ± 3.2	76.6 ± 4.7	0.07	0.923	80.2 ± 5.1	89.8 ± 2.4§	2.40	0.083	78.7 ± 5.1	87.6 ± 2.6 §	2.19	0.118	92.0 ± 2.5 §	94.6 ± 1.3 §	1.30	0.322
Sports	47.0 ± 5.2	42.3 ± 2.8	1.12	0.439	43.6 ± 3.8	50.5 ± 4.5	1.65	0.229	45.1 ± 4.7	44.0 ± 2.8	0.28	0.835	45.2 ± 3.8	45.6 ± 4.9	0.09	0.936
QoL	56.6 ± 4.7	60.0 ± 6.3	0.61	0.696	64.6 ± 7.3	71.4 ± 6.0	1.01	0.474	63.4 ± 5.8	72.8 ± 4.3 §	1.84	0.193	68.1 ± 5.8	91.0 ± 2.8 §	5.02	0.001 *

Data are presented as mean ± SD. *: A significant difference between two groups. §: Significant difference compared to baseline within the individual group ($p < 0.05$). Abbreviations: CON: control; WT: walk training; Sym: symptom; ADL: activity of daily livings; QoL: quality of life; ES: effect size. The scores of the QoL subscale in the WT group exhibited a significant increase in the post-exercise and follow-up assessments compared with the baseline.

Table 7. The effect sizes and *p*-values of comparison of KOOS subscales between different time points and baselines within the WT and CON groups.

		ES			*p*		
	Variable	Mid-Exercise	Post-Exercise	Follow-Up	Mid-Exercise	Post-Exercise	Follow-Up
WT	Sym	0.97	1.51	3.34	0.521	0.321	0.002
	Pain	1.93	3.22	5.42	0.074	0.024	0.002
	ADL	3.54	2.90	5.22	0.032	0.008	0.002
	Sports	2.19	0.61	0.83	0.658	0.713	0.072
	QoL	1.85	2.37	6.36	0.067	0.012	0.001
CON	Sym	1.71	1.80	4.63	0.047	0.051	0.035
	Pain	1.91	1.57	3.97	0.514	0.624	0.041
	ADL	0.92	0.56	5.47	0.727	0.694	0.037
	Sports	0.75	0.38	0.40	0.865	0.792	0.698
	QoL	1.30	1.29	2.18	0.476	0.523	0.331

Abbreviation: CON: control; WT: walk training; Sym: symptom; ADL: activity of daily livings; QoL: quality of life; ES: effect size.

4. Discussion

The most important findings of the present study were the statistical increase in the QoL subcategory of KOOS and the knee muscle strength (extensors and flexors) in the WT group compared to the CON group at follow-up (Tables 2 and 6). While the comparison was performed within the individual group in a temporal manner (Tables 3 and 5), the knee extensors and functional fitness test (6 MWT: ambulatory capacity, 8UG: motor agility) in the WT group were improved at the mid-exercise, post-exercise, and follow-up measurements compared with the baseline, suggesting that WT enhanced the recovery of functional mobility after TKA [30–33]. These increases were based on the concomitant improvements in the pain, symptoms, and ADL subcategories of KOOS after TKA, quadriceps training, and range of motion exercises [10,34,35]. In addition to the range of motion of knee and muscle strength on which the TKA standard exercises focus, WT may provide extra endurance and proprioception.

Many studies have shown that an increase in the quadriceps strength of TKA patients is associated with superior outcomes and performance [36–40]. Previous studies have also revealed that muscle strengthening improves quadriceps strength and functional fitness test scores (timed up-and-go, stair-climbing test, and 6MWT) [41,42]. Our results were similar to other findings in the literature, and showed that the knee extensor and flexor strength in the WT group significantly improved compared with the CON group (Table 2). In the temporal comparison within individual groups, the knee extensor improvement in the WT group occurred simultaneously with the improvements in the 6MWT and 8UG functional tests (Tables 4 and 5), as well as in the ADL domains of KOOS (Tables 6 and 7). These results imply that lower leg muscle strength is a determinant physical factor and plays a pivotal role in functional outcomes in TKA patients [43,44]. However, no significant difference in the 6MWT, 8UG, or the ADL subscales of KOOS was observed between the two groups. The lack of a significant difference might have resulted from the small sample size.

All patients in the WT group were supervised by a trained physical therapist during training, and no patient in the WT group reported discomfort or needed further medical management. Our results suggest that WT is beneficial for patients who have undergone TKA. Considering the ease of standardization and performance of WT, it can be effectively applied during rehabilitation following TKA in community-dwelling patients. However, a trained physical therapist also plays an important role in the intervention. In this study, a therapist monitored the status of the participants and gave real-time assistance when the participants had an issue during the exercise. Therefore, the deployment of a trained physical therapist would be recommended.

Several limitations of this study should be acknowledged. First, evidence regarding the entire effect of WT in TKA was restricted because of the small sample size in this retrospective study. Nevertheless, improvements in muscle strength and function over time indicated the benefits of WT in TKA. Second, our results (12 months after surgery) did not fully demonstrate the long-term effects of WT. The results showed that multiple metrics gradually improved in the CON group. Third, there was no random allocation, and certain participants could be more motivated, which could influence the results. Thus, a larger randomized controlled trial is needed to verify the findings of this study, as it was limited to only 28 who completed the study out of 46 total patients. Fourth, neuromuscular activation also plays a critical role in muscle function. Electromyography of the lower limbs might be included for greater clarity regarding the effect of WT on the neuromuscular system in further study.

5. Conclusions

During WT, no patient in the WT exercise group reported discomfort or injury or required further medical management. The 9 km/week WT protocol is feasible for TKA patients. The WT could significantly improve muscle strength (knee flexion and extension), functional outcome (6WT and 8UG), and QoL of patients. The simplicity of WT could reduce the limitations of poor supervision, impaired exercise adherence, and low compliance with exercise regimens, which cause poor rehabilitation outcomes. The authors believe that WT should be broadly applied in the rehabilitation of patients who have undergone TKA.

Supplementary Materials: The following supporting information can be downloaded at: https://www.mdpi.com/article/10.3390/healthcare11030356/s1, Figure S1: Schematic diagram of the experimental time course; Table S1: The coefficient of validity and reliability in the WT and CON groups; Table S2: Shapiro–Wilk test of normality.

Author Contributions: Conceptualization, Z.-R.L. and W.-W.R.H.; methodology, W.-H.H., W.-B.H. and W.-W.R.H.; software, S.-H.C. and C.-H.F.; validation, W.-H.H., W.-B.H., Z.-R.L. and W.-W.R.H.; formal analysis, S.-H.C. and C.-H.F.; investigation, W.-H.H., W.-B.H. and W.-W.R.H.; resources, W.-H.H., W.-B.H. and W.-W.R.H.; data curation, S.-H.C. and C.-H.F.; writing—original draft preparation, W.-H.H. and W.-B.H.; writing—review and editing, W.-H.H., W.-B.H. and W.-W.R.H.; supervision, L.-T.K., Z.-R.L. and W.-W.R.H.; project administration, W.-W.R.H.; funding acquisition, W.-W.R.H. All authors have read and agreed to the published version of the manuscript.

Funding: This research was funded by Chang Gung Memorial Hospital Grant, grant number CORPG6C0021-23, CORPG6C0011-13, CORPG6G0251-52, CORPG6G0261-62, CMRPG6K0161-3, CMRPG6G0311, CMRPG6I0011, and CMRPG6K0131-3.

Institutional Review Board Statement: The study was conducted in accordance with the Declaration of Helsinki and approved by the Institutional Review Board of Chang Gung Memorial Hospital (Approval No. 102-0979B, Approval Date: 26 April 2013).

Informed Consent Statement: Informed consent was obtained from all subjects involved in the study. Written informed consent has been obtained from the patient(s) to publish this paper.

Data Availability Statement: Remaining data are available from the corresponding author upon reasonable request.

Acknowledgments: The authors thank Chia-Fang Chang for collecting the participants' data.

Conflicts of Interest: The authors declare no conflict of interest.

References

1. Hussain, S.M.; Neilly, D.W.; Baliga, S.; Patil, S.; Meek, R. Knee osteoarthritis: A review of management options. *Scott. Med. J.* **2016**, *61*, 7–16. [CrossRef] [PubMed]
2. Schiavone Panni, A.; Falez, F.; D'Apolito, R.; Corona, K.; Perisano, C.; Vasso, M. Long-term follow-up of a non-randomised prospective cohort of one hundred and ninety two total knee arthroplasties using the NexGen implant. *Int. Orthop.* **2017**, *41*, 1155–1162. [CrossRef] [PubMed]
3. Singh, J.A.; Yu, S.; Chen, L.; Cleveland, J.D. Rates of Total Joint Replacement in the United States: Future Projections to 2020–2040 Using the National Inpatient Sample. *J. Rheumatol.* **2019**, *46*, 1134–1140. [CrossRef] [PubMed]
4. Paravlic, A.H.; Kovac, S.; Pisot, R.; Marusic, U. Neurostructural correlates of strength decrease following total knee arthroplasty: A systematic review of the literature with meta-analysis. *Bosn. J. Basic Med. Sci.* **2020**, *20*, 1–12. [CrossRef]
5. Paravlic, A.H.; Meulenberg, C.J.; Drole, K. The Time Course of Quadriceps Strength Recovery After Total Knee Arthroplasty Is Influenced by Body Mass Index, Sex, and Age of Patients: Systematic Review and Meta-Analysis. *Front. Med.* **2022**, *9*, 865412. [CrossRef]
6. Pozzi, F.; Snyder-Mackler, L.; Zeni, J. Physical exercise after knee arthroplasty: A systematic review of controlled trials. *Eur. J. Phys. Rehabil. Med.* **2013**, *49*, 877–892.
7. Hsu, W.H.; Hsu, W.B.; Shen, W.J.; Lin, Z.R.; Chang, S.H.; Hsu, R.W. Circuit training enhances function in patients undergoing total knee arthroplasty: A retrospective cohort study. *J. Orthop. Surg. Res.* **2017**, *12*, 156. [CrossRef]
8. Hsu, W.H.; Hsu, W.B.; Shen, W.J.; Lin, Z.R.; Chang, S.H.; Hsu, R.W. Twenty-four-week hospital-based progressive resistance training on functional recovery in female patients post total knee arthroplasty. *Knee* **2019**, *26*, 729–736. [CrossRef]
9. Henderson, K.G.; Wallis, J.A.; Snowdon, D.A. Active physiotherapy interventions following total knee arthroplasty in the hospital and inpatient rehabilitation settings: A systematic review and meta-analysis. *Physiotherapy* **2018**, *104*, 25–35. [CrossRef]
10. Hiyama, Y.; Kamitani, T.; Wada, O.; Mizuno, K.; Yamada, M. Effects of Group-Based Exercise on Range of Motion, Muscle Strength, Functional Ability, and Pain During the Acute Phase After Total Knee Arthroplasty: A Controlled Clinical Trial. *J. Orthop. Sports Phys. Ther.* **2016**, *46*, 742–748. [CrossRef]

11. Han, A.S.; Nairn, L.; Harmer, A.R.; Crosbie, J.; March, L.; Parker, D.; Crawford, R.; Fransen, M. Early rehabilitation after total knee replacement surgery: A multicenter, noninferiority, randomized clinical trial comparing a home exercise program with usual outpatient care. *Arthritis Care Res.* **2015**, *67*, 196–202. [CrossRef] [PubMed]

12. Chen, M.; Li, P.; Lin, F. Influence of structured telephone follow-up on patient compliance with rehabilitation after total knee arthroplasty. *Patient Prefer. Adherence* **2016**, *10*, 257–264. [CrossRef] [PubMed]

13. Sipila, S.; Tirkkonen, A.; Hanninen, T.; Laukkanen, P.; Alen, M.; Fielding, R.A.; Kivipelto, M.; Kokko, K.; Kulmala, J.; Rantanen, T.; et al. Promoting safe walking among older people: The effects of a physical and cognitive training intervention vs. physical training alone on mobility and falls among older community-dwelling men and women (the PASSWORD study): Design and methods of a randomized controlled trial. *BMC Geriatr.* **2018**, *18*, 215. [CrossRef]

14. Monteagudo, P.; Roldan, A.; Cordellat, A.; Gomez-Cabrera, M.C.; Blasco-Lafarga, C. Continuous Compared to Accumulated Walking-Training on Physical Function and Health-Related Quality of Life in Sedentary Older Persons. *Int. J. Environ. Res. Public Health* **2020**, *17*, 6060. [CrossRef] [PubMed]

15. Les, A.; Guszkowska, M.; Piotrowska, J.; Rutkowska, I. Changes in perceived quality of life and subjective age in older women participating in Nordic Walking classes and memory training. *J. Sports Med. Phys. Fit.* **2019**, *59*, 1783–1790. [CrossRef]

16. Whitlock, K.G.; Piponov, H.I.; Shah, S.H.; Wang, O.J.; Gonzalez, M.H. Gender Role in Total Knee Arthroplasty: A Retrospective Analysis of Perioperative Outcomes in US Patients. *J. Arthroplast.* **2016**, *31*, 2736–2740. [CrossRef]

17. Deshpande, P.R.; Rajan, S.; Sudeepthi, B.L.; Abdul Nazir, C.P. Patient-reported outcomes: A new era in clinical research. *Perspect. Clin. Res.* **2011**, *2*, 137–144. [CrossRef]

18. Roos, E.M.; Toksvig-Larsen, S. Knee injury and Osteoarthritis Outcome Score (KOOS)—Validation and comparison to the WOMAC in total knee replacement. *Health Qual. Life Outcomes* **2003**, *1*, 17. [CrossRef]

19. Ahlback, S. Osteoarthrosis of the knee. A radiographic investigation. *Acta Radiol. Diagn.* **1968**, *227*, 7–72.

20. Chun, E.H.; Kim, Y.J.; Woo, J.H. Which is your choice for prolonging the analgesic duration of single-shot interscalene brachial blocks for arthroscopic shoulder surgery? intravenous dexamethasone 5 mg vs. perineural dexamethasone 5 mg randomized, controlled, clinical trial. *Medicine* **2016**, *95*, e3828. [CrossRef]

21. Savkin, R.; Buker, N.; Gungor, H.R. The effects of preoperative neuromuscular electrical stimulation on the postoperative quadriceps muscle strength and functional status in patients with fast-track total knee arthroplasty. *Acta Orthop. Belg.* **2021**, *87*, 735–744. [CrossRef] [PubMed]

22. Schimpl, M.; Moore, C.; Lederer, C.; Neuhaus, A.; Sambrook, J.; Danesh, J.; Ouwehand, W.; Daumer, M. Association between walking speed and age in healthy, free-living individuals using mobile accelerometry—A cross-sectional study. *PLoS ONE* **2011**, *6*, e23299. [CrossRef] [PubMed]

23. Serwe, K.M.; Swartz, A.M.; Hart, T.L.; Strath, S.J. Effectiveness of long and short bout walking on increasing physical activity in women. *J. Women's Health* **2011**, *20*, 247–253. [CrossRef] [PubMed]

24. Riebe, D.; Liguori, G.; Magal, M. *ACSM's Guidelines for Exercise Testing and Prescription*, 10th ed.; Wolters Kluwer/Lippincott Williams & Wilkins: Philadelphia, PA, USA, 2018.

25. Bazett-Jones, D.M.; Cobb, S.C.; Joshi, M.N.; Cashin, S.E.; Earl, J.E. Normalizing hip muscle strength: Establishing body-size-independent measurements. *Arch. Phys. Med. Rehabil.* **2011**, *92*, 76–82. [CrossRef]

26. Rikli, R.E.; Jones, C.J. Development and validation of criterion-referenced clinically relevant fitness standards for maintaining physical independence in later years. *Gerontologist* **2013**, *53*, 255–267. [CrossRef] [PubMed]

27. Peer, M.A.; Lane, J. The Knee Injury and Osteoarthritis Outcome Score (KOOS): A review of its psychometric properties in people undergoing total knee arthroplasty. *J. Orthop. Sports Phys. Ther.* **2013**, *43*, 20–28. [CrossRef]

28. Liang, K.-Y.; Zeger, S.L. Longitudinal data analysis using generalized linear models. *Biometrika* **1986**, *73*, 13–22. [CrossRef]

29. Sawilowsky, S. New effect size rules of thumb. *J. Mod. Appl. Stat. Methods* **2009**, *8*, 467–474. [CrossRef]

30. Unver, B.; Kalkan, S.; Yuksel, E.; Kahraman, T.; Karatosun, V. Reliability of the 50-foot walk test and 30-sec chair stand test in total knee arthroplasty. *Acta Ortop. Bras.* **2015**, *23*, 184–187. [CrossRef]

31. Ko, V.; Naylor, J.M.; Harris, I.A.; Crosbie, J.; Yeo, A.E. The six-minute walk test is an excellent predictor of functional ambulation after total knee arthroplasty. *BMC Musculoskelet. Disord.* **2013**, *14*, 145. [CrossRef]

32. Rossi, M.D.; Hasson, S.; Kohia, M.; Pineda, E.; Bryan, W. Mobility and perceived function after total knee arthroplasty. *J. Arthroplasty* **2006**, *21*, 6–12. [CrossRef] [PubMed]

33. Calatayud, J.; Casana, J.; Ezzatvar, Y.; Jakobsen, M.D.; Sundstrup, E.; Andersen, L.L. High-intensity preoperative training improves physical and functional recovery in the early post-operative periods after total knee arthroplasty: A randomized controlled trial. *Knee Surg. Sports Traumatol. Arthrosc.* **2017**, *25*, 2864–2872. [CrossRef] [PubMed]

34. Fransen, M.; McConnell, S.; Harmer, A.R.; Van der Esch, M.; Simic, M.; Bennell, K.L. Exercise for osteoarthritis of the knee. *Cochrane Database Syst. Rev.* **2015**, *1*, CD004376. [CrossRef] [PubMed]

35. Taniguchi, M.; Sawano, S.; Kugo, M.; Maegawa, S.; Kawasaki, T.; Ichihashi, N. Physical Activity Promotes Gait Improvement in Patients with Total Knee Arthroplasty. *J. Arthroplasty* **2016**, *31*, 984–988. [CrossRef] [PubMed]

36. Mizner, R.L.; Petterson, S.C.; Stevens, J.E.; Axe, M.J.; Snyder-Mackler, L. Preoperative quadriceps strength predicts functional ability one year after total knee arthroplasty. *J. Rheumatol.* **2005**, *32*, 1533–1539.

37. Mizner, R.L.; Petterson, S.C.; Snyder-Mackler, L. Quadriceps strength and the time course of functional recovery after total knee arthroplasty. *J. Orthop. Sports Phys. Ther.* **2005**, *35*, 424–436. [CrossRef]

38. Mizner, R.L.; Petterson, S.C.; Stevens, J.E.; Vandenborne, K.; Snyder-Mackler, L. Early quadriceps strength loss after total knee arthroplasty. The contributions of muscle atrophy and failure of voluntary muscle activation. *J. Bone Joint Surg. Am.* **2005**, *87*, 1047–1053. [CrossRef]

39. Mizner, R.L.; Snyder-Mackler, L. Altered loading during walking and sit-to-stand is affected by quadriceps weakness after total knee arthroplasty. *J. Orthop. Res.* **2005**, *23*, 1083–1090. [CrossRef]

40. Saleh, K.J.; Lee, L.W.; Gandhi, R.; Ingersoll, C.D.; Mahomed, N.N.; Sheibani-Rad, S.; Novicoff, W.M.; Mihalko, W.M. Quadriceps strength in relation to total knee arthroplasty outcomes. *Instr. Course Lect.* **2010**, *59*, 119–130.

41. Petterson, S.C.; Mizner, R.L.; Stevens, J.E.; Raisis, L.; Bodenstab, A.; Newcomb, W.; Snyder-Mackler, L. Improved function from progressive strengthening interventions after total knee arthroplasty: A randomized clinical trial with an imbedded prospective cohort. *Arthritis Rheum.* **2009**, *61*, 174–183. [CrossRef]

42. Johnson, A.W.; Myrer, J.W.; Hunter, I.; Feland, J.B.; Hopkins, J.T.; Draper, D.O.; Eggett, D. Whole-body vibration strengthening compared to traditional strengthening during physical therapy in individuals with total knee arthroplasty. *Physiother. Theory Pract.* **2010**, *26*, 215–225. [CrossRef] [PubMed]

43. Accettura, A.J.; Brenneman, E.C.; Stratford, P.W.; Maly, M.R. Knee Extensor Power Relates to Mobility Performance in People with Knee Osteoarthritis: Cross-Sectional Analysis. *Phys. Ther.* **2015**, *95*, 989–995. [CrossRef] [PubMed]

44. Pua, Y.H.; Seah, F.J.; Clark, R.A.; Lian-Li Poon, C.; Tan, J.W.; Chong, H.C. Factors associated with gait speed recovery after total knee arthroplasty: A longitudinal study. *Semin. Arthritis Rheum.* **2017**, *46*, 544–551. [CrossRef] [PubMed]

Article

Influence of Transcranial Direct Current Stimulation and Exercise on Fatigue and Quality of Life in Multiple Sclerosis

Inés Muñoz-Paredes [1,*], Azael J. Herrero [2,3], Natalia Román-Nieto [4], Alba M. Peña-Gomez [4,5] and Jesús Seco-Calvo [6,7,*]

1 Faculty of Health Sciences, University of León, 24071 León, Spain
2 Department of Health Sciences, European University Miguel de Cervantes, 47012 Valladolid, Spain
3 Research Center on Physical Disability, ASPAYM Castilla y León, 47008 Valladolid, Spain
4 Multiple Sclerosis Association of Palencia, 34004 Palencia, Spain
5 Physiotherapy Department, Hospital of Cabueñes, University of Oviedo, 33394 Gijón, Spain
6 Faculty of Physiotherapy and Nursing, University of Leon, 24071 León, Spain
7 Physiology Department, University of the Basque Country, 48940 Leioa, Spain
* Correspondence: inesmunozparedes@gmail.com (I.M.-P.); jesus.seco@unileon.es (J.S.-C.)

Abstract: Background: Multiple sclerosis (MS) is a chronic autoimmune disease of the central nervous system that leads to a great deterioration in the quality of life. Objective: We aimed to assess the effectiveness of two individual programs, one based on transcranial direct current stimulation (tDCS) and another based on the effect of physical exercise on fatigue and quality of life in patients with MS. Methods: A total of 12 patients with relapsing–remitting and progressive secondary MS participated. Fatigue and quality of life were assessed before and after intervention. The exercise program and tDCS were carried out over a 4-week period, with a washout period of 5 months. Results: The results show significant improvements in the different quality of life subscales after the application of tDCS, activities of daily living (r = 0.625; p = 0.037) (g = 0.465), psychological well-being (r = 0.856; p = 0.004) (g = 0.727) and coping (r = 0.904; p = 0.18) (g = 0.376), and in those after the application of exercise, activities of daily living (r = 0.853; p = 0.003) (g = 0.570) and psychological well-being (r = 0.693; p = 0.041) (g = 0.417). After the application of both therapies, more than 50% of the subjects did not have a positive fatigue score on the MFIS scale. Conclusion: The major findings suggest that the application of both therapies produces a beneficial effect with significant improvements in the quality of life of this sample.

Keywords: physical training; activities of daily living; Modified Fatigue Impact Scale

Citation: Muñoz-Paredes, I.; Herrero, A.J.; Román-Nieto, N.; Peña-Gomez, A.M.; Seco-Calvo, J. Influence of Transcranial Direct Current Stimulation and Exercise on Fatigue and Quality of Life in Multiple Sclerosis. *Healthcare* **2023**, *11*, 84. https://doi.org/10.3390/healthcare11010084

Academic Editors: Rafael Oliveira and João Paulo Brito

Received: 23 November 2022
Revised: 19 December 2022
Accepted: 21 December 2022
Published: 28 December 2022

1. Introduction

Multiple sclerosis (MS) is a chronic autoimmune disease of the central nervous system, with effects including demyelination, axonal loss and inflammatory episodes, making MS the most common cause of neurologic disability in young adults in the Western world [1,2]. Its distribution and prevalence are heterogeneous, and the global median prevalence is 33 per 100,000 people. Generally, young adults are the most affected, and the disease is more frequent in the female sex (3:1) [3,4].

Motor and cognitive dysfunctions are frequent in this type of patients. The limitation of activities, the onset of psychiatric disorders and fatigue cause severe disability, which worsens the physical condition, mobility and quality of life. Quality of life is more limited than in the general population or in those with other chronic diseases, and the decline is progressive in at least one-third of patients after diagnosis [5–7].

The worsening of symptoms in this population is associated with lower levels of physical activity, independently of other factors such as depression, neurological disability or the course of the disease, which have a serious impact on quality of life [8]. Therefore, the application of a concurrent physical exercise program, such as the one applied by

Grazioli [9], improves the quality of life as well as other factors that affect it, such as the severity of the illness, depression and fatigue.

Another beneficial therapy for MS symptoms is the use of transcranial direct current stimulation (tDCS). The application of anodic-type tDCS produces an increase in cortical excitability, changes in the neurotransmission system of glial cells and changes in the state of the cerebral microvasculature and inflammatory processes [10]. In fact, the application of anodic-type tDCS on the left dorsolateral prefrontal cortex (DLPFC) shows a significant decrease in fatigue, but only the Mortezanejad [11] study has assessed its impact on quality of life, reporting improvements that were maintained up to 4 weeks after intervention.

The current evidence on MS shows a tendency toward improvement in quality of life after the application of an exercise program or anodic tDCS on the DLPFC area. However, such evidence is insufficient to prove these treatments' effectiveness, and it is important to consider that most studies apply a very small number of sessions of tDCS to an MS population. From this point of view, it is important and necessary to carry out a study with objectives that focuse on the analysis of quality of life after the application of the two interventions. Our hypothesis is that subjects receiving these treatments will show an improvement in quality of life and in the variable of fatigue.

2. Materials and Methods

2.1. Study Design

A cross-over design was used to carry out this study. On 2 March 2020, the tDCS was applied to the first participant; however, due to health measures imposed by COVID-19, the study was discontinued. Data collection was resumed on 8 June 2020, and this first period ended on 28 August 2020. After a washout period of 5 months, the exercise program was implemented following the guidelines of Muñoz et al. (2022) [12], and data collection was completed on 19 April 2021.

2.2. Participants

A sample size calculation was carried out by calculating the difference between dependent groups utilizing the G*Power-3.1.9.2 software (G*Power©, Dusseldorf University, Dusseldorf, Germany). The predetermination of sample size was calculated for α-error of 0.05 and B error of 0.20. However, after the COVID-19 pandemic, many of the people with MS who attended the rehabilitation centers where the sample was recruited stopped attending due to caution or fear, which made recruitment difficult. For this reason, a crossover design was chosen. In addition, our sample is very small; for this reason, the findings from the patients we present must be considered as preliminary results, and the study should be considered as a pilot study.

In this study, 15 patients participated (6 females and 9 males), but 3 of them were excluded. The reasons for exclusion were as follows: the first one had a surgical intervention, the second presented with COVID-19 and the third was hospitalized due to exacerbation of the disease. The sample was recruited at the Palencia head office of Aspaym Castilla y León and at the Multiple Sclerosis Association of the city of Palencia.

The inclusion criteria followed in this study were a diagnosis of multiple sclerosis (no type of MS was excluded), a score of 38 points or more indicating the presence of fatigue on the Modified Fatigue Impact Scale (MFIS) [13], ability to walk at least 20 m without rest and age of older than 18 years. However, patients were excluded if they had any disease that could affect muscle function, along with those with a cardiovascular risk profile, respiratory disease or other disturbances that could interfere with the exercise program. Those who were pregnant or who did not have a good understanding of and good writing skills in Spanish were also excluded.

2.3. Ethical Considerations

The study was approved by the legal ethical committee of the University of León ULE-010-2020. The principles described in the Declaration of Helsinki were followed. An information sheet was given describing the study, including its risks and benefits. This was also explained verbally, in case there were any doubts, including that participants may abandon the intervention at any time. All participants agreed to and signed the written informed consent form prior to enrollment into the study. All participants consented to the publication of identifiable details, which may include photograph(s), video, case history and/or personal details in an online open-access publication.

2.4. Interventions

2.4.1. Transcranial Direct Current

The HDCstim stimulator (Newronika, Milán, Italy) was used for the application of tDCS: #HS0042/01-13; HDcel: #HE0021/02-13. Direct current stimulation was distributed with 2mA of current intensity in 35cm2 sponge electrodes. During the session, the current was increased during the first 15 s to a maximum of 2mA that was maintained throughout the 20 min stimulation session.

An adjustable head mesh was used to fix the stimulation electrodes in place, and the points of application were determined with the 10–20 EEG system, using the protocol described by DaSilva [14] (Figure 1), as this has been proven to be a useful and cost-effective method to localize the cortical areas. The cathode was laid in the right supraorbital cortex, while the anode was laid in the left DLPFC region (F3 according to the 10–20 EEG system). A total of 10 sessions, 20 min in duration, over the course of 4 weeks were applied by a specialized physiotherapist.

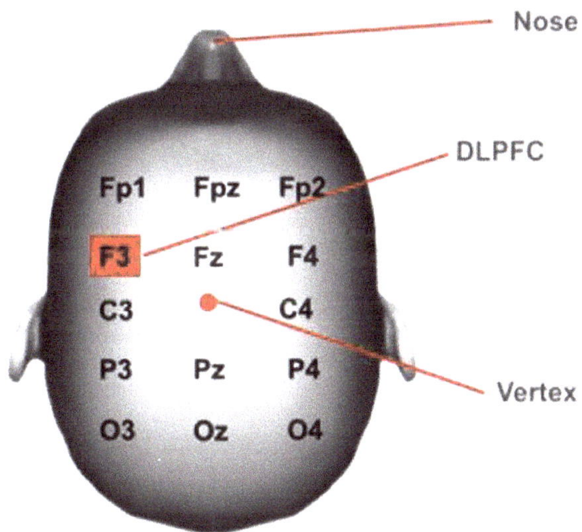

Figure 1. DLPFC (dorsolateral prefrontal cortex) position.

2.4.2. Exercise Program

The exercise program, applied by a specialized physiotherapist, consisted of both strength and aerobic training. Current evidence shows that this type of training is more effective than others for the treatment of fatigue in MS [15,16]. The exercise program was distributed over 4 weeks and conformed to the recommendations that exist for the development of exercise programs in this group of patients [16–19]. The exercise plan was carried out over 4 weeks, where the sessions progressively increased in intensity (Tables A1 and A2). The strength and aerobic training sessions were performed on different days according to the recommendations established for this population group.

The strength training was carried out in a circuit. In each circuit, there were 6 exercises of pushing and pulling exercises of the upper and lower limbs, trunk and pelvic girdle.

Two circuits were designed that require working the same muscle groups, but in different positions, thus adapting the circuit to facilitate its execution for subjects with some functional limitation and involving the same muscular activity. Moreover, the repetitions and the rest time between exercise and circuits were set in the program. The intensity used for the aerobic training was moderate, which corresponds to a level 3–5 on the rating scale of perceived exertion (RPE). Motomed kinesio therapeutic equipment® (RECK-Technik GmbH & Co. KG, Medizintechnik, Reckstraße 1–5, D-88422 Betzenweiler, Germany) or a static bike were used, depending on the participant's preference.

2.5. Outcome Measures

The primary outcome is the assessment of quality of life after the application of the two treatments. The secondary outcome is to study the relationship between quality of life and sociodemographic, clinical and anthropometric variables, as well as to evaluate fatigue after the application of the two treatments.

Quality of life was assessed before and after each intervention using the Spanish version of the Multiple Sclerosis International Quality of Life (MusiQoL) questionnaire. It consists of 31 items describing 9 dimensions: activities of daily living, psychological well-being, symptomatology, social and family relationships, relationship with the health care system, intimate/sexual life, coping and rejection. The questionnaire uses a 6-point Likert scale, and, for each patient, the score of each dimension is obtained by taking the mean of the scores of the items of each dimension. All dimension scores are linearly translated to a scale from 0 to 100, and negatively marked scores are inverted so that higher scores indicate a higher level of quality of life. The global index score is computed as the mean of the dimension scores. The scale shows good construct validity, internal consistency, reproducibility and reliability [20,21].

Physical activity was evaluated at baseline using the Spanish version of the International Physical Activity Questionnaire Short Form (IPAQ-SF). The reliability and validity have been investigated and analyzed in different countries and contexts and with different types of populations, including in MS patients. Frequency and duration of vigorous, moderate and walking physical activity is measured by this questionnaire during a 7-day period. The respective frequencies and durations were initially multiplied, and the resulting volumes were later multiplied to obtain the METs, by 8 in the case of vigorous activity, 4 in the case of moderate activity and 3.3 in the case of walking [22,23].

The Kurtzke Expanded Disability Status Scale (EDSS) was used to assess neurological involvement and disability. This variable was investigated in the first data collection. It consists of neurological examination findings and includes 20 grades on a scale from 0 (normal examination) to 10 (death due to MS), with 0.5-point intervals. Patients are assessed with regard to the neurological examination and clinical history for each functional system, and, then, taking into account the ability to walk [24,25], an overall score is obtained.

The Spanish version of the Modified Fatigue Impact Scale (MFIS) was used to assess fatigue before and after each intervention. This questionnaire uses a multidimensional approach and consists of 21 items distributed in 3 subscales. According to the frequency with which the symptom has occurred, the patient will respond to each item regarding the last week. The final scores ranged from 0 to 84, and a score of 38 was established as the cutoff point to indicate the presence of fatigue [13,26].

2.6. Procedure

First of all, the study information sheet was distributed, answering any queries presented by the participants. Then, all participants signed the informed consent form.

Afterwards, each participant was given a registration form with basic information to check compliance with the inclusion and exclusion criteria of the study and, thus, determine their ability to complete the self-administered questionnaires. This part had an estimated duration of 15 min.

Consecutively, tDCS was applied over 10 sessions of 20 min in duration over the course of 4 weeks. All participants were stimulated from the beginning. Fatigue and quality of life were measured again with the same questionnaires after the application of the intervention.

A stabilization period of 5 months was used to avoid affecting results obtained after the first intervention. In our case, we selected 5 months because the literature determines that the benefits of the treatment last up to 3 weeks after the application of tDCS, when it is applied for 5 days. In a study by Ferrucci [27], an intensity of 1.5mA was applied for 15 min in the primary motor region (M1) for 5 days. Since it has been shown that the effect of tDCS is cumulative and that this effect is necessary to generate the appropriate adaptations for the desired objective, a long washout period was chosen to prevent these effects from interfering with those of the other therapy.

After this period, fatigue and quality of life data were gathered again, and a concurrent training program was performed for a period of 4 weeks. After this time, fatigue and quality of life data were gathered for the last time. Compliance with attendance at the interventions was assessed by means of an attendance control chart.

2.7. Statistical Analysis

For statistical analysis, the Statistical Package for Social Sciences (SPSS 21, SPSS Inc., Chicago, IL USA) software package was used. Descriptive statistics were used in the data analysis to show the data for continuous variables presented as ± standard deviation (SD) and relative frequency (percentage) for categorical variables. The Shapiro–Wilk test was used to evaluate the normality, and the result indicated that not all of them met normality, so non-parametric tests were used for the statistical calculation. In addition, a Wilcoxon signed-rank test was used to analyze the results in fatigue and quality of life after the application of exercise and tDCS.

The results for the indicators of fatigue and quality of life were analyzed by Friedman post hoc Dunn test.

The effect size was calculated to express the magnitude of the differences between the samples. For this purpose, Hedges' g (scale: 0–1) was used, which sets the effect size as small (0.2), medium (0.5) or large (0.8) [28].

The significance level for all tests was set at $p < 0.05$.

3. Results

The initial characteristics of the participants are shown in Table 1, and the flowchart is shown in Figure 2.

Table 1. Participant's baseline demographic and clinical characteristics.

VARIABLES	N [MIN–MAX]; MEAN ± (SD)	FREQUENCY (%)
Age	12 [35–66]; 48.08 ± 8.55	
Years of diagnosis	12 [0.8–28]; 16.65 ± 7.44	
Outbreaks per year	11 [0–2]; 0.36 ± 0.67	
Walking time (minutes)	9 [0.0–120]; 51.11 ± 41.06	
Sitting time (minutes)	9 [0–960]; 466.667 ± 304.13	
	Type of sclerosis	
Relapsing–remitting		7 (58.3%)
Progressive secondary		5 (41.7%)
	Outbreak intensity	
Mild		2 (18.2%)
Moderate		1 (9.1%)
Intense		1 (9.1%)
No outbreaks		7 (63.6%)
	Medical recommendation	
Physical activity		6 (11.3%)
Other		1 (8.3%)

Table 1. *Cont.*

VARIABLES	N [MIN–MAX]; MEAN ± (SD)	FREQUENCY (%)
No recommendation		5 (41.7%)
Fatigue medication		
Yes		5 (41.7%)
No		7 (58.3%)
Type of medication fatigue		
LioresalLioresal + Avonex		1 (9.1%)
Lioresal + Rebif 44		1 (9.1%)
Other medication		1 (9.1%)
No medication		2 (16.7%)
		7 (58.3%)
Rehabilitation		
Yes		11 (91.7%)
No		1 (8.3%)
Intensity rehabilitation		
Occasional		5 (41.7%)
Periodic		7 (58.3%)
Exercise habits		
Occasional		2 (16.7%)
Regularly		10 (83.3%)
Education level		
Primary education		1 (8.3%)
Secondary studies		1 (8.3%)
Vocational training		6 (50%)
University studies		4 (33.3%)
Employment situation		
Homemaker		1 (8.3%)
Part-time employee		1 (8.3%)
Full-time employee		5 (41.7%)
Retired		3 (25%)
Permanently disabled		2 (16.7%)

Our study shows that, after the implementation of both therapies, there is a significant improvement in the fatigue and quality of life scales. This suggests that tDCS and exercise may be effective in improving quality of life and fatigue in this sample (Table 2).

Table 2. Pre–post tDCS and exercise through Wilcoxon test: quality of life and fatigue.

	Pre tDCS Median [Range]	Post tDCS Median [Range]	p	Size Effect Hedges' g	Pre-Exercise Median [Range]	Post-Exercise Median [Range]	p	Size Effect Hedges' g
MQOL	68.04 [25.28]	75.19 [32.23]	0.015 *	0.646	70.07 [28.99]	74.94 [27.2]	0.003 **	0.56
MQOLADL	58.9 [6.25]	64.80 [6.55]	0.037 *	0.465	57.73 [1.5]	68.09 [6.95]	0.003 **	0.570
MQOLPWB	66.24 [3]	80.42 [6.82]	0.004 **	0.727	65 [2]	72.5 [4.33]	0.41 *	0.417
MQOLSYM	72.5 [4]	77.5 [5.8]	0.438	0.258	72.5 [6.17]	74.58 [4.42]	0.625	0.122
MQOLSOREL	76.93 [4.88]	78.31 [5.92]	0.413	0.055	84.98 [2]	87.2 [3.67]	0.336	0.120
MQOLRFAREL	81.65 [6.25]	83.32 [3.92]	0.44	0.066	88.88 [2.3]	90.53 [3.4]	0.102	0.104
MQOLSEXLIFE	73.75 [5]	75 [3.25]	0.865	0.045	73.33 [2]	77.5 [2.67]	0.257	0.129
MQOLCOP	50.83 [3]	61.67 [5.25]	0.018 *	0.376	51.67 [5.75]	56.66 [4.79]	0.177	0.266
MQOLREJEC	68.33 [3]	79.17 [6]	0.103	0.475	75 [3.5]	80 [4.08]	0.058	0.221
MQOLREHEALTH	79.41 [3.38]	78.88 [4.83]	0.933	0.036	78.87 [2.3]	82.07 [2.67]	0.269	0.152
MFIS	39.5 [31]	38.5 [45]	0.028 *	0.525	43 [33]	36 [52]	0.003 **	0.742

Non-parameter statistics. Wilcoxon signed-rank test. MQOL: Multiple Sclerosis International Quality of Life. MQoLADL: MusiQoL subscale activities of daily life. MQoLPWB: MusiQoL subscale psychological well-being. MQoLSYM: MusiQoL subscale symptomatology. MQoLSOREL: MusiQoL subscale social relationship. MQoL-FAREL: MusiQoL subscale family relationship. MQoLSEXLIFE: MusiQoL subscale sexual life. MQoLCOP: MusiQoL subscale coping. MQoLREJEC: MusiQoL subscale rejection. MQoLREHEALTH: MusiQoL subscale relationship with the health care. MFIS: Modified Fatigue Impact Scale. tDCS: current direct transcranial stimulation. Pre tDCS: treatment before transcranial direct current stimulation (tDCS). Post tDCS: treatment after transcranial direct current stimulation (tDCS). * $p < 0.05$ and ** $p < 0.001$.

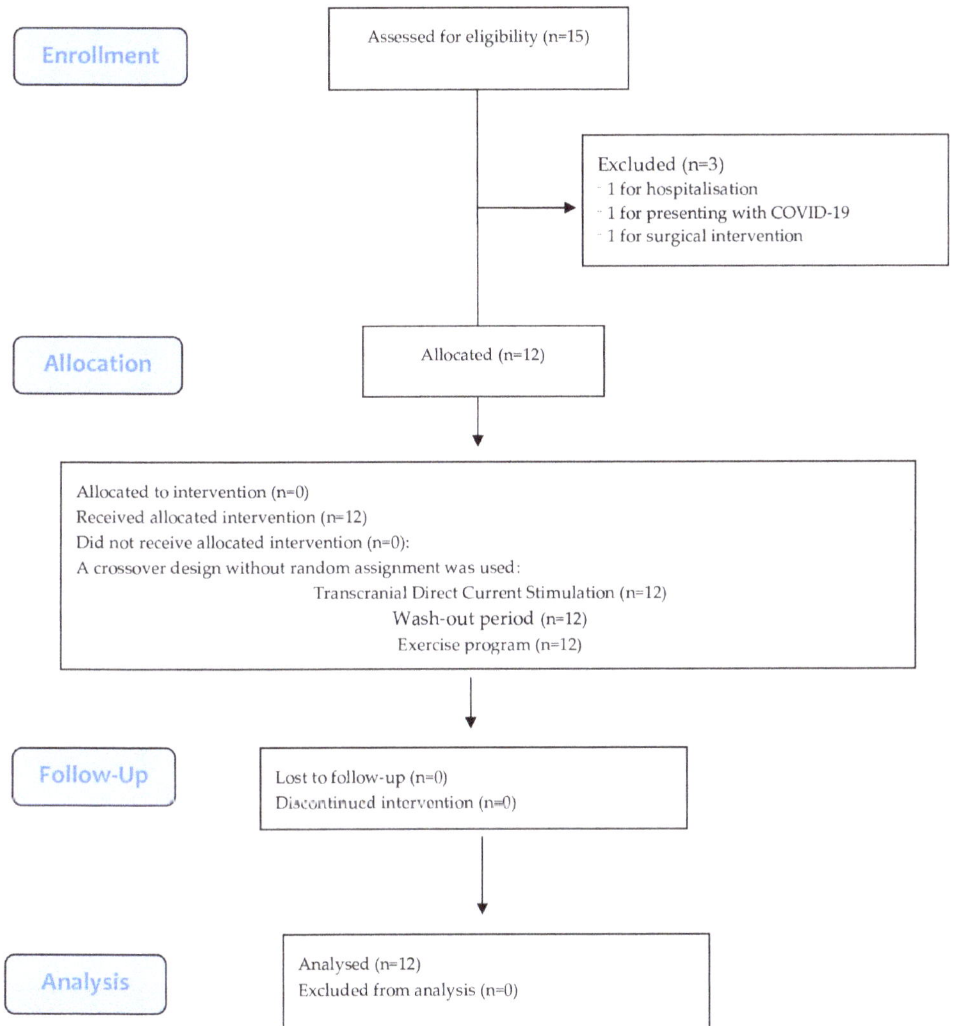

Figure 2. CONSORT flowchart diagram.

Furthermore, the effect size for these outcomes is moderate, which indicates the clinical relevance of the interventions.

Likewise, for the quality-of-life scale, we found significant improvements and clinical changes after the application of tDCS ($p = 0.015$) (g = 0.646) and after exercise ($p = 0.003$) (g = 0.56). Moreover, statistical analysis shows that there have been significant changes after the application of the tDCS in the subscales of activities ADL ($p = 0.037$) (g = 0.465), PWB ($p = 0.004$) (g = 0.727) and COP ($p = 0.18$) (g = 0.376). After the exercise program, significant changes were found in the subscales of ADL ($p = 0.003$) (g = 0.570) and PWB ($p = 0.041$) (g = 0.417).

Furthermore, the Friedman results showed that all variables improved in the tDCS group (Friedman´s test $X^2 = 8.33$, $p = 0.002$, N = 12), with a significant difference between after tDCS treatment and before treatment tDCS ($p = 0.009$). They also improved in the exercise group (Friedman´s test $X^2 = 13.44$, $p = 0.004$, N = 12) with a significant difference between after exercise treatment and before treatment exercise (p= 0.021), as shown in Figure 3.

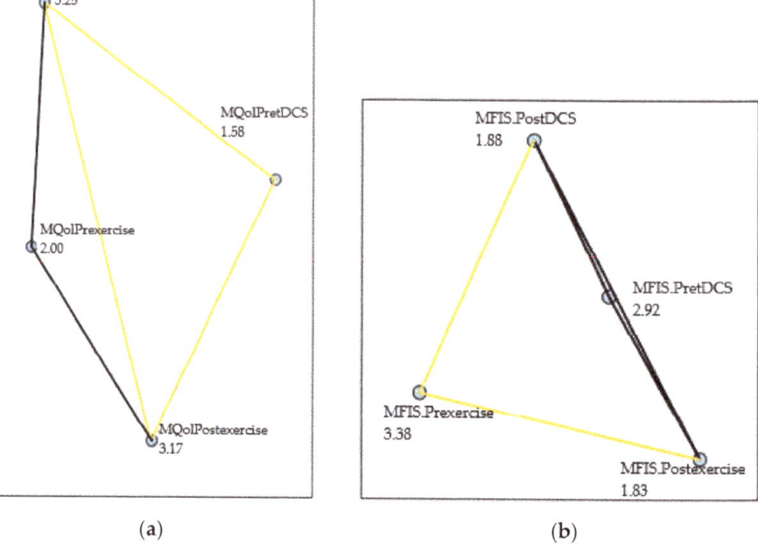

Figure 3. Pairwise comparisons Friedman's test. (**a**) Multiple Sclerosis International Quality of Life (MusiQoL). (**b**) Modified Fatigue Impact Scale (MFIS).

Finally, we observed that there were significant changes in fatigue when applying the tDCS ($p = 0.028$) (g = 0.525) and when applying the exercise program ($p = 0.003$) (g = 0.742). Moreover, more than 50% of the subjects did not have a positive fatigue score on the MFIS scale after the application of the therapies.

4. Discussion

The results of this study suggest that tDCS and an exercise program may be effective in improving quality of life and fatigue in subjects with MS. Moreover, both variables improved in the groups, with significant differences over time. However, it is important to note that given the sample size and the type of design, our results have to be interpreted with caution.

Quality of life is a multidimensional construct and has special importance in the assessment of MS because it is considered more impaired than in other chronic diseases [29]. In addition, because of the impact of fatigue on quality of life, the studies that focus their therapy on improving quality of life orient the therapies toward the treatment of fatigue symptomatology. For this reason, we thought it was important to analyze the improvement in fatigue after the application of the treatments, so that our results indicate that both the application of the exercise and the tDCS produce beneficial effects.

The current pharmacological treatments are unsatisfactory in many patients because they have modest benefits and numerous side effects. This is perhaps why the combination of pharmacological and other types of therapies may provide the optimal management of the disease and the symptomatology of fatigue [30–34]. Accordingly, the use of physical exercise has the capacity to reverse the consequences of inactivity in these patients and can also modify the anti-inflammatory effect of the disease. This demystifies the belief that the practice of physical activity and the corresponding rise in body temperature could be detrimental for this disease, resulting in a relapse in symptoms [18,35,36].

Despite the benefits of exercise, there is sedentary behavior present in our sample because the mean sitting time spent by the subjects analyzed is 7.77 h, while the mean walking time on a normal day is reduced to 51 min; according to the IPAQ questionnaire, only four of the participants maintain high levels of activity. Therefore, it is important to take into account that sedentary behavior is associated with high levels of mortality and morbidity in the general population and with disability and fatigue in MS [37–39].

In our study, the practice of concurrent training shows significant improvements in fatigue and a higher score on the quality of life scale and its various subscales. In particular, in the subscales of activities of daily living and psychological well-being, significant changes were noted. Along the same lines, we found improvements in the same subscales of the MusiQoL questionnaire as in Tarkci's study [40], comparing a 12-week training program including stretching, balance, core exercises and functional activities with a control group. Performing calisthenics also reportedly brings a significant improvement on the MusiQoL questionnaire [41]. Both studies performed a 12-week exercise program, although we obtained similar results with a 4-week program. It has been shown that practicing a concurrent training program is more effective than strength training and task-oriented training. If we focus on resistance training, several authors support that it is not a consistent measure of fatigue reduction on its own, and, therefore, it is not a consistent measure of quality of life improvement. In addition, one of the reasons why our 4-week exercise program has shown similar results may be due to a better type of training being chosen. Moreover, the intensity of our program is better delimited, since it has been shown that a moderate intensity is ideal in this population. To delimit the intensity of the exercise program, we have different options: the application of 50–70% of the maximum volume of oxygen (VO2 max), 80–60% of the maximum cardiac capacity or a submaximal effort that would correspond to a score of 11–14 of the Borg scale [15,16,19]. However, we must bear in mind that these studies, like ours, include subjects with different types of MS and of different ages. Even so, the beneficial effect of physical exercise on this population has been demonstrated, although in the future it could be interesting to study whether there is a specific type of training that benefits each type of MS.

According to our data, a significant improvement in quality of life and a clinically relevant improvement in the fatigue variable are observed after the application of tDCS. There are several areas of application of this therapy that have been shown to be effective for improvement in fatigue in MS. For example, Cancelli's group [42] applied the stimulation on the primary somatosensory area, reducing fatigue by 42% after stimulation. Meanwhile, Chalah's group [43] applied the stimulation on the right posterior parietal lobe, with beneficial effects on fatigue, anxiety and depression. It is true that the area of preference for stimulation is the left DLPFC area, which is the one chosen in this study. One of the reasons for this is that it has been shown to be dose-dependent, which is why we have applied a total of 10 sessions. On the other hand, this stimulation area has been shown to have significant effects both in the decrease in fatigue and in the increase in the quality of life, maintained even 4 weeks after the intervention [11,44,45].

Few articles assessed the quality-of-life scale after the application of tDCS, and none of them used the MusiQoL questionnaire. Only Marzieh's [11] study evaluated it after application in the left DLPFC area; the other authors used stimulation in the primary motor area, as their objectives were focused on pain. Marzieh's group used the Multiple Sclerosis Specific Quality of Life Questionnaire (MSQOL-54), obtaining significant improvements that are maintained after stimulation.

Currently, there is no research where tDCS is applied over the left DLPFC area, compared or combined with an exercise program to assess quality of life in MS or other neurological pathologies. Our results indicate that both therapies are effective in improving quality of life, so the application of tDCS in the DLPFC area may have interesting results in MS, without forgetting the cumulative effect of tDCS in order to induce reliable and persistent changes. However, we cannot suggest that one therapy is superior to another.

4.1. Practical Application

Quality of life is a complex assessment due to its multifactorial nature, where clinical, psychological and social data converge and where the patient's perception takes on special relevance. It is, therefore, of particular importance to assess whether the treatments applied in this population benefit this aspect. In this case, both the application of the exercise program and the application of tDCS have shown beneficial effects in quality of life and in several of its subscales. Therefore, we could suggest that the application of concurrent training, which not only focuses its work on the knee musculature but also on the secondary and stabilizing muscles of the trunk, hip and upper limb, could be beneficial in these subjects, regardless of their age and type of sclerosis, as all of them significantly increased their score on the quality of life scale. On the other hand, the benefits suggested by the application of tDCS on the left DLPFC area on quality of life may be interesting to study in the future, along with its effectiveness and potential in larger samples, and to check, if appropriate, if the application on this area shows better results than in other areas, such as the primary motor cortex. Nevertheless, in line with the suggestions of other studies, such as the review by Giuseppina et al. [46], the combination of tDCS and exercise with prolonged treatment with multiple sessions would be optimal to generate measurable benefits.

4.2. Limitations

One of the limitations of our study is the small sample size; for this reason, the findings from the patients we present must be considered as preliminary results, and the study should be considered as a pilot study. In addition, our data should be treated with caution as the sample is limited, and we did not have a long follow-up period. Future research with larger samples is necessary; similarly, randomized clinical trials are needed to accurately determine the efficacy and effectiveness of the treatment.

5. Conclusions

Major findings suggest that both the implementation of an exercise program and the application of tDCS produce a beneficial effect with significant improvements in quality of life in this sample. In the same way, it could be suggested that adding tDCS could provide further improvements to physical therapy, but further research with a larger sample and sham stimulation as a control is needed.

Author Contributions: Conceptualization: I.M.-P., A.J.H. and J.S.-C.; data curation: I.M.-P.; formal analysis: I.M.-P.; investigation: I.M.-P., A.J.H. and J.S.-C.; data collection: I.M.-P., N.R.-N. and A.M.P.-G.; methodology: I.M.-P. and J.S.-C.; project administration: I.M.-P. and J.S.-C.; resources: I.M.-P. and A.J.H.; software: I.M.-P.; supervision: I.M.-P., A.J.H. and J.S.-C.; visualization: I.M.-P., N.R.-N. and A.M.P.-G.; writing—original draft: I.M.-P. and J.S.-C.; writing—review and editing: I.M.-P., A.J.H. and J.S.-C. All authors have read and agreed to the published version of the manuscript.

Funding: This research was funded by a grant from the Professional Association of Physiotherapists of Castilla y León (grant number INV2020-15).

Institutional Review Board Statement: The study was approved by the legal ethical committee of the University of León, ULE-010-2020. The study followed the principles described in the Declaration of Helsinki. All participants agreed to and signed the written informed consent form prior to enrollment in the study.

Informed Consent Statement: Written informed consent was obtained from all the subjects involved in the study.

Data Availability Statement: The datasets used and/or analyzed in the current study are available from the corresponding author on reasonable request.

Acknowledgments: The authors acknowledge Laura Varas López for proofreading the manuscript and the Esclerosis Mútiple and Aspaym of Palencia associations for providing administrative and material support.

Conflicts of Interest: The authors declare no conflict of interest.

Appendix A

Table A1. Exercise program schedule.

	Week 1	Week 2	Week 3	Week 4
Endurance	1 Session 10 min 3–5 RPE MOTOmed/static bike	2 Sessions 15 min 3–5 RPE MOTOmed/static bike	2 Sessions 10 min + 10 min (5 min of rest) 3–5 RPE MOTOmed/static bike	2-3 Sessions 15 min + 15 min (5 min of rest) 3–5 RPE MOTOmed/static bike
Strength	2 Sessions (A, B) 6 Exercise in each circuit 15 Rep (rest: 2 min) 2 Circuits (rest: 3 min)	2 Sessions (A, B) 6 Exercise in each circuit 15 Rep (rest: 2 min) 2 Circuits (rest: 3 min)	3 Sessions (circuit to choose) 6 Exercise in each circuit 10 Rep (rest: 3 min) 3 Circuits (rest: 5 min)	3 Sessions (circuit to choose) 6 Exercise in each circuit 10 Rep (rest: 3 min) 3 Circuits (rest: 5 min)

Exercise program. A, B refers to exercises specified in the table. RPE = rate of perceived exertion. Min = minutes. MOTOmed = aerobic kinesiotherapy equipment. Rep = repetitions.

Table A2. Exercises are used in circuits A and B of the exercise program.

Exercise	Circuit A	Circuit B
MMSS PULL	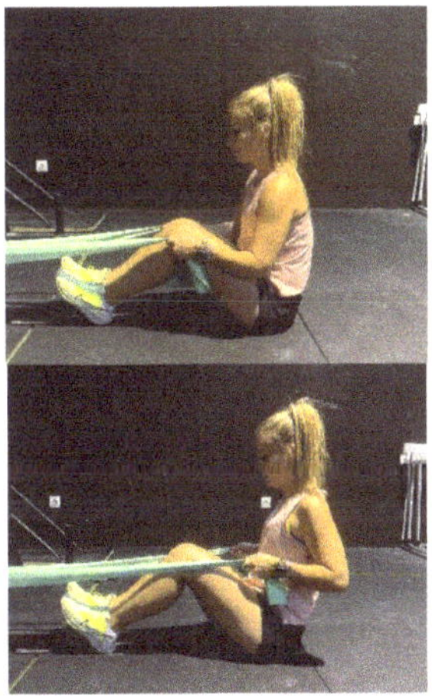 	 
MMSS PUSH		

Table A2. *Cont.*

Exercise	Circuit A	Circuit B
KNEE EXTENSION	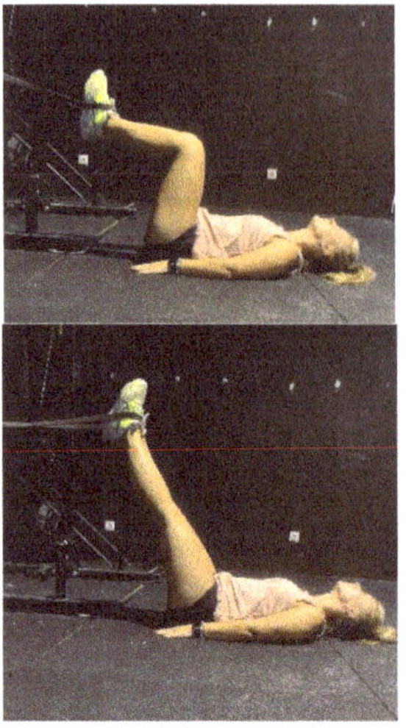	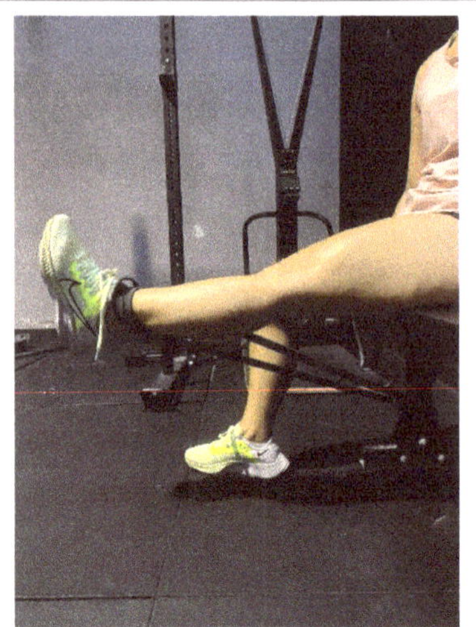
KNEE FLEXION	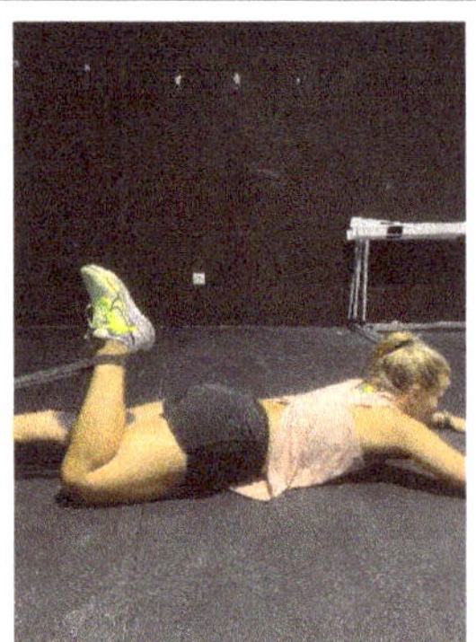	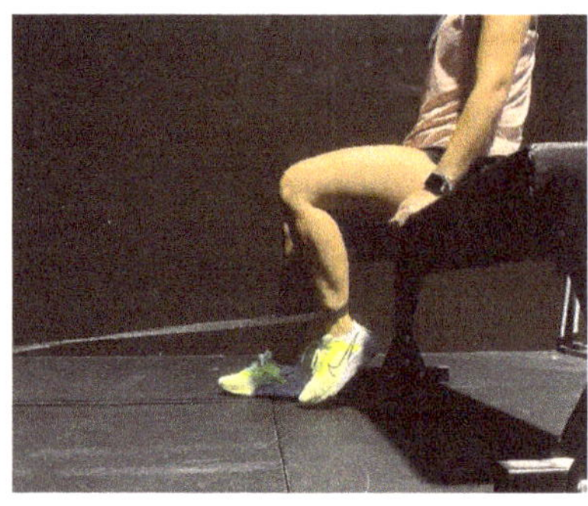
HIP EXTENSION	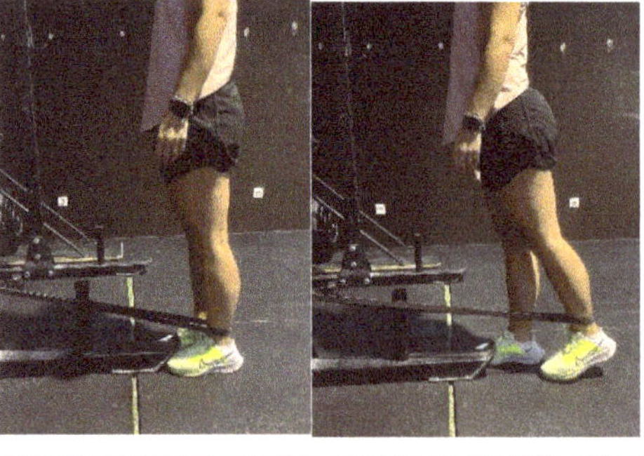	

Table A2. *Cont.*

Exercise	Circuit A	Circuit B
HIP FLEXION		
CORE	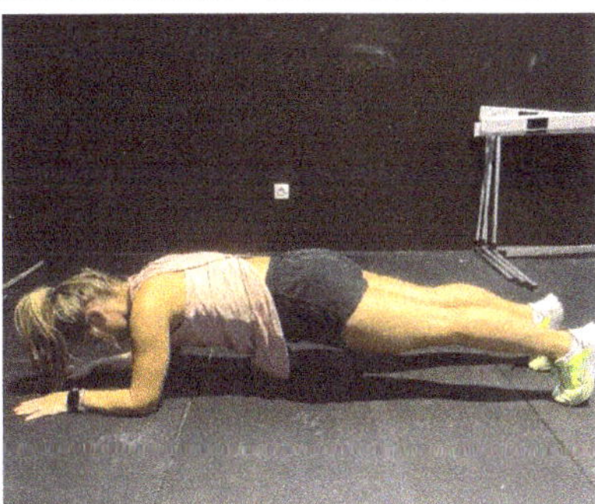	

Images of the exercises developed according to the objectives to be achieved. Two different exercises are shown for each objective.

References

1. Fernández Fernández, Ó.; Arroyo González, R.V.; Rodríguez Antigüedad, A.; García Merino, J.A.; Comabella López, M.; Villar, L.M.; Izquierdo Ayuso, G.; Tintoré Subirana, M.d.M.; Oreja Guevara, C.; Álvarez Cermeño, J.C.; et al. Biomarcadores En Esclerosis Múltiple. *Rev. Neurol.* **2013**, *56*, 375. [CrossRef]
2. Tremlett, H.; Zhao, Y.; Rieckmann, P.; Hutchinson, M. New Perspectives in the Natural History of Multiple Sclerosis. *Neurology* **2010**, *74*, 2004–2015. [CrossRef] [PubMed]
3. Abad, P.; Pérez, M.; Castro, E.; Alarcón, T.; Santibáñez, R.; Díaz, F. Prevalence of Multiple Sclerosis in Ecuador. *Neurología* **2010**, *25*, 309–313. [CrossRef] [PubMed]
4. Vidal-Jordana, A.; Montalban, X. Multiple Sclerosis: Epidemiologic, Clinical, and Therapeutic Aspects. *Neuroimaging Clin. N. Am.* **2017**, *27*, 195–204. [CrossRef] [PubMed]
5. Pilutti, L.A.; Platta, M.E.; Motl, R.W.; Latimer-Cheung, A.E. The Safety of Exercise Training in Multiple Sclerosis: A Systematic Review. *J. Neurol. Sci.* **2014**, *343*, 3–7. [CrossRef]
6. Mitchell, A.J.; Benito-León, J.; González, J.M.M.; Rivera-Navarro, J. Quality of Life and Its Assessment in Multiple Sclerosis: Integrating Physical and Psychological Components of Wellbeing. *Lancet Neurol.* **2005**, *4*, 556–566. [CrossRef]
7. Tallner, A.; Waschbisch, A.; Hentschke, C.; Pfeifer, K.; Mäurer, M. Mental Health in Multiple Sclerosis Patients without Limitation of Physical Function: The Role of Physical Activity. *Int. J. Mol. Sci.* **2015**, *16*, 14901–14911. [CrossRef]
8. Motl, R.W.; Arnett, P.A.; Smith, M.M.; Barwick, F.H.; Ahlstrom, B.; Stover, E.J. Worsening of Symptoms Is Associated with Lower Physical Activity Levels in Individuals with Multiple Sclerosis. *Mult. Scler.* **2008**, *14*, 140–142. [CrossRef]
9. Grazioli, E.; Tranchita, E.; Borriello, G.; Cerulli, C.; Minganti, C.; Parisi, A. The Effects of Concurrent Resistance and Aerobic Exercise Training on Functional Status in Patients with Multiple Sclerosis. *Curr. Sports Med. Rep.* **2019**, *18*, 452–457. [CrossRef]

10. Akcali, A.; Zengin, F.; Aksoy, S.N.; Zengin, O. Fatigue in Multiple Sclerosis: Is It Related to Cytokines and Hypothalamic-Pituitary-Adrenal Axis? *Mult. Scler. Relat. Disord.* **2017**, *15*, 37–41. [CrossRef]
11. Mortezanejad, M.; Ehsani, F.; Masoudian, N.; Zoghi, M.; Jaberzadeh, S. Comparing the Effects of Multi-Session Anodal Trans-Cranial Direct Current Stimulation of Primary Motor and Dorsolateral Prefrontal Cortices on Fatigue and Quality of Life in Patients with Multiple Sclerosis: A Double-Blind, Randomized, Sham-Controlled Trial. *Clin. Rehabil.* **2020**, *34*, 1103–1111. [CrossRef] [PubMed]
12. Muñoz-Paredes, I.; Herrero, A.J.; Llamas-Ramos, R.; Rodríguez-Pérez, V.; Seco-Calvo, J. The Effect of Combining Transcranial Direct Current Stimulation Treatment and an Exercise Program on Fragility in a Population with Multiple Sclerosis: Cross-Over Design Trial. *Int. J. Environ. Res. Public Health* **2022**, *19*, 12747. [CrossRef] [PubMed]
13. Kos, D.; Kerckhofs, E.; Carrea, I.; Verza, R.; Ramos, M.; Jansa, J. Evaluation of the Modified Fatigue Impact Scale in Four Different European Countries. *Mult. Scler.* **2005**, *11*, 76–80. [CrossRef] [PubMed]
14. DaSilva, A.F.; Volz, M.S.; Bikson, M.; Fregni, F. Electrode Positioning and Montage in Transcranial Direct Current Stimulation. *J. Vis. Exp.* **2011**, *51*, e2744. [CrossRef] [PubMed]
15. Heine, M.; van de Port, I.; Rietberg, M.B.; van Wegen, E.E.; Kwakkel, G. Exercise Therapy for Fatigue in Multiple Sclerosis. *Cochrane Database Syst. Rev.* **2015**, *9*, CD009956. [CrossRef]
16. Latimer-Cheung, A.E.; Pilutti, L.A.; Hicks, A.L.; Martin Ginis, K.A.; Fenuta, A.M.; MacKibbon, K.A.; Motl, R.W. Effects of Exercise Training on Fitness, Mobility, Fatigue, and Health-Related Quality of Life among Adults with Multiple Sclerosis: A Systematic Review to Inform Guideline Development. *Arch. Phys. Med. Rehabil.* **2013**, *94*, 1800–1828.e3. [CrossRef]
17. White, L.J.; McCoy, S.C.; Castellano, V.; Gutierrez, G.; Stevens, J.E.; Walter, G.A.; Vandenborne, K. Resistance Training Improves Strength and Functional Capacity in Persons with Multiple Sclerosis. *Mult. Scler.* **2004**, *10*, 668–674. [CrossRef]
18. Dalgas, U.; Stenager, E.; Ingemann-Hansen, I. Multiple Sclerosis and Physical Exercise: Recommendations for the Application of Resistance-, Endurance- and Combined Training. *Mult. Scler.* **2008**, *14*, 35–53. [CrossRef]
19. Halabchi, F.; Alizadeh, Z.; Sahraian, M.A.; Abolhasani, M. Exercise Prescription for Patients with Multiple Sclerosis; Potential Benefits and Practical Recommendations. *BMC Neurol.* **2017**, *17*, 185. [CrossRef]
20. Simeoni, M.C.; Auquier, P.; Fernandez, O.; Flachenecker, P.; Stecchi, S.; Constantinescu, C.S.; Idiman, E.; Boyko, A.; Beiske, A.G.; Vollmer, T.; et al. Validation of the Multiple Sclerosis International Quality of Life Questionnaire. *Mult. Scler.* **2008**, *14*, 219–230. [CrossRef]
21. Fernández, O.; Fernández, V.; Baumstarck-Barrau, K.; Muñoz, L.; Gonzalez Alvarez, M.D.M.; Arrabal, J.C.; León, A.; Alonso, A.; López-Madrona, J.C.; Bustamante, R.; et al. Validation of the Spanish Version of the Multiple Sclerosis International Quality of Life (Musiqol) Questionnaire. *BMC Neurol.* **2011**, *11*, 127. [CrossRef] [PubMed]
22. Gosney, J.L.; Scott, J.A.; Snook, E.M.; Motl, R.W. Physical Activity and Multiple Sclerosis: Validity of Self-Report and Objective Measures. *Fam. Community Health* **2007**, *30*, 144–150. [CrossRef] [PubMed]
23. Craig, C.L.; Marshall, A.L.; Sjöström, M.; Bauman, A.E.; Booth, M.L.; Ainsworth, B.E.; Pratt, M.; Ekelund, U.; Yngve, A.; Sallis, J.F.; et al. International Physical Activity Questionnaire: 12-Country Reliability and Validity. *Med. Sci. Sports Exerc.* **2003**, *35*, 1381–1395. [CrossRef] [PubMed]
24. Hobart, J.; Freeman, J.; Thompson, A. Kurtzke Scales Revisited: The Application of Psychometric Methods to Clinical Intuition. *Brain* **2000**, *123*, 1027–1040. [CrossRef] [PubMed]
25. Meyer-Moock, S.; Feng, Y.S.; Maeurer, M.; Dippel, F.W.; Kohlmann, T. Systematic Literature Review and Validity Evaluation of the Expanded Disability Status Scale (EDSS) and the Multiple Sclerosis Functional Composite (MSFC) in Patients with Multiple Sclerosis. *BMC Neurol.* **2014**, *14*, 58. [CrossRef] [PubMed]
26. Learmonth, Y.C.; Dlugonski, D.; Pilutti, L.A.; Sandroff, B.M.; Klaren, R.; Motl, R.W. Psychometric Properties of the Fatigue Severity Scale and the Modified Fatigue Impact Scale. *J. Neurol. Sci.* **2013**, *331*, 102–107. [CrossRef] [PubMed]
27. Ferruci, R.; Vergari, M.; Cogiamanian, F.; Bocci, T.; Ciocca, M.; Tomasini, E.; De Riz, M.; Scarpin, E.; Priori, A. Transcranial Direct Current Stimulation (TDCS) for Fatigue in Multiple Sclerosis. *NeuroRehabilitation* **2014**, *34*, 121–127. [CrossRef]
28. Cohen, J. *Statistical Power Analysis for the Behavioral Sciences*, 2nd ed.; Associated, L.E., Ed.; Elsevier Science: Morristown, NJ, USA, 1988; ISBN 9781483276489.
29. Olascoaga, J. Quality of Life and Multiple Sclerosis. *Rev. Neurol.* **2010**, *51*, 279–288. [CrossRef]
30. Brenner, P.; Piehl, F. Fatigue and Depression in Multiple Sclerosis: Pharmacological and Non-Pharmacological Interventions. *Acta Neurol. Scand.* **2016**, *134*, 47–54. [CrossRef]
31. Benito-León, J.; Morales, J.M.; Rivera-Navarro, J.; Mitchell, A.J. A Review about the Impact of Multiple Sclerosis on Health-Related Quality of Life. *Disabil. Rehabil.* **2003**, *25*, 1291–1303. [CrossRef]
32. Ayache, S.S.; Chalah, M.A. Fatigue and Affective Manifestations in Multiple Sclerosis—A Cluster Approach. *Brain Sci.* **2019**, *10*, 10. [CrossRef] [PubMed]
33. Ayache, S.S.; Chalah, M.A. Fatigue in Multiple Sclerosis–Insights into Evaluation and Management. *Neurophysiol. Clin.* **2017**, *47*, 139–171. [CrossRef] [PubMed]
34. Ayache, S.S.; Chalah, M.A. Fatigue in Multiple Sclerosis: Pathophysiology and Emergent Interventions. *Arch. Ital. Biol.* **2018**, *156*, 149–152. [CrossRef] [PubMed]
35. Dalgas, U.; Stenager, E. Exercise and Disease Progression in Multiple Sclerosis: Can Exercise Slow down the Progression of Multiple Sclerosis? *Ther. Adv. Neurol. Disord.* **2012**, *5*, 81–95. [CrossRef]

36. Tallner, A.; Waschbisch, A.; Wenny, I.; Schwab, S.; Hentschke, C.; Pfeifer, K.; Mäurer, M. Multiple Sclerosis Relapses Are Not Associated with Exercise. *Mult. Scler. J.* **2012**, *18*, 232–235. [CrossRef]
37. Alpa, P.; Leslie, B.; Anusila, D.; Heather, F.; Peter, C.; Susan, G.; Graham, C.; Michael, T. Leisure Time Spent Sitting in Relation to Total Mortality in a Prospective Cohort of US Adults. *Am. J. Epidemiol.* **2010**, *172*, 419–429. [CrossRef]
38. Cavanaugh, J.; Gappmaier, V.; Dibble, L.; Gappmaier, E. Ambulatory Activity in Individuals with Multiple Sclerosis. *J. Neurol. Phys. Ther.* **2011**, *35*, 26–33. [CrossRef]
39. Owen, N. Sedentary Behavior: Understanding and Influencing Adults' Prolonged Sitting Time. *Prev. Med.* **2012**, *55*, 535–539. [CrossRef]
40. Tarakci, E.; Yeldan, I.; Huseyinsinoglu, B.; Zenginler, Y.; Eraksoy, M. Group Exercise Training for Balance, Functional Status, Spasticity, Fatigue and Quality of Life in Multiple Sclerosis: A Randomized Controlled Trial. *Clin. Rehabil.* **2013**, *27*, 813–822. [CrossRef]
41. Aydin, T.; Taşpinar, Ö.; Sariyildiz, M.A.; Güneşer, M.; Keskin, Y.; Canbaz, N.; Gök, M.; Camli, A.; Kiziltan, H.; Eris, A.H. Evaluation of the Effectiveness of Home Based or Hospital Based Calisthenic Exercises in Patients with Ankylosing Spondylitis. *J. Back Musculoskelet. Rehabil.* **2016**, *29*, 723–730. [CrossRef]
42. Cancelli, A.; Cottone, C.; Giordani, A.; Migliore, S.; Lupoi, D.; Porcaro, C.; Mirabella, M.; Rossini, P.M.; Filippi, M.M.; Tecchio, F. Personalized, Bilateral Whole-Body Somatosensory Cortex Stimulation to Relieve Fatigue in Multiple Sclerosis. *Mult. Scler.* **2018**, *24*, 1366–1374. [CrossRef] [PubMed]
43. Chalah, M.A.; Riachi, N.; Ahdab, R.; Mhalla, A.; Abdellaoui, M.; Créange, A.; Lefaucheur, J.P.; Ayache, S.S. Effects of Left DLPFC versus Right PPC TDCS on Multiple Sclerosis Fatigue. *J. Neurol. Sci.* **2017**, *372*, 131–137. [CrossRef] [PubMed]
44. Chalah, M.A.; Grigorescu, C.; Padberg, F.; Kümpfel, T.; Palm, U.; Ayache, S.S. Bifrontal Transcranial Direct Current Stimulation Modulates Fatigue in Multiple Sclerosis: A Randomized Sham-Controlled Study. *J. Neural Transm.* **2020**, *127*, 953–961. [CrossRef] [PubMed]
45. Palm, U.; Chalah, M.; Padberg, F.; Al-Ani, T.; Abdellaoui, M.; Sorel, M.; Dimitri, D.; Créange, A.; Lefaucheur, J.; SS, A. Effects of Transcranial Random Noise Stimulation (TRNS) on Affect, Pain and Attention in Multiple Sclerosis. *Restor. Neurol. Neurosci.* **2016**, *34*, 189–199. [CrossRef] [PubMed]
46. Pilloni, G.; Choi, C.; Shaw, M.T.; Coghe, G.; Krupp, L.; Moffat, M.; Cocco, E.; Pau, M.; Charvet, L. Walking in Multiple Sclerosis Improves with TDCS: A Randomized, Double-Blind, Sham-Controlled Study. *Ann. Clin. Transl. Neurol.* **2020**, *7*, 2310–2319. [CrossRef] [PubMed]

 *healthcare*

Article

Physical Activity Reduces the Risk of Developing Diabetes and Diabetes Medication Use

Ángel Denche-Zamorano [1], David Manuel Mendoza-Muñoz [1,*], Sabina Barrios-Fernandez [2,*], Carolina Perez-Corraliza [3], Juan Manuel Franco-García [4], Jorge Carlos-Vivas [5], Raquel Pastor-Cisneros [1] and María Mendoza-Muñoz [6,7]

[1] Promoting a Healthy Society Research Group (PHeSO), Faculty of Sport Sciences, University of Extremadura, 10003 Caceres, Spain

[2] Occupation, Participation, Sustainability and Quality of Life (Ability Research Group), Nursing and Occupational Therapy College, University of Extremadura, 10003 Caceres, Spain

[3] Research Group on Physical and Health Literacy and Health-Related Quality of Life, Centro Universitario de Plasencia, Universidad de Extremadura, 10600 Plasencia, Spain

[4] Health, Economy, Motricity and Education (HEME) Research Group, Faculty of Sport Sciences, University of Extremadura, 10003 Caceres, Spain

[5] Physical Activity for Education, Performance and Health, Faculty of Sport Sciences, University of Extremadura, 10003 Caceres, Spain

[6] Research Group on Physical and Health Literacy and Health-Related Quality of Life (PHYQOL), Faculty of Sport Sciences, University of Extremadura, 10003 Caceres, Spain

[7] Departamento de Desporto e Saúde, Escola de Saúde e Desenvolvimento Humano, Universidade de Évora, 7004-516 Évora, Portugal

* Correspondence: dmendozaj@alumnos.unex.es (D.M.M.-M.); sabinabarrios@unex.es (S.B.-F.)

Citation: Denche-Zamorano, Á.; Mendoza-Muñoz, D.M.; Barrios-Fernandez, S.; Perez-Corraliza, C.; Franco-García, J.M.; Carlos-Vivas, J.; Pastor-Cisneros, R.; Mendoza-Muñoz, M. Physical Activity Reduces the Risk of Developing Diabetes and Diabetes Medication Use. *Healthcare* **2022**, *10*, 2479. https://doi.org/10.3390/healthcare10122479

Academic Editor: Herbert Löllgen

Received: 18 November 2022
Accepted: 5 December 2022
Published: 8 December 2022

Publisher's Note: MDPI stays neutral with regard to jurisdictional claims in published maps and institutional affiliations.

Abstract: Diabetes is a global public health challenge, exerting a large socioeconomic burden on healthcare systems. This study aimed to explore Diabetes prevalence and Diabetes medication use in diabetics regarding sex, age group, Physical Activity Level (PAL) and Body Mass Index (BMI) by studying possible differences and calculating the risks of developing Diabetes and Diabetes medication use in the population according to their PAL. A cross-sectional study was conducted using data extracted from the Spanish National Health Survey (ENSE2017). The sample was finally composed of 17,710 participants. A descriptive analysis was performed to characterise Diabetes prevalence and Diabetes medication use (Chi-square test and a z-test for independent proportions). Odds Ratios (OR) and 95% Confidence Intervals (CI) were calculated for Diabetes prevalence and Diabetes medication use according to the participants' PAL. Both the Diabetes and Diabetes medication use was higher in men than in women, increasing with age and BMI, and decreasing with increasing PAL ($p < 0.001$). Higher prevalence levels were observed in the inactive group versus very active or active people ($p < 0.001$). Inactive people had a higher risk of Diabetes and use of Diabetes medication risk compared to the very active and active groups. Prevalence decreased the higher the PAL both in men and women.

Keywords: Diabetes; medication; physical activity; physical heath

1. Introduction

Diabetes mellitus is a chronic disease which has emerged as a significant public health challenge worldwide, both in developed and emerging countries [1]. In 2015, 415 million people were estimated to be living with Diabetes (8.8% worldwide). The has data doubled in the instance of those with Diabetes since 2000 (4.6%, 151 million people), which is an estimated increase to 10.4% (642 million) by 2040 [2]. Furthermore, in 2019, this disease was the ninth leading cause of death, responsible for 1.5 million deaths, 48% in people under 70. Between 2000 and 2016, premature mortality due to Diabetes increased by 5% [3]. As a result, Diabetes was considered a major socio-economic burden for many countries [4]. In

Spain, the total annual direct cost of Diabetes amounts to EUR 5809 million, 8.2% of the total healthcare expenditure; pharmacological costs exerted the greatest influence on the total direct cost (38%) with EUR 2232 million per year [5]. Moreover, Diabetes has become a major cause of cardiovascular disease, non-traumatic lower limb amputations, blindness, kidney failure and death worldwide [6]. Furthermore, an association between Diabetes and cancer has been demonstrated, with Diabetes considered a risk factor for cancer in all locations, with a stronger impact on men than women [7].

There is a significant number of people with Diabetes undiagnosed. In Europe, 37.9% of individuals belong to this group, which could mean that around 22 million individuals with an increased risk of developing cardiovascular disease are unaware of their condition [8]. In Spain, it was found that almost half of the cases detected were undiagnosed [9]. Although there are several types of Diabetes, Diabetes mellitus types 1 (DM1) and 2 (DM2) are the most common [10]. DM1 is associated with deficits in insulin production and requires daily administration of insulin [11], while DM2 results from a decrease in insulin production due to insulin resistance. Symptoms are similar in both types of Diabetes (thirst, excessive excretion of urine, constant hunger, visual disturbances, etc.), but are less intense in DM2 [12]. Between 5–10% of cases have DM1, while the remaining 90–95% have DM2 which can be controllable and/or improved by physical activity and healthy lifestyle promotion [13].

Non-pharmacological interventions such as physical activity (PA) performance and a healthy diet are considered promising methods in the prevention and control of this disease, reducing the socio-economic cost associated with its treatment [10]. Exercise improves blood glucose control in DM2, influences weight loss, reduces cardiovascular risk factors and improves well-being [14,15]. Performing regular exercise can prevent or delay the development of the disease [16] and, in the case of DM1, lead to improvements in insulin sensitivity, muscle strength and cardiovascular fitness [17]. In terms of types of exercise training, moderate/high-volume aerobic is associated with lower cardiovascular and mortality risks in both types of Diabetes [18]. Thus, regular PA increases cardiorespiratory fitness, reduces insulin resistance, and improves lipid levels, and endothelial function [19], while DM2 decreases A1C, blood pressure, insulin resistance and serum triglycerides [20]. Resistance exercise reports positive effects in reducing the exercise-induced hypoglycemia risk in DM1 [21], producing improvements in glycemic control, insulin resistance, strength, and BMI, and reducing blood pressure [22]. Nevertheless, current evidence shows that PA and diet adherence in diabetic patients is still lower than adherence to medication [16].

Therefore, this research aims (1) to explore Diabetes prevalence and Diabetes medication use related to sex, age group, Physical Activity Level (PAL) and Body Mass Index (BMI); (2) to study potential differences in the proportions of Diabetes prevalence and Diabetes medication use by the Physical Activity Level (PAL) according to sex, age and Body Mass Index (BMI); and (3) to estimate the Diabetes and Diabetes medication use risk probability risks in the population according to their Physical Activity Level (PAL).

2. Materials and Methods

2.1. Recruitment and Data Source

A cross-sectional study was conducted using data from the Spanish National Health Survey (ENSE2017), based on Diabetes prevalence and Diabetes medication use according to the PAL and by socio-demographic characteristics such as sex, age and BMI. The ENSE is a survey conducted every five years by the Ministry of Health, Consumer Affairs and Social Welfare (MSCBS) and the Spanish National Statistics Institute (INE), which aims to identify health status, indicators and socioeconomic factors of the population residing in Spain [23]. Trained and accredited interviewers conducted the surveys between October 2016 and October 2017 and published the data in June 2018. The ENSE2017 was the last one conducted before the COVID-19 pandemic. Moreover, the ENSE2017 follows a three-stage random sampling system by strata among individuals aged 15 years and older residents in Spain. Before the first stage, Spanish municipalities were grouped by strata, based on population size. In this first stage, municipalities were randomly selected from the strata. In the second stage, a random selection of dwellings was made and, in the third stage, a random selection of one of the residents of the dwelling, among those aged 15 and over, was performed. Subsequently, the selected sample was informed and asked for volunteers, keeping confidentiality and anonymous treatment of data. Finally, 23,089 participants responded to the ENSE2017 adult questionnaire [24]. These data were previously submitted and published as public and non-confidential files on the website of the MSCBS: https://www.sanidad.gob.es/estadisticas/microdatos.do (accessed on 10 May 2022).

2.2. Participants

The ENSE2017 complied with 23,089 participants' data aged 15 years and older. Inclusion criteria were being younger than 70 years (as they were not questioned about their PA), providing all data on PA items (Q.113–Q.117) and Diabetes status (Q.25a.12). Thus, 5,312 participants were excluded for being older than 70 years, 60 for not having fulfilled items on PA and 7 for not completing Diabetes status data. Then, 17,710 participants' responses were analysed, 8,486 men and 9,224 women. Figure 1 shows the flow chart with this process. For analyses that included Diabetes medication use, one participant who did not submit data was excluded (Q.87a.19). A total of 497 participants did not present data on the BMI Group variable; therefore, they were excluded in the analyses that included this variable.

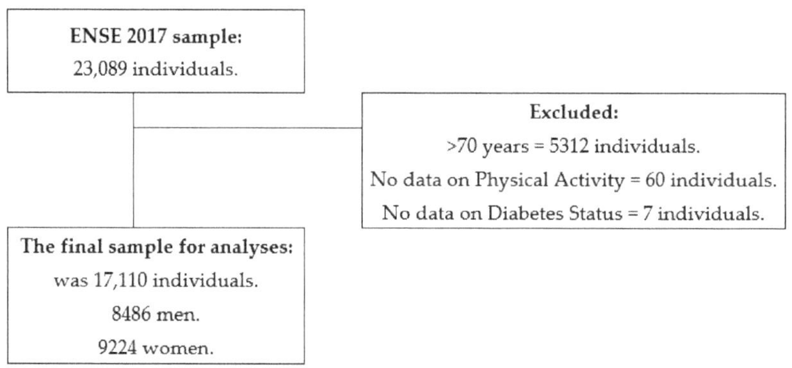

Figure 1. Chart outlining the study sample's eligibility criteria.

2.3. Variables

<u>Sex</u>: male or female.

<u>Age</u>: from which the variable "Age groups" was created with the following groups: 15–39, 35–49, 50–64 and 65–69 years.

<u>BMI Groups</u>: The ENSE2017 grouped participants according to their BMI into underweight (<18.5 kg/m^2), normal weight (18.5 to 24.9 kg/m^2), overweight (25 to 29.9 kg/m^2), and obese (≥30 kg/m^2).

<u>Diabetes Status</u>: extracted from the answers provided to item Q.25a.12 ("do you suffer, or have you ever suffered from Diabetes?". Possible answers were "yes", "no", "I don't know" or "no answer".

<u>Diabetes Medication Use</u>: extracted from the answers to item Q.85 ("during the past two weeks, have you taken any medicines prescribed by a doctor?". Possible answers were "yes" or "no"; and Q.87a.19 ("I will read a list of drugs, please tell me which one(s) you have taken in the last two weeks, and which ones have been prescribed by the doctor: Diabetes medicines?" Answers: "yes", "no", "I don't know" or "not answer". Participants were labelled as YES if they answered "yes" to items Q.85 and Q.85a.19, and NO if they answered "no" to item Q.85, "yes" to item Q.85 and "no" to item Q.87a.19.

Physical Activity Level (PAL): this variable grouped participants according to their PA level. For this purpose, the answers to items Q.113–Q.117 correspond to the International Physical Activity Questionnaire (IPAQ short form) in its Spanish version into a Physical Activity Index (PAI) [25]. The PAI could take values between 0 and 67.5, and this formula was described by Denche et al. [26] adapting the Nes et al. PAI, applying factors to the intensity (vigorous: 10; moderate: 5; light: 0), frequency (0 days/week: 0; One day/week: 1; Two or three days/week: 2; More than 3 days/week: 3) and duration (Less than 30 min: 1; 30 or more minutes: 1.5) that participants performed physical activity and calculating the sum [27]. Participants were grouped into four levels: Inactive (PAI = 0 and Q.117 = 0); Walkers (PAI = 0 and Q.117 > 0); Active (PAI between 1 and 30) and Very Active (PAI > 30); these groupings followed the indications of previous research [28].

2.4. Statistical analysis

The variables distribution was tested with the Kolmogorov–Smirnov test. The sample was characterised according to their Diabetes status and Diabetes medication use according to the general population, by sex, BMI, age and PAL group, reporting data in absolute and relative frequencies. Possible dependency relationships between Diabetes and medication use and socio-demographic variables were analysed with a Chi-square test. The relationship intensity was assessed using the contingency coefficient, interpreted according to Schubert [29]. Differences between Diabetes and prevalence of Diabetes medication use by the PAL were analysed with a pairwise z-test for independent proportions. Odds Ratios (OR) and their 95% confidence intervals (CI) for the Diabetes status and Diabetes medication use according to the PAL were calculated, taking the inactive group as a reference. Two multiple binary logistic regressions were performed taking as dependent variables the Diabetes status and the Diabetes medication use and sex, age, BMI and PAL as independent variables, analysing the predictor effects of these variables. The statistical software IBM SPSS Statistics for Windows, Version 25.0 (IBM Corp., Armonk, NY, USA) was used, considering two-sided p-values ≤ 0.05 as statistically significant.

3. Results

Table 1 shows the associations between Diabetes prevalence and sex, age group, PAL, and BMI ($p < 0.001$). There were more diabetic men than women (6.8% vs. 4.7%, $p < 0.05$). The Diabetes prevalence increased with age, being 0.6% in those under 34, and 18.3% in the 65–69 age group, with differences between all group proportions ($p < 0.05$). According to the PAL, the highest prevalence was found in the inactive and walker groups (8.0% and 7.4%) with no significant differences between them, decreasing to 2.2% in the very active group, $p < 0.05$. Diabetes was found in 0.5% of underweight people compared to 6.6% in overweight and 13.6% with obesity, with differences in the proportions in these groups ($p < 0.05$) and between them ($p < 0.05$).

Same dependency relationships were found between Diabetes medication use and sex, age, PAL and BMI ($p < 0.001$). Again, men showed higher Diabetes medication use than women (6.2% vs. 3.9%, $p < 0.05$). The under-35-year-olds had the lowest prevalence among the age groups (0.4%), with the 65–69-year-old group having the highest (16.3%), $p < 0.05$. Similarly, the inactive and very active groups had the highest and lowest prevalence according to PAL (7.2% vs. 1.7%, $p < 0.05$), while people with obesity (12.3%) had the highest among the BMI groups (Table 2).

Figure 2 shows Diabetes and the prevalence of Diabetes medication use according to the PAL in the general population, being, in both cases, higher in the inactive group, with trends to lower prevalence as the population presents higher PAL.

Table 3 displays the Diabetes prevalence related to PAL in men and women from 50–64 years normal and overweight ($p < 0.001$). In all cases, the highest prevalence was found in the Inactive group. A decreasing trend in the Diabetes prevalence could be discerned as higher PAL was in sex, age, and BMI groups, but with no significant differences between the PALs. In men, the highest prevalence were found in inactive and walker (9.6%)

groups with no differences between them. In contrast, the active (4.3%) and very active (2.3%) groups showed differences in proportions regarding the previous groups and with each other, $p < 0.05$. In females, the inactive and walker groups (6.6% and 5.8%) also had the highest Diabetes prevalence, with no significant differences between them, while there were significant differences between the active and very actives (2.6% and 2.2%), $p < 0.05$.

Table 1. Population Characteristics by their Diabetes Prevalence in the ENSE2017.

Characteristic	Overall	Diabetes		No Diabetes		X^2	df	p	CC
	n	n	%	n	%				
Overall	17,710	1016	(5.7)	16,694	(94.3)	n.a	n.a.	n.a	n.a
Sex									
Men	8486	578 a	(6.8)	7908 a	(93.2)	34.8	1	<0.001	0.044
Women	9224	438 b	(4.7)	8786 b	(95.3)				
Age (years)									
15–34	3872	25 a	(0.6)	3847 a	(99.4)				
35–49	6176	137 b	(2.2)	6039 b	(97.8)	950.6	3	<0.001	0.226
50–64	5953	541 c	(9.1)	5412 c	(90.9)				
65–69	1709	313 d	(18.3)	1396 d	(81.7)				
PAL Group									
Inactive	2531	203 a	(8.0)	2328 a	(92.0)				
Walkers	8063	593 a	(7.4)	7470 a	(92.6)	160.3	3	<0.001	0.095
Actives	4888	171 b	(3.5)	4717 b	(96.5)				
Very actives	2228	49 c	(2.2)	2179 c	(97.8)				
BMI (kg/m^2)	n = 17,213	n = 998		n = 16,225					
	n	n	%	n	%				
<18.5	415	2a	(0.5)	413	(99.5)				
[18.5–25)	7765	193a	(2.5)	7572 a	(97.5)	501.9	3	<0.001	0.168
[25–30)	6192	408b	(6.6)	5784 b	(93.4)				
>=30	2841	385c	(13.6)	2456 c	(86.4)				

Data presented in absolute and relative values; n: Participants; %: Percentage; PAL: Physical Activity Level; BMI: Body Mass Index; X^2: Pearson's Chi-square; df: degrees of freedom; p: p-value; CC: Contingency Coefficient; abcd: Different letters indicate significant differences between people with Diabetes proportions according to Sex, Age, BMI and Physical Activity Level groups; with $p < 0.05$ from pairwise z-test for independent proportions; n.a. not applicable.

Table 2. Population Characteristics by Diabetes Medication Use in ENSE2017.

Characteristic	Overall	Medication		No Medication		X^2	df	p	CC
	n	n	%	n	%				
Overall	17709	886	(5.0)	16823	(95.0)	n.a	n.a.	n.a	n.a
Sex									
Men	8485	524 a	(6.2)	7961	(93.8)	47.1	1	<0.001	0.052
Women	9224	362 b	(3.9)	8862	(96.1)				
Age (years)									
15–34	3872	17 a	(0.4)	3855	(99.6)				
35–49	6176	110 b	(1.8)	6066	(98.2)	879.5	3	<0.001	0.218
50–64	5952	481 c	(8.1)	5741	(91.9)				
65–69	1709	278 d	(16.3)	1431	(83.7)				

Table 2. *Cont.*

Characteristic	Overall	Medication		No Medication		X^2	df	p	CC
	n	n	%	n	%				
PAL Group									
Inactive	2531	181 a	(7.2)	2350	(92.8)				
Walkers	8062	516 a	(6.4)	7546	(93.6)	146.9	3	<0.001	0.091
Actives	4888	152 b	(3.1)	4736	(96.9)				
Very actives	2228	37 c	(1.7)	2191	(98.3)				
BMI (kg/m²)	n = 17,212	n = 860		n = 16,352					
	n	n	%	n	%				
<18.5	415	2 a	(0.5)	413	(99.5)				
[18.5–25)	7765	153 a	(2.0)	7612	(98.0)	493.2	3	<0.001	0.167
[25–30)	6192	356 b	(5.7)	5836	(94.3)				
>=30	2840	349 c	(12.3)	2491	(87.7)				

Data presented in absolute and relative values; n: Participants; %: Percentage; PAL: Physical Activity Level; BMI: Body Mass Index; X^2: Pearson's Chi-square; df: degrees of freedom; *p*: *p*-value; CC: Contingency Coefficient; abcd: Different letters indicate significant differences between People with Diabetes proportions according to Sex, Age, BMI and Physical Activity Level groups; with $p < 0.05$ from pairwise z-test for independent proportions; n.a. not applicable.

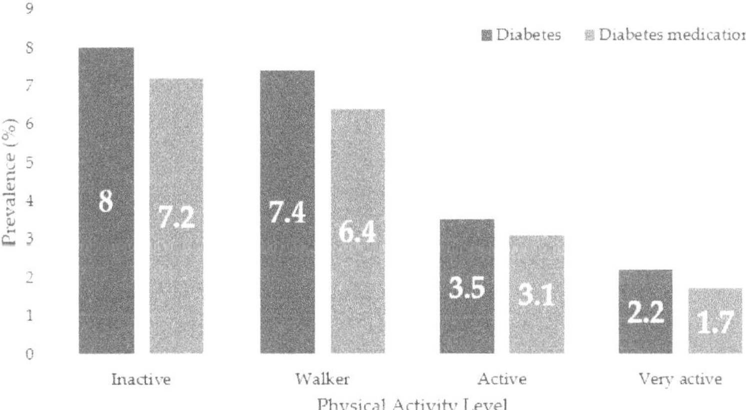

Figure 2. Diabetes and Diabetes Medication use Prevalence According to the Physical Activity Level (PAL) in the ENSE2017.

Figure 3 shows the Diabetes prevalence in males and females according to the PAL.

Table 4 shows the associations between PAL and Diabetes prevalence by sex, age, and BMI groups. Significant dependence relationships were found between these variables in men, women, adults between 35–49, and in normal and overweight, $p < 0.001$. Prevalence were higher the lower the PAL, although no differences were found between the groups. Significant differences were found in women between the inactive/walker groups (5.5% and 4.8%, respectively) and the active/very active groups (2.1% and 1.4%) prevalence, $p < 0.05$. The same significant differences were found in men between the inactive/walker (9.0% and 8.6%), the active (4.1%) and the very active (1.8%) groups ($p < 0.05$), with significant differences also found between the latter two groups.

Figure 4 shows the Diabetes medication use prevalence according to the Physical Activity Level (PAL) in men and women.

Table 5 shows the Diabetes risk according to the PAL based on the inactive group. Significantly reduced risks were found in the active and very active groups compared to the inactive group in the general population, in both sexes, in the 35–49 and 50–64 years groups, and most BMI groups.

Table 3. Diabetes prevalence through the Physical Activity Level (PAL) according to Sex, Age, and Body Mass Index (BMI).

	Physical Activity Level											
Variables	**Inactive**		**Walkers**		**Active**		**Very Active**					
Sex	**n**	**%**	**n**	**%**	**n**	**%**	**n**	**%**	**X^2**	**df**	**p**	**CC**
Male	113 a	9.6	324 a	9.6	107 b	4.3	34 c	2.3	125.0	3	<0.001	0.120
Female	90 a	6.6	269 a	5.8	64 b	2.6	15 b	2.2	58.1	3	<0.001	0.079
Age (years)												
15–34	6 a	1.3	9 a	0.6	6 a	0.5	4 a	0.5	3.6	3	0.310	0.030
35–49	26 ab	2.8	73 b	2.8	29 a	1.6	9 a	1.1	14.5	3	0.002	0.048
50–64	115 a	13.1	318 a	10.2	84 b	5.8	24 b	4.7	52.5	3	<0.001	0.093
65–69	56 a	22.6	193 a	19.5	52 b	13.3	12 ab	15.0	11.0	3	0.012	0.080
BMI (kg/m^2)												
<18.5	0 a	0.0	2 a	0.3	0 a	0.0	0 a	0.0	2.3	3	0.506	0.075
[18.5–25)	35 a	3.9	107 a	3.4	40 b	1.6	11 b	0.9	37.6	3	<0.001	0.069
[25–30)	67 a	7.9	242 a	8.3	74 b	4.4	25 b	3.3	42.0	3	<0.001	0.082
>=30	91 a	15.2	224 a	14.8	57 ab	10.7	13 b	6.5	15.7	3	0.001	0.074

n: participants; %: percentage; X^2: Pearson's Chi-square; df: degrees of freedom; p: p-value; CC: Contingency Coefficient; abc: different letters indicate significant differences between the Diabetes proportions according to their Physical Activity Level (PAL), $p < 0.05$ from pairwise z-test for independent proportions.

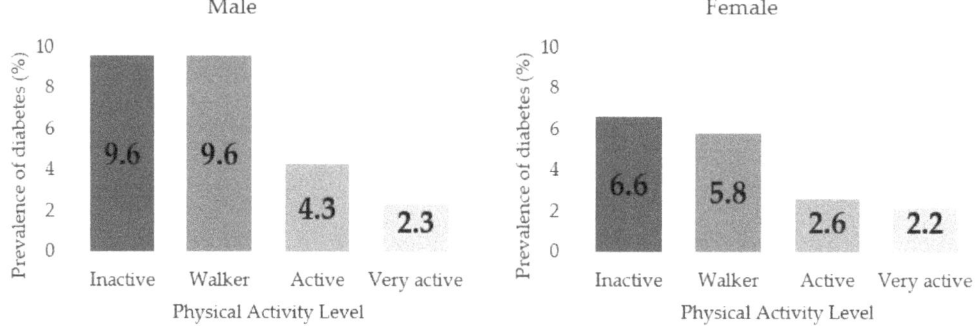

Figure 3. Diabetes Prevalence by Sex according to their Physical Activity Level (PAL).

Table 4. Diabetes Medication Use prevalence by the Physical Activity Level (PAL) and Sex, Age and Body Mass Index (BMI).

	Physical Activity Level											
Variables	**Inactive**		**Walkers**		**Active**		**Very Active**					
Sex	**n**	**%**	**n**	**%**	**n**	**%**	**n**	**%**	**X^2**	**df**	**p**	**CC**
Male	106 a	9.0	292a	8.6	100 b	4.1	26 c	1.8	119.3	3	<0.001	0.118
Female	75 a	5.5	224a	4.8	52 b	2.1	11 b	1.4	51.5	3	<0.001	0.075
Age (years)												
15–34	4 a	0.8	5a	0.4	5 a	0.4	3 a	0.4	2.1	3	0.553	0.023
35–49	23 ab	2.5	60b	2.3	22 ac	1.2	5 c	0.6	17.3	3	0.001	0.053
50–64	102 a	11.6	283a	9.1	78 b	5.4	18 b	3.5	47.4	3	<0.001	0.089
65–69	52 a	21.0	168ab	17.0	47 b	12.1	11 ab	13.8	9.8	3	0.020	0.076

Table 4. *Cont.*

	Physical Activity Level											
Variables	**Inactive**		**Walkers**		**Active**		**Very Active**					
BMI (kg/m^2)												
<18.5	0 a	0.0	2a	1.0	0 a	0.0	0 a	0.0	2.3	3	0.506	0.075
[18.5–25)	30 a	3.3	82a	2.6	34 b	1.4	7 b	0.6	31.4	3	<0.001	0.063
[25–30)	57 a	6.7	216a	7.4	66 b	3.9	17 b	2.3	43.2	3	<0.001	0.084
>=30	84 a	14.1	201a	13.3	52 ab	9.8	12 b	6.0	13.7	3	0.003	0.069

n: participants; %: percentage; X^2: Pearson's Chi-square; df: degrees of freedom; *p*: *p*-value; CC: Contingency Coefficient; abc: different letters indicate significant differences between the Diabetes proportions according to their Physical Activity Level (PAL), $p < 0.05$ from pairwise z-test for independent proportions.

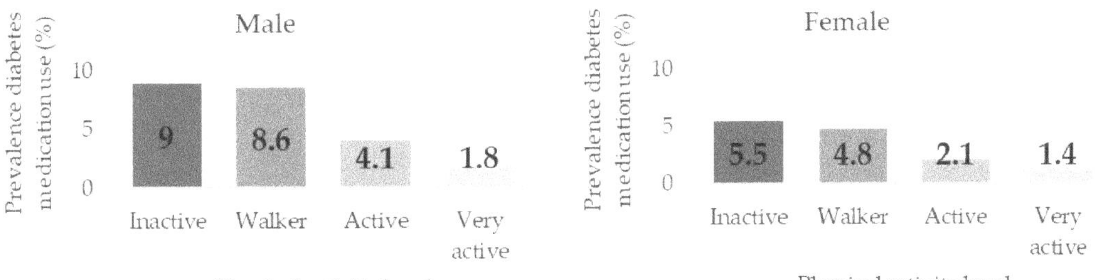

Figure 4. Diabetes Medication Use prevalence according to their Physical Activity Level by sex.

Table 5. Diabetes Risk According to the Physical Activity Level.

	Physical Activity Level								
	Inactive	**Walkers**			**Active**			**Very Active**	
Variables		**OR**	**CI95%**		**OR**	**CI95%**		**OR**	**CI95%**
Overall	Ref.	0.91	0.77	1.07	0.42	0.34	0.51	0.26	0.19	0.35
Sex										
Male	Ref.	0.99	0.70	1.24	0.42	0.32	0.56	0.22	0.15	0.33
Female	Ref.	0.86	0.67	1.10	0.38	0.28	0.53	0.28	0.16	0.49
Age Group										
Young	Ref.	0.51	0.18	1.43	0.39	0.12	1.21	0.39	0.11	1.38
Young adults	Ref.	1.02	0.65	1.61	0.56	0.33	0.95	0.39	0.18	0.83
Older adults	Ref.	0.75	0.00	0.94	0.41	0.31	0.55	0.32	0.21	0.51
Older	Ref.	0.83	0.59	1.16	0.53	0.35	0.80	0.61	0.31	1.20
BMI										
<18.5	Ref.	n.a.	n.a.	n.a.	n.a.	n.a.	n.a.	n.a.	n.a.	n.a.
[18.5–25)	Ref.	0.87	0.59	1.28	0.40	0.26	0.64	0.23	0.12	0.46
[25–30)	Ref.	1.06	0.80	1.40	0.54	0.38	0.76	0.40	0.25	0.64
>=30	Ref.	0.97	0.74	1.26	0.67	0.47	0.95	0.38	0.21	0.70

Ref: reference; OR: Odds Ratio, >1 higher risk of reporting Diabetes; CI95%: 95% OR Confidence Interval; n.a. not applicable.

The Diabetes medication use risk was similarly reduced in the active and very active groups compared to the inactive ones (Table 6).

According to the binary multiple regression analysis on the Diabetes status and Diabetes medication use, older, male, inactive and obese people showed increased Diabetes and Diabetes medication use risks. These models explained 20.5% and 21.4% of the variance (Nagelkerke R^2) in the Diabetes status and Diabetes medication use, respectively (Table 7).

Table 6. Diabetes Medication Use Risk according to the Physical Activity Level.

		Physical Activity Levels								
	Inactive	Walkers			Active			Very Active		
Variables		**OR**	**CI95%**		**OR**	**CI95%**		**OR**	**CI95%**	
Overall	Ref.	0.89	0.74	1.06	0.42	0.33	0.52	0.22	0.15	0.31
Sex										
Male	Ref.	0.95	0.75	1.20	0.43	0.32	0.57	0.18	0.12	0.28
Female	Ref.	0.86	0.66	1.13	0.37	0.26	0.54	0.25	0.13	0.47
Age Group										
Young	Ref.	0.42	0.11	1.59	0.49	0.13	1.82	0.44	0.10	1.96
Young adults	Ref.	0.95	0.58	1.55	0.48	0.26	0.86	0.24	0.09	0.64
Older adults	Ref.	0.76	0.60	0.96	0.43	0.32	0.59	0.27	0.16	0.46
Older	Ref.	0.77	0.54	1.09	0.52	0.34	0.80	0.60	0.30	1.22
BMI										
<18.5	Ref.	n.a.	n.a.	n.a.	n.a.	n.a.	n.a.	n.a.	n.a.	n.a.
[18.5–25)	Ref.	0.77	0.51	1.18	0.40	0.25	0.66	0.17	0.07	0.39
[25–30)	Ref.	1.11	0.82	1.50	0.57	0.40	0.82	0.32	0.19	0.56
>=30	Ref.	0.94	0.71	1.23	0.66	0.46	0.96	0.39	0.21	0.73

Ref: reference; OR: Odds Ratio, >1 higher risk of reporting Diabetes; CI95%: 95% OR Confidence Interval; n.a. not applicable.

Table 7. Logistic Binary Regression Model for Diabetes and Diabetes Medication Use Risk Factor.

							Diabetes	
							95% CI for EXP(B)	
	B	SE	Wald	df	Sig	Exp(B)	Lower	Upper
Sex (Women)	−0.358	0.071	25.764	1	0.000	0.699	0.609	0.803
Age	0.085	0.004	537.721	1	0.000	1.089	1.081	1.097
Inactive			42.017	3	0.000			
Walker	−0.167	0.092	3.317	1	0.069	0.846	0.706	1.013
Active	−0.587	0.114	26.535	1	0.000	0.556	0.445	0.695
Very active	−0.773	0.169	20.802	1	0.000	0.462	0.331	0.644
Underweight			187.363	3	0.000			
Normal	1.247	0.717	3.020	1	0.082	3.479	0.853	14.192
Overweight	1.782	0.716	6.192	1	0.013	5.939	1.460	24.162
Obesity	2.487	0.716	12.059	1	0.001	12.025	2.954	48.943
Constant	−8.609	0.744	133.738	1	0.000	0.000		

							Diabetes Medication Use	
							95% CI for EXP(B)	
	B	SE	Wald	df	Sig	Exp(B)	Lower	Upper
Sex (Women)	−0.467	0.076	38.116	1	0.000	0.627	0.540	0.727
Age	0.090	0.004	493.403	1	0.000	1.094	1.085	1.103
Inactive			39.358	3	0.000			
Walker	−0.184	0.097	3.582	1	0.058	0.832	0.687	1.007
Active	−0.560	0.121	21.563	1	0.000	0.571	0.451	0.723
Very active	−0.947	0.192	24.208	1	0.000	0.388	0.266	0.566
Underweight			185.194	3	0.000			
Normal	0.967	0.719	1.808	1	0.179	2.630	0.643	10.761
Overweight	1.564	0.717	4.762	1	0.029	4.780	1.173	19.483
Obesity	2.305	0.717	10.337	1	0.001	10.027	2.459	40.878
Constant	−8.755	0.751	135.743	1	0.000	0.000		

B: understandarized beta; SE: standard error of the regression; Wald: Wald Chi-Squared Test; Df: degrees of freedom; Sig: statistical significance; Exp: exponential regression; CI: Confidence Interval).

4. Discussion

This study found associations between Diabetes prevalence and sex, with men having a higher prevalence than women, whose differences in proportions were statistically significant. One study carried out in the general population was in the same line, with a higher Diabetes prevalence in men than in women and increased prevalence in older people [30]. Similar results were found in the north-western European population, where the Diabetes mean prevalence was 5.9% for women and 7.9% for men [31]. In the study carried out by Aregbesola et al., a 61% higher risk in men rather than in women was found [32]. This higher prevalence in men could be justified by a higher accumulation of body iron in men and a limited capacity for subcutaneous fat expansion in men [33], so males would accumulate more fat in visceral (liver, spleen, and pancreas) and skeletal muscles, generating greater oxidation of the accumulated fats, leading to increased insulin resistance and glucose homeostasis disturbance [34,35]. Dependent relationships were also found between Diabetes prevalence and age groups. Diabetes prevalence increased with increasing age, with the highest prevalence (18.3%) in the older age group (65–69 years). The study performed by Bullard in the USA population, based on data from the 2016 National Health Interview Survey (NHIS) showed that DM2 prevalence increased progressively in the older age group; the highest prevalence was found among adults over 65 years and older [36].

Diabetes percentage was higher in the BMI groups, with the lowest percentage (0.5%) of diabetic patients in the underweight group and the highest percentage (13.6%) in the obesity group. These data were consistent with those of Zhao with the same group classification. However, percentages were established within the Diabetes group, with 0.54% in the low weight group and 61.94% in the obesity group [37]. The research conducted by Glovaci with adults with BMI above 40 kg/m^2 showed a higher risk (OR = 7.37, 95 % CI: 6.39–8.5) of Diabetes diagnosis than those with a BMI within the normal range [38]. Obesity may be one of the most important predictors of DM2 as excess body fat and unfavourable body fat distribution lead to a state of chronic inflammation and insulin resistance, as well as impaired insulin secretion [39]. The most sedentary individuals had higher BMI, waist circumference, and increased systolic blood pressure was found [40]. Among the individuals with DM1, the active group had a lower BMI with a low obesity prevalence, a lower fat mass index and a lower waist circumference [41]. Our study obtained that Diabetes prevalence in obese subjects (BMI $\geq$ 30) was lower in the groups with higher PAL (Active 10.7 % and Very Active 6.5%) than in those with lower PAL (Inactive 15.2%), being these differences between proportions were significant. Therefore, there could be a strong association between sedentary behaviours (lower PAL) and higher BMI in diabetic and non-diabetic patients. Among the most sedentary individuals, generally, a higher BMI, waist circumference, and increased systolic blood pressure were found [42]. In individuals with DM1, the active group had lower BMI with a low obesity prevalence, lower fat mass index and a lower waist circumference [43]. The Diabetes prevalence in our study in Obese subjects (BMI $\geq$ 30) was lower in the groups with higher PAL (active 10.7% and very active 6.5%) than in those with lower PAL (inactive 15.2%), with these differences being significant proportions [42]. Therefore, there could be a strong association between sedentary behaviours (lower PAL) and higher BMI in both diabetic and non-diabetic individuals [42,43].

The highest Diabetes prevalence were found in the participants with the lowest PAL (inactive and walking groups) and the lowest in those with the highest PAL (active and very active groups). In this line, Colak used the IPAQ-SF Questionnaire [44] to assess PA and found that in 129 patients with DM2, 39.5% were Inactive and only 8.5% showed high PAL [45]. Oguntibeju found [46] reduced Diabetes risks in the active and very active groups compared to the inactive group in the general population, in both sexes, in the age groups 35–49 and 50–64 years, and most BMI groups. Other research has shown this inverse relationship between PA and Diabetes risk [47–50]. However, these studies used different instruments to assess PA (quantitatively, qualitatively, or mixed), measured different domains (occupational PA, leisure time PA, etc.) and dimensions (type of PA, intensity, duration, etc.), which could be problematic in making comparisons [51,52].

Diabetes medication use prevalence were higher in subjects with lower PAL, with prevalence decreasing the higher the PAL. In this regard, another study showed an association between a daily dose of vigorous exercise and lower odds of Diabetes medication use [53]. These findings were in line with those of the current study: the higher the PAL, the lower the medication prevalence. Therefore, diabetic individuals taking more medication have more severe symptoms of the disease and, consequently, experience greater challenges or barriers to physical activity or even prevent physical activity, possibly related to the adverse effects of the medication itself.

The importance of this study lies in the analysis of the associations between the physical activity level and Diabetes prevalence and Diabetes medication use in the Spanish population during the last period before the COVID-19 pandemic, serving as a frame of reference for future research examining post-pandemic periods, as the ENSE is addressed every 5 years. This research showed the relationship between PAL and Diabetes. Hence, introducing exercise programs in Diabetes could be considered as a potential tool for its benefits on patients' health such as improvements in glucose metabolism and insulin sensitivity [54], though this should be confirmed by studies that allow cause-effect relationships to be established.

This study has some limitations. On the one hand, a cross-sectional design does not allow for establishing cause–effect relations. Thus, further research using designs which allow causal relations to be established would be advisable. This work was based on self-reported Diabetes without any medical history or medical judgement. Additionally, this study lacked data on the type of Diabetes, which could influence the results. In future research, it would be interesting to use means of collecting objective data for Diabetes diagnosis. Moreover, additional sociocultural, socio-demographic, and socio-economic variables that could influence the results of Diabetes prevalence were also not considered [55–57].

5. Conclusions

Diabetes and Diabetes medication prevalence use was higher in men than in women and increased with age groups, with the highest prevalence levels among those aged 65–69 years and the lowest prevalence levels among those under the age of 34 years. These prevalence increased with the higher the BMI, with underweight people having the lowest prevalence levels and obese people the highest. The prevalence in men and women were highest in the inactive and walking groups and lowest in the active and very active groups, i.e., prevalence levels decreased the higher the PAL. Therefore, Diabetes risk and Diabetes medication use could be reduced the more active and very active a person is compared to inactive people in the general population in both sexes, in age groups of 35–49 and 50–64, and in most BMI groups.

Author Contributions: Conceptualization, D.M.M.-M. and J.M.F.-G.; Data curation, Á.D.-Z.; Formal analysis, Á.D.-Z.; Funding acquisition, M.M.-M.; Investigation, Á.D.-Z. and J.M.F.-G.; Methodology, D.M.M.-M.; Resources, R.P.-C.; Software, J.C.-V.; Supervision, S.B.-F.; Validation, J.C.-V.; Writing—original draft, D.M.M.-M., C.P.-C. and M.M.-M.; Writing—review and editing, S.B.-F., R.P.-C. and M.M.-M. All authors have read and agreed to the published version of the manuscript.

Funding: The authors Á.D.-Z. (FPU20/04201) and J.M.F-G. (FPU20/04143) were supported by a grant from the Spanish Ministry of Education, Culture, and Sport. Grants FPU20/04201 and FPU20/04143 funded by MCIN/AEI/ 10.13039/501100011033 and, as appropriate, by "European Social Found Investing in your future" or by "European Union NextGenerationEU/PRTR". The author M.M.-M. was supported by a grant from the Universities Ministry and the European Union (NextGenerationUE) (MS-12).

Institutional Review Board Statement: Not applicable.

Informed Consent Statement: Not applicable.

Data Availability Statement: Datasets are available through the corresponding author upon reasonable request.

Conflicts of Interest: The authors declare no conflict of interest.

References

1. Koye, D.N.; Magliano, D.J.; Nelson, R.G.; Pavkov, M.E. The Global Epidemiology of Diabetes and Kidney Disease. *Adv. Chronic Kidney Dis.* **2018**, *25*, 121–132. [CrossRef]
2. Ogurtsova, K.; da Rocha Fernandes, J.D.; Huang, Y.; Linnenkamp, U.; Guariguata, L.; Cho, N.H.; Cavan, D.; Shaw, J.E.; Makaroff, L.E. IDF Diabetes Atlas: Global Estimates for the Prevalence of Diabetes for 2015 and 2040. *Diabetes Res. Clin. Pract.* **2017**, *128*, 40–50. [CrossRef] [PubMed]
3. Organización Mundial de la Salud Diabetes. Available online: https://www.who.int/es/news-room/fact-sheets/detail/Diabetes (accessed on 28 July 2022).
4. Abdulah, D.M.; Hassan, A.B.; Saadi, F.S.; Mohammed, A.H. Impacts of Self-Management Education on Glycaemic Control in Patients with Type 2 Diabetes Mellitus. *Diabetes Metab. Syndr. Clin. Res. Rev.* **2018**, *12*, 969–975. [CrossRef] [PubMed]
5. Crespo, C.; Brosa, M.; Soria-Juan, A.; Lopez-Alba, A.; López-Martínez, N.; Soria, B. Costes Directos de La Diabetes Mellitus y de Sus Complicaciones En España (Estudio SECCAID: Spain Estimated Cost Ciberdem-Cabimer in Diabetes). *Av. Diabetol.* **2013**, *29*, 182–189. [CrossRef]
6. Rojo-Martínez, G.; Valdés, S.; Soriguer, F.; Vendrell, J.; Urrutia, I.; Pérez, V.; Ortega, E.; Ocón, P.; Montanya, E.; Menéndez, E.; et al. Incidence of Diabetes Mellitus in Spain as Results of the Nation-Wide Cohort Di@bet.Es Study. *Sci. Rep.* **2020**, *10*, 1–9. [CrossRef] [PubMed]
7. Ohkuma, T.; Peters, S.A.E.; Woodward, M. Sex Differences in the Association between Diabetes and Cancer: A Systematic Review and Meta-Analysis of 121 Cohorts Including 20 Million Individuals and One Million Events. *Diabetologia* **2018**, *61*, 2140–2154. [CrossRef]
8. Cho, N.H.; Shaw, J.E.; Karuranga, S.; Huang, Y.; da Rocha Fernandes, J.D.; Ohlrogge, A.W.; Malanda, B. IDF Diabetes Atlas: Global Estimates of Diabetes Prevalence for 2017 and Projections for 2045. *Diabetes Res. Clin. Pract.* **2018**, *138*, 271–281. [CrossRef]
9. Soriguer, F.; Goday, A.; Bosch-Comas, A.; Bordiú, E.; Calle-Pascual, A.; Carmena, R.; Casamitjana, R.; Castaño, L.; Castell, C.; Catalá, M.; et al. Prevalence of Diabetes Mellitus and Impaired Glucose Regulation in Spain: The Di@bet.Es Study. *Diabetologia* **2012**, *55*, 88–93. [CrossRef]
10. Tan, S.Y.; Mei Wong, J.L.; Sim, Y.J.; Wong, S.S.; Mohamed Elhassan, S.A.; Tan, S.H.; Ling Lim, G.P.; Rong Tay, N.W.; Annan, N.C.; Bhattamisra, S.K.; et al. Type 1 and 2 Diabetes Mellitus: A Review on Current Treatment Approach and Gene Therapy as Potential Intervention. *Diabetes Metab. Syndr. Clin. Res. Rev.* **2019**, *13*, 364–372. [CrossRef]
11. Gharravi, A.M.; Jafar, A.; Ebrahimi, M.; Mahmodi, A.; Pourhashemi, F.; Haseli, N.; Talaie, N.; Hajiasgarli, P. Current Status of Stem Cell Therapy, Scaffolds for the Treatment of Diabetes Mellitus. *Diabetes Metab. Syndr. Clin. Res. Rev.* **2018**, *12*, 1133–1139. [CrossRef]
12. American Diabetes Association. 2. Classification and Diagnosis of Diabetes: Standards of Medical Care in Diabetes—2022. *Diabetes Care* **2022**, *45*, S17–S38. [CrossRef]
13. Colberg, S.R.; Sigal, R.J.; Yardley, J.E.; Riddell, M.C.; Dunstan, D.W.; Dempsey, P.C.; Horton, E.S.; Castorino, K.; Tate, D.F. Physical Activity/Exercise and Diabetes: A Position Statement of the American Diabetes Association. *Diabetes Care* **2016**, *39*, 2065–2079. [CrossRef] [PubMed]
14. Chen, L.; Pei, J.H.; Kuang, J.; Chen, H.M.; Chen, Z.; Li, Z.W.; Yang, H.Z. Effect of Lifestyle Intervention in Patients with Type 2 Diabetes: A Meta-Analysis. *Metabolism* **2015**, *64*, 338–347. [CrossRef] [PubMed]
15. Lin, X.; Zhang, X.; Guo, J.; Roberts, C.K.; McKenzie, S.; Wu, W.C.; Liu, S.; Song, Y. Effects of Exercise Training on Cardiorespiratory Fitness and Biomarkers of Cardiometabolic Health: A Systematic Review and Meta-Analysis of Randomized Controlled Trials. *J. Am. Heart Assoc.* **2015**, *4*, e002014. [CrossRef] [PubMed]
16. Sumamo Schellenberg, E.; Dryden, D.M.; Vandermeer, B.; Ha, C.; Korownyk, C. Lifestyle Interventions for Patients with and at Risk for Type 2 Diabetes: A Systematic Review and Meta-Analysis. *Ann. Intern. Med.* **2013**, *159*, 543–551. [CrossRef] [PubMed]
17. Yardley, J.E.; Hay, J.; Abou-Setta, A.M.; Marks, S.D.; McGavock, J. A Systematic Review and Meta-Analysis of Exercise Interventions in Adults with Type 1 Diabetes. *Diabetes Res. Clin. Pract.* **2014**, *106*, 393–400. [CrossRef]
18. Sluik, D.; Buijsse, B.; Muckelbauer, R.; Kaaks, R.; Teucher, B.; Johnsen, N.F.; Tjønneland, A.; Overvad, K.; Østergaard, J.N.; Amiano, P.; et al. Physical Activity and Mortality in Individuals With Diabetes Mellitus: A Prospective Study and Meta-Analysis. *Arch. Intern. Med.* **2012**, *172*, 1285–1295. [CrossRef]
19. Chimen, M.; Kennedy, A.; Nirantharakumar, K.; Pang, T.T.; Andrews, R.; Narendran, P. What Are the Health Benefits of Physical Activity in Type 1 Diabetes Mellitus? A Literature Review. *Diabetologia* **2012**, *55*, 542–551. [CrossRef]
20. Snowling, N.J.; Hopkins, W.G. Effects of Different Modes of Exercise Training on Glucose Control and Risk Factors for Complications in Type 2 Diabetic PatientsA Meta-Analysis. *Diabetes Care* **2006**, *29*, 2518–2527. [CrossRef]
21. Yardley, J.E.; Kenny, G.P.; Perkins, B.A.; Riddell, M.C.; Balaa, N.; Malcolm, J.; Boulay, P.; Khandwala, F.; Sigal, R.J. Resistance Versus Aerobic ExerciseAcute Effects on Glycemia in Type 1 Diabetes. *Diabetes Care* **2013**, *36*, 537–542. [CrossRef]

22. Gordon, B.A.; Benson, A.C.; Bird, S.R.; Fraser, S.F. Resistance Training Improves Metabolic Health in Type 2 Diabetes: A Systematic Review. *Diabetes Res. Clin. Pract.* **2009**, *83*, 157–175. [CrossRef] [PubMed]
23. Ministerio De Sanidad Consumo Y Bienestar Social. *Encuesta Nacional de Salud 2017*; Metodología;: Madrid, Spain, 2017; 64p, Available online: https://www.sanidad.gob.es/estadEstudios/estadisticas/encuestaNacional/encuestaNac2017/ENSE1 7_Metodologia.pdf (accessed on 30 August 2022).
24. Ministerio De Sanidad Consumo Y Bienestar Social. *Encuesta Nacional de Salud 2017, Cuestionario de Adultos*; Instituto Nacional de Estadística: Madrid, Spain, 2017; pp. 1–61. Available online: https://www.sanidad.gob.es/estadEstudios/estadisticas/ encuestaNacional/encuestaNac2017/ENSE17_ADULTO_.pdf (accessed on 30 August 2022).
25. Craig, C.L.; Marshall, A.L.; Sjo¨stro¨m, M.; Sjo¨stro, S.; Sjo¨stro¨m, S.; Bauman, A.E.; Booth, M.L.; Ainsworth, B.E.; Pratt, M.; Ekelund, U.; et al. International Physical Activity Questionnaire: 12-Country Reliability and Validity. *Med. Sci. Sports Exerc.* **2003**, *35*, 1381–1395. [CrossRef]
26. Denche-Zamorano, Á.; Franco-García, J.M.; Carlos-Vivas, J.; Mendoza-Muñoz, M.; Pereira-Payo, D.; Pastor-Cisneros, R.; Merellano-Navarro, E.; Adsuar, J.C. Increased Risks of Mental Disorders: Youth with Inactive Physical Activity. *Healthcare* **2022**, *10*, 237. [CrossRef] [PubMed]
27. Nes, B.M.; Janszky, I.; Vatten, L.J.; Nilsen, T.I.L.; Aspenes, S.T.; WislØff, U. Estimating V·O 2peak from a Nonexercise Prediction Model: The HUNT Study, Norway. *Med. Sci. Sports Exerc.* **2011**, *43*, 2024–2030. [CrossRef] [PubMed]
28. Myint, K.; Denche-Zamorano, Á.; Pérez-Gómez, J.; Mendoza-Muñoz, M.; Carlos-Vivas, J.; Oliveira, R.; Brito, J.P. Risk of Hypertension and Use of Antihypertensive Drugs in the Physically Active Population Under-70 Years Old—Spanish Health Survey. *Healthcare* **2022**, *10*, 1283. [CrossRef]
29. Schubert, P.; Leimstoll, U. Importance and Use of Information Technology in Small and Medium-Sized Companies. *Electron. Mark.* **2007**, *17*, 38–55. [CrossRef]
30. Xu, G.; Liu, B.; Sun, Y.; Du, Y.; Snetselaar, L.G.; Hu, F.B.; Bao, W. Prevalence of Diagnosed Type 1 and Type 2 Diabetes among US Adults in 2016 and 2017: Population Based Study. *Bmj* **2018**, *362*, k1497. [CrossRef]
31. Zhou, B.; Lu, Y.; Hajifathalian, K.; Bentham, J.; di Cesare, M.; Danaei, G.; Bixby, H.; Cowan, M.J.; Ali, M.K.; Taddei, C.; et al. Worldwide Trends in Diabetes since 1980: A Pooled Analysis of 751 Population-Based Studies with 4.4 Million Participants. *Lancet* **2016**, *387*, 1513–1530. [CrossRef]
32. Aregbesola, A.; Voutilainen, S.; Virtanen, J.K.; Mursu, J.; Tuomainen, T.P. Gender Difference in Type 2 Diabetes and the Role of Body Iron Stores. *Ann. Clin. Biochem.* **2017**, *54*, 113–120. [CrossRef]
33. Koster, A.; Stenholm, S.; Alley, D.E.; Kim, L.J.; Simonsick, E.M.; Kanaya, A.M.; Visser, M.; Houston, D.K.; Nicklas, B.J.; Tylavsky, F.A.; et al. Body Fat Distribution and Inflammation Among Obese Older Adults With and Without Metabolic Syndrome. *Obesity* **2010**, *18*, 2354–2361. [CrossRef]
34. Varlamov, O.; Bethea, C.L.; Roberts, C.T. Sex-Specific Differences in Lipid and Glucose Metabolism. *Front. Endocrinol.* **2014**, *5*, 241. [CrossRef] [PubMed]
35. Geer, E.B.; Shen, W. Gender Differences in Insulin Resistance, Body Composition, and Energy Balance. *Gend. Med.* **2009**, *6*, 60–75. [CrossRef] [PubMed]
36. Bullard, K.M.; Cowie, C.C.; Lessem, S.E.; Saydah, S.H.; Menke, A.; Geiss, L.S.; Orchard, T.J.; Rolka, D.B.; Imperatore, G. Prevalence of Diagnosed Diabetes in Adults by Diabetes Type—United States, 2016. *Morb. Mortal. Wkly. Rep.* **2018**, *67*, 359. [CrossRef] [PubMed]
37. Zhao, F.; Wu, W.; Feng, X.; Li, C.; Han, D.; Guo, X.; Lyu, J. Physical Activity Levels and Diabetes Prevalence in US Adults: Findings from NHANES 2015–2016. *Diabetes Ther.* **2020**, *11*, 1303–1316. [CrossRef] [PubMed]
38. Glovaci, D.; Fan, W.; Wong, N.D. Epidemiology of Diabetes Mellitus and Cardiovascular Disease. *Curr. Cardiol. Rep.* **2019**, *21*, 1–8. [CrossRef] [PubMed]
39. Hauner, H. Obesity and Diabetes. Available online: https://onlinelibrary.wiley.com/doi/10.1002/9781118924853.ch16 (accessed on 20 October 2022).
40. Hernández Rodríguez, J.; Domínguez, Y.A.; Mendoza Choqueticlla, J. Efectos Benéficos Del Ejercicio Físico En Las Personas Con Diabetes Mellitus Tipo 2. *Rev. Cuba. Endocrinol.* **2018**, *29*, 1–18.
41. Brazeau, A.S.; Leroux, C.; Mircescu, H.; Rabasa-Lhoret, R. Physical Activity Level and Body Composition among Adults with Type1 Diabetes. *Diabet. Med.* **2012**, *29*, e402–e408. [CrossRef] [PubMed]
42. Hankinson, A.L.; Daviglus, M.L.; Bouchard, C.; Carnethon, M.; Lewis, C.E.; Schreiner, P.J.; Liu, K.; Sidney, S. Maintaining a High Physical Activity Level Over 20 Years and Weight Gain. *Jama* **2010**, *304*, 2603–2610. [CrossRef] [PubMed]
43. Aadahl, M.; Kjær, M.; Jørgensen, T. Associations between Overall Physical Activity Level and Cardiovascular Risk Factors in an Adult Population. *Eur. J. Epidemiol.* **2007**, *22*, 369–378. [CrossRef] [PubMed]
44. Lee, P.H.; Macfarlane, D.J.; Lam, T.H.; Stewart, S.M. Validity of the International Physical Activity Questionnaire Short Form (IPAQ-SF): A Systematic Review. *Int. J. Behav. Nutr. Phys. Act.* **2011**, *8*, 1–11. [CrossRef]
45. Çolak, T.K.; Acar, G.; Elçin Dereli, E.; Özgül, B.; Demirbüken, İ.; Alkaç, Ç.; Polat, M.G. Association between the Physical Activity Level and the Quality of Life of patients with Type 2 Diabetes Mellitus. *J. Phys. Ther. Sci.* **2016**, *28*, 142. [CrossRef] [PubMed]
46. Oguntibeju, O.; Odunaiya, N.; Oladipo, B.; Truter, E. Health Behaviour and Quality of Life of Patients with Type 2 Diabetes Attending Selected Hospitals in South Western Nigeria. *West Indian Med. J.* **2012**, *61*, 619–626. [PubMed]

47. Villegas, R.; Shu, X.O.; Li, H.; Yang, G.; Matthews, C.E.; Leitzmann, M.; Li, Q.; Cai, H.; Gao, Y.T.; Zheng, W. Physical Activity and the Incidence of Type 2 Diabetes in the Shanghai Women's Health Study. *Int. J. Epidemiol.* **2006**, *35*, 1553–1562. [CrossRef] [PubMed]

48. Aune, D.; Norat, T.; Leitzmann, M.; Tonstad, S.; Vatten, L.J. Physical Activity and the Risk of Type 2 Diabetes: A Systematic Review and Dose-Response Meta-Analysis. *Eur. J. Epidemiol.* **2015**, *30*, 529–542. [CrossRef] [PubMed]

49. Honda, T.; Kuwahara, K.; Nakagawa, T.; Yamamoto, S.; Hayashi, T.; Mizoue, T. Leisure-Time, Occupational, and Commuting Physical Activity and Risk of Type 2 Diabetes in Japanese Workers: A Cohort Study Chronic Disease Epidemiology. *BMC Public Health* **2015**, *15*, 1004. [CrossRef]

50. Mutie, P.M.; Drake, I.; Ericson, U.; Teleka, S.; Schulz, C.A.; Stocks, T.; Sonestedt, E. Different Domains of Self-Reported Physical Activity and Risk of Type 2 Diabetes in a Population-Based Swedish Cohort: The Malmö Diet and Cancer Study. *BMC Public Health* **2020**, *20*, 261. [CrossRef]

51. Macera, C.A.; Hootman, J.M.; Sniezek, J.E. Major Public Health Benefits of Physical Activity. *Arthritis Care Res.* **2003**, *49*, 122–128. [CrossRef]

52. Williams, P.T.; Franklin, B. Vigorous Exercise and Diabetic, Hypertensive, and Hypercholesterolemia Medication Use. *Med. Sci. Sports Exerc.* **2007**, *39*, 1933–1941. [CrossRef]

53. Correia, M.A.; Silva, G.O.; Longano, P.; Trombetta, I.C.; Consolim-Colombo, F.; Puech-Leão, P.; Wolosker, N.; Cucato, G.G.; Ritti-Dias, R.M. In Peripheral Artery Disease, Diabetes Is Associated with Reduced Physical Activity Level and Physical Function and Impaired Cardiac Autonomic Control: A Cross-Sectional Study. *Ann. Phys. Rehabil. Med.* **2021**, *64*, 101365. [CrossRef]

54. Shrivastava, U.; Misra, A.; Gupta, R.; Viswanathan, V. Socioeconomic Factors Relating to Diabetes and Its Management in India. *J. Diabetes* **2016**, *8*, 12–23. [CrossRef]

55. Reisig, V.; Reitmeir, P.; Döing, A.; Rathmann, W.; Mielck, A. Social Inequalities and Outcomes in Type 2 Diabetes in the German Region of Augsburg. A Cross-Sectional Survey. *Int. J. Public Health* **2007**, *52*, 158–165. [CrossRef] [PubMed]

56. Lopez, J.M.S.; Bailey, R.A.; Rupnow, M.F.T. Demographic Disparities Among Medicare Beneficiaries with Type 2 Diabetes Mellitus in 2011: Diabetes Prevalence, Comorbidities, and Hypoglycemia Events. *Popul. Health Manag.* **2015**, *18*, 283–289. [CrossRef] [PubMed]

57. Emoto, N.; Okajima, F.; Sugihara, H.; Goto, R. A Socioeconomic and Behavioral Survey of Patients with Difficult-to-Control Type 2 Diabetes Mellitus Reveals an Association between Diabetic Retinopathy and Educational Attainment. *Patient Prefer. Adherence* **2016**, *10*, 2151. [CrossRef] [PubMed]

Protocol

Women's Involvement in Steady Exercise (WISE): Study Protocol for a Randomized Controlled Trial

Irene Ferrando-Terradez [1], Lirios Dueñas [1,2,*], Ivana Parčina [3], Nemanja Ćopić [3], Svetlana Petronijević [3], Gianfranco Beltrami [4], Fabio Pezzoni [5], Constanza San Martín-Valenzuela [1,6,7], Maarten Gijssel [8], Stefano Moliterni [9], Panagiotis Papageorgiou [10] and Yelko Rodríguez-Carrasco [11]

[1] Department of Physiotherapy, University of Valencia, 46010 Valencia, Spain
[2] Physiotherapy in Motion, Multi-Specialty Research Group (PTinMOTION), Department of Physiotherapy, University of Valencia, 46010 Valencia, Spain
[3] Faculty of Sport, University "Union—Nikola Tesla", 11000 Belgrade, Serbia
[4] University of Parma, 43121 Parma, Italy
[5] Federazione Italiana Triathlon, 00135 Roma, Italy
[6] Unit of Personal Autonomy, Dependency and Mental Disorder Assessment, INCLIVA Biomedical Research Institute, 46010 Valencia, Spain
[7] Centro de Investigación Biomédica en Red de Salud Mental (CIBERSAM), Instituto de Salud Carlos III, 28029 Madrid, Spain
[8] Kinetic Analysis, Jheronimus Academy of Data Science, 5211 DA 's-Hertogenbosh, The Netherlands
[9] European Culture and Sport Organization (ECOS), 00183 Roma, Italy
[10] European Platform for Sport Innovation (EPSI), 1000 Brussels, Belgium
[11] Department of Food Science and Toxicology, Faculty of Pharmacy, University of Valencia, Av. Vicent Andrés Estellés s/n, 46100 Burjassot, Spain
* Correspondence: lirios.duenas@uv.es

Citation: Ferrando-Terradez, I.; Dueñas, L.; Parčina, I.; Ćopić, N.; Petronijević, S.; Beltrami, G.; Pezzoni, F.; San Martín-Valenzuela, C.; Gijssel, M.; Moliterni, S.; et al. Women's Involvement in Steady Exercise (WISE): Study Protocol for a Randomized Controlled Trial. *Healthcare* **2023**, *11*, 1279. https://doi.org/10.3390/healthcare11091279

Academic Editors: Rafael Oliveira and João Paulo Brito

Received: 15 March 2023
Revised: 24 April 2023
Accepted: 27 April 2023
Published: 29 April 2023

Abstract: Background: Physical inactivity is a serious public health problem for people of all ages and is currently the fourth highest global risk factor for mortality. The transition period from adolescence to adulthood coincides with a marked reduction in participation in physical activity, with more than 50% (and up to 80%) of young adults stopping physical activity. This decrease in physical activity is more evident in women than in men. Despite efforts, existing programs face challenges in effectively initiating and maintaining physical activity among individuals, particularly women, for extended durations. To address these limitations, the Women's Involvement in Steady Exercise (WISE) randomized controlled trial (RCT) seeks to assess the efficacy of a digital high-intensity training intervention complemented by nutritional plans and other health-related advice. Methods: The study will be a three-center, randomized (1:1), controlled, parallel-group trial with a six-month intervention period. A total of 300 participants will be recruited at three study sites in Spain, Serbia and Italy. The participants will be randomized to one of the two groups and will follow a six-month program. The primary outcome of the study is the daily step count. Self-reported physical activity, the adherence to the exercise program, body composition, physical activity enjoyment, quality of sleep and physical capacities will also be evaluated.

Keywords: health promotion; exercise; patient adherence; sedentary behavior; mobile applications

1. Introduction

Physical inactivity is a serious public health problem for people of all ages and is currently the fourth highest global risk factor for mortality [1]. A lack of physical activity is related to pathologies such as type 2 diabetes, coronary heart disease and certain types of cancer, among others [2]. In addition, in the case of children and adolescents, several studies have shown that those who perform greater physical activity have better physical and mental health and better psychosocial well-being than those who lead a sedentary lifestyle [3].

It is estimated that 23% of the adult population and 81% of adolescents (between 11–17 years) do not meet the physical activity recommendations of the World Health Organization (WHO) [4]. These indications suggest that children and adolescents between 5 and 17 should perform at least 60 min a day of moderate-to-vigorous-intensity exercise, mostly aerobic, as well as incorporating strength/impact activities at least three days a week. On the other hand, the recommendations for adults between the ages of 18 and 64 are that they should perform at least 150–300 min of moderate-intensity aerobic physical activity, or at least 75–150 min of vigorous-intensity aerobic physical activity, combined with muscle-strengthening activities at moderate intensity or greater on two or more days a week [5].

Statistics from the European Union (EU) member states reveal that 60% of individuals aged 15 and above rarely or never participate in exercise or sports, and over 50% seldom or never engage in other forms of physical activity such as cycling, walking, gardening or household chores [6]. The transition period from adolescence to adulthood coincides with a marked reduction in participation in physical activity, with more than 50% (and up to 80%) of young adults stopping physical activity [7]. This decrease in physical activity is more evident in women than in men. According to the data provided in 2020 by the Survey of Sports Habits in Spain, in annual terms, only 53.9% of women practiced sport in 2020 compared to 65.5% of men. The difference between men and women is most evident in younger age groups: 15% of men aged 15–24 never exercise or play sports compared to 33% of women in the same age range [8].

Evidence reports that the best way to combat the consequences of a sedentary lifestyle is to exercise. However, one of the great problems of sports practice is the lack of adherence [9], that is, the lack of continuity to perform some type of physical activity. Barriers to physical activity and sports participation for women include a lack of time, lack of interest or low motivational level, societal norms and expectations related to gender roles and appearance, lack of social support, fear of injury, limited access to exercise facilities or prohibitive costs of training programs, cultural norms and stereotypes and personal beliefs and attitudes, among other factors. These barriers may vary by age, ethnicity and socio-economic status [10–14].

In order to overcome these barriers, different strategies have been described, such as group training, supervised training and different, more "dynamic" exercise modalities such as those that combine moderate aerobic exercise with strength exercise or high-intensity functional training [15]. One of these interventions is High Intensity Interval Training (hereinafter HIIT). This training modality is acquiring great acceptance as one of its main advantages is that it is a form of exercise of short duration (the sessions usually last between 20 and 30 min). However, despite its short execution time, its effects are maintained in the long term (up to 48 h after having completed the training [16]). Furthermore, it has been seen in some studies that participants prefer this type of exercise to a traditional exercise program because they enjoy it more and find it more motivating [17]. A review of HIIT programs has shown that a range of HIIT protocols, including those that use cycling, running and bodyweight exercises, can enhance cardiorespiratory fitness and other health markers in both healthy individuals and those with chronic diseases such as diabetes and hypertension [18]. Other studies have suggested that HIIT programs involving Tabata-style intervals (i.e., exercise periods that involve performing an exercise at maximum intensity for 20 s, followed by 10 s of rest, and repeating this cycle for a total of four minutes) can be effective for improving fitness in inexperienced populations [19,20]. However, it is worth noting that a gradual progression and individualized programming are crucial to prevent injury and ensure long-term success with a HIIT program.

The evidence also reports that new technologies could be a great tool to combat the aforementioned adhesion problems. Some studies have shown that new technologies, such as phone apps or smartwatches, can be a useful support to improve exercise adherence [21,22]. The Xiaomi Mi Band smartwatches have shown good accuracy in the measurement of steps [23,24], although more studies are needed in this regard as there is a lack of literature on the subject. Leveraging behavior change theories and techniques (BCTs) is a critical factor in enhancing the effectiveness of e-health interventions as they enable the targeting of key components for behavior change [25]. The Consolidated Standards of Reporting Trials (CONSORT) statement [26] and the World Health Organization (WHO) [27] have stressed the importance of incorporating a theory-based approach in the creation of digital interventions, according to their recommendations.

On the other hand, nutrition is well-recognized as a central component of a healthy lifestyle. Maintaining a healthy diet is a challenge for adolescents and young adults [28]. Dietary behaviors also tend to worsen during early adulthood, when young individuals transition into independent living [29]. Dietary behavior change programs are needed that appeal to large numbers and diverse types of people who could benefit from the educational/behavior change procedures. Innovative approaches to dietary change are needed to engage participants in enjoyable experiences to reach the largest number of participants.

For these reasons, in order to engage young women in physical activity habits, it is necessary to create a new perspective of promoting exercise. New ways of encouraging physical activity should be included. These should include a combination of new technologies with shorter and more varied exercise sessions that facilitate women's enjoyment of physical activity and appear as a new way of improving their adherence to exercise within a healthier lifestyle.

Study Aims

The primary goal of the Women's Involvement in Steady Exercise (WISE) randomized controlled trial (RCT) is to evaluate the adherence to a six-month HIIT program, accompanied by nutritional plans and other health-related advice, delivered via a mobile application, among sedentary young women aged 15 to 24 years. This trial aims to determine the effectiveness of the exercise intervention by assessing the changes in the participants' daily step count over the six-month period. The secondary objectives include analyzing the medium- and long-term effects of the program on various variables, including physical activity, anthropometric measurements, body composition, physical capacities, well-being and psychological mediators.

2. Materials and Methods

2.1. Study Design

The WISE experimental RCT consists of three components embedded into a smartphone app: (1) a remote HIIT program with video sessions; (2) an interface that includes health information; (3) an activity monitoring tool. The study will be a three-center, randomized (1:1), controlled, parallel-group trial with a six-month intervention period. Young women will be recruited from the communities at three study sites in Spain (University of Valencia, Valencia), Serbia (University Nikola Tesla, Belgrade) and Italy (SPORTLAB, Parma). This study protocol will be reported in accordance with the Standard Protocol Items: Recommendations for Interventional Trials (SPIRIT) guidelines [30], and will follow the CONSORT (Consolidated Standards for Reporting Trials) guidelines for the transparent reporting of parallel group randomized trials [31]. To gain a comprehensive understanding of the study design, refer to Figure 1.

2.2. Participants

2.2.1. Eligibility Criteria

Participants eligible for the trial must comply with all of the following:

1. Young women aged between 15 and 24 years.
2. Sedentary young women who do not comply with the WHO recommendations for physical exercise and with a low international physical activity questionnaire (IPAQ) score, which means that they do not perform at least one of these:
   - Three or more days of vigorous activity for at least 20 min a day.

- Five or more days of moderate-intensity activity.
- Walk at least 30 min a day every day.
- Five or more days of combining activities of moderate or vigorous intensity or walking reached a minimum of 600 MET (min/week).

The presence of one of the following criteria will lead to the exclusion of the participant:

1. Young women with diabetes.
2. Young women with possible heart problems or another type of contraindication that does not allow physical exercise (for this item the physical activity readiness questionnaire (PAR-Q) survey is used).
3. Young women who are not willing to wear the smartwatch during the six-month intervention.
4. Young women who have contracted severe COVID-19 prior to the intervention. Severe COVID-19 is defined as dyspnea, a respiratory rate of 30 or more breaths per minute, a blood oxygen saturation of 93% or less, a ratio of the partial pressure of arterial oxygen to the fraction of inspired oxygen (Pao2:Fio2) of less than 300 mm Hg, or infiltrates in more than 50% of the lung field [32].

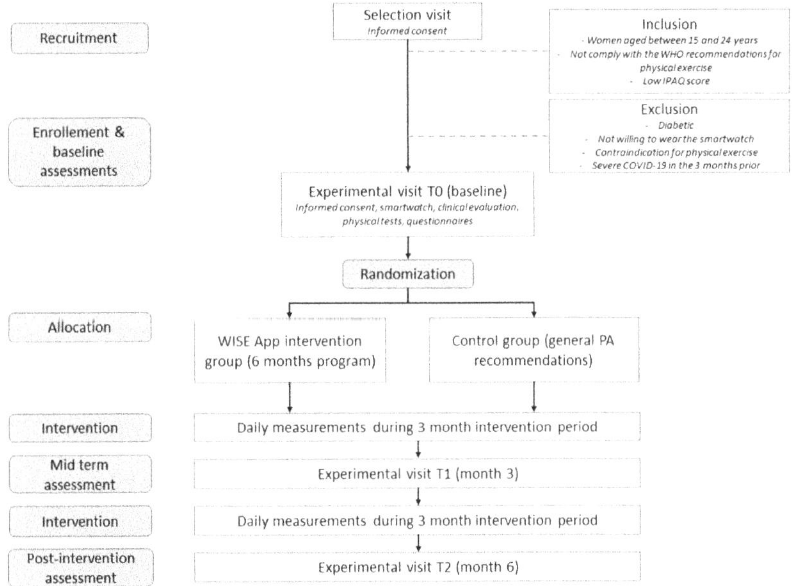

Figure 1. Study flow chart. WHO, World Health Organization; IPAQ, international physical activity questionnaire; PA, physical activity.

2.2.2. Recruitment

A total of 300 participants (150 per group) will be recruited between the three countries. Each country will recruit 100 young women for the study. Dissemination of the project will be conducted via email and posters in close contact with universities, schools and local authorities. Through the information of the posters and emails, interested participants will contact the investigators and will be informed about the study in order to decide if they finally want to participate. Those willing to participate will be summoned for a personal interview to check if they meet the inclusion criteria. Participants who meet the inclusion criteria will be invited to participate in the study and will sign a written consent form before being included in the study. At the start of the study, a wearable device (Xiaomi Mi Band 5) will be distributed to all participants.

2.2.3. Sample Size and Power Analysis

Sample size estimations were derived from the primary outcome measure of daily step count, measured using the Xiaomi Mi Band 5. A priori sample size calculation was conducted based on a previous meta-analysis [33], which reported an effect size of d = 0.51 (95% CI 0.12 to 0.91, I^2 = 90%), compared with control groups, for physical activity interventions that included wearables and smartphone apps. Our sample size estimation was based on an intermediate effect size of d = 0.50.

To detect a difference equivalent to an effect size of 0.5 on the primary outcome with 90% power and a two-sided type I error of 0.05, a sample size of 240 is required. Specifically, assuming that the statistical unit is an individual day and considering an intraclass correlation coefficient of 0.5 to account for interindividual and intraindividual variability, a total of 2002 individual days per group (equivalent to 22 participants per group) are necessary. We suggest including a total of 300 participants (150 per group) in order to protect against a potential dropout of 20%, inherent to such a trial.

2.2.4. Randomization, Allocation and Blinding

After the first experimental visit (T0), the participants will be randomized using a computer-generated randomization list into one of the two conditions with a 1:1 allocation. The allocation will be transmitted using numbered (0 = control/1 = Intervention), opaque and sealed envelopes, which will be given to the young women and then registered by the data collectors. The assessors will be blinded to the treatment allocation, although double blinding is not applicable in interventions of this nature as it is impossible to conceal the allocation from the participants. The participants will be instructed not to reveal their allocation to the assessors to ensure unbiased assessment.

2.3. Intervention

2.3.1. Procedure

All participants will undergo a total of four visits, including the initial inclusion visit and three experimental visits (T0, T1 and T2), as detailed in Table 1. During the selection visit, qualified personnel will make sure that the young women fulfil the inclusion criteria and that they (or their parent/legal representative in case the girl was a minor) sign the informed consent form after explaining the project itself and after responding to all of their questions. One week after the initial visit, the T0 experimental visit will be conducted to measure all of the baseline assessments before the start of the intervention. At the halfway mark of the program (i.e., three months in), a mid-term measurement will be carried out, denoted as T1. The T2 experimental visit will be scheduled at the end of the program (i.e., after six months of intervention). The three experimental sessions will be identical.

Table 1. Timeline for enrolment, interventions and assessments.

	Selection Visit	T0	Study Period				
				Intervention			
TIMEPOINT	M_{-1}	0	M_1	M_2	M_3/T1	M_4	M_5	M_6/T2
ENROLMENT:							
Eligibility screen	X						
Informed consent	X						
Randomization		X					
INTERVENTIONS:							
Intervention group		●━━━━━━━━━━━━━━━━━━━━●					
Control group		●━━━━━━━━━━━━━━━━━━━━●					
ASSESSMENTS:							

Table 1. *Cont.*

	Selection Visit	T0	Intervention
			Study Period
Step count		●━━━━━━━━━━━━━━━━━━━━━━━●	
Adherence		●━━━━━━━━━━━━━━━━━━━━━━━●	
IPAQ		X	X	X
Height		X		
Weight		X	X	X
Body composition		X	X	X
Plank test		X	X	X
HLPCQ		X	X	X
PSQI		X	X	X
PACES		X	X	X
Period pain		X	X	X
CPM		X	X	X
6MWT		X	X	X

IPAQ, International Physical Activity Questionnaire; HLPCQ, Healthy Lifestyle and Personal Control Questionnaire; PSQI, Pittsburgh Sleep Quality Index; PACES, Physical Activity Enjoyment Scale; CPM, Conditioned Pain Modulation; 6MWT, 6 Minutes Walking Test. Continue line: duration over time.

2.3.2. Intervention Group

In order to promote behavior change, we implemented 12 BCTs within the intervention program. Table 2 displays how the BCTs have been implemented within the WISE intervention group.

Table 2. BCT implementation in the intervention group following Michie et al.'s taxonomy [34].

BCT	Implementation in the WISE RCT
Goal setting behavior (1.1)	Set monthly step goal
Feedback on behavior (2.2)	Feedback on daily steps via the activity monitoring tool included in the application with weekly graphs
Self-monitoring of behavior (2.3)	Weekly diary used to document whether participants performed the exercise video and if they were engaged in any additional exercise during the study period
Feedback on outcome(s) of behavior (2.7)	Bioimpedance sheet with all the information of the changes in their body composition after 3 and 6 months of exercise
Social support (practical) (3.2)	Social days created so the participants can exercise together in real life
Social support (emotional) (3.3)	Promote social interaction through social media groups
Instruction on how to perform a behavior (4.1) Behavioral practice/rehearsal (8.1) Demonstration of the behavior (6.1)	Exercise videos
Information about health consequences (5.1)	Tips and information about the benefits of exercise and a healthy lifestyle given through the application
Prompts/cues (7.1)	Push notification through the application
Graded tasks (8.7)	Exercise program with increased difficulty

BCTs, behaviour change techniques; WISE, Women's Involvement in Steady Exercise; RCT, randomized controlled trial.

The WISE intervention is composed of four main features:

1. Exercise videos: participants in the intervention group will receive two HIIT video sessions a week via the WISE app (Figure 2A). These sessions will be carried out at their own home without the need of sport equipment. The exercise program will be developed by two experts in physical activity and sports sciences. Each session will consist of four exercises, and, from session 10, with two different intensities so that the

participants can choose one according to their physical condition. The exercises will be different every week. The second session will always have a greater intensity than the first, and the intensity will be increased from week to week. Each session will last between 20 and 30 min and will follow the following structure: 5–7 min of warm-up; 10–15 min of HIIT training; 5–7 min of stretching. Table 3 shows the progression for the WISE exercise protocol. The complete WISE exercise protocol with links to the videos of the sessions can be seen in Table S1.

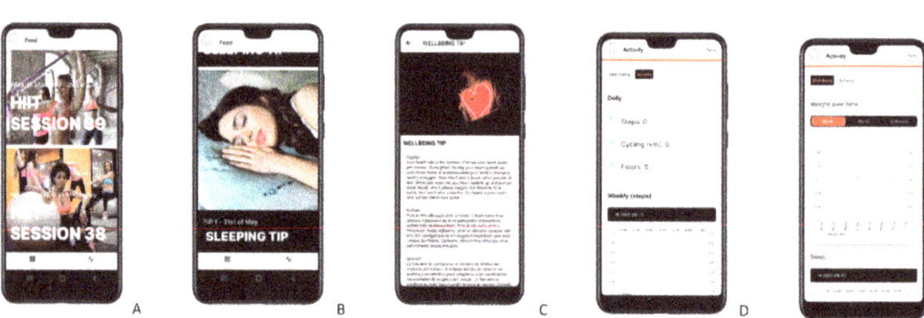

Figure 2. Screenshots examples of the WISE application. (**A**) HIIT video sessions. (**B**) Sleeping tips. (**C**) Well-being tips. (**D**) The activity monitoring tool. (**E**) The weight and sleep monitoring tool.

Table 3. WISE exercise protocol progression.

Month	Session	Work Time	Recovery Time	Level
1	1 2	20″	20″	Initiation and no impact Medium-advanced with impact
2	1 2	20″	15″	Initiation and no impact Medium-advanced with impact
3	1 2	20″	10″	Initiation and no impact Medium-advanced with impact
From week 12th, we introduce 2 different intensity levels in each session (one without impact or less demanding) so the participants can choice between the 2 levels in order to better adapt their effort				
4	1 2	30″	20″	Medium Advanced
5 and 6	1 2	40″	15″	Medium Advanced

2. Education in healthy habits: once a week, the participants will receive reading material, with an approximate duration of 2 to 5 min, on general advice on nutrition, sleep (Figure 2B) and well-being (Figure 2C), written by experts in the field. Dietary guidelines will also be created by a team of professionals specialized in nutrition to stimulate the participants and give them basic indications on proper eating behavior. These reading materials will be delivered to the participants through the WISE application and will appear by means of a push notification. The complete list of messages for the WISE participants can be seen in Table S2.
3. Activity tracker: participants will be able to check their activity at any time of the day through the Xiao Mi Band 5 smartwatch (Figure 2D). The data will also be displayed in the app with weekly charts so participants can see their progression. The participants will also be able to track their weight and their sleep over time (Figure 2E) through the application. Moreover, their heart rate will be recorded. This tool aims to give feedback on behavior and promote self-monitoring of physical activity.
4. Motivational activities: in order to motivate the participants to perform the videos, social media, email, WhatsApp groups and/or other channels, such as Viber (depending on the availability/popularity in each country), will be created for the participants to

communicate with each other. In addition, social meetings will be organized through "Open Days" held in each country and transmitted through streaming so all the participants can attend the event. During the "Open Days", talks on nutrition and physical activity will be held, as well as in-person meetings to perform the exercises. Moreover, the participants can expose their doubts and concerns, which can be solved in a personal manner during these events.

2.3.3. Control Condition

The participants in the control group will benefit from general physical activity recommendations (at the start of the intervention: general physical activity guidelines) and will also have access to the activity monitoring tool supplied by the smartwatch. The content of both groups is summarized in Table 4.

Table 4. Summary of the groups' content.

Intervention Group (Videos)	Control Group
50 HIIT video exercise sessions	Physical activity recommendations (at the start of the intervention: general physical activity guidelines) *
Communication group	–
Activity monitoring tool (mobile app + Xiaomi Mi Band 5)	Activity monitoring tool (Xiaomi Mi Band 5)
Education in healthy habits	–

HIIT, High Intensity Interval Training. * At the end of the program, the participants will have access to the HIIT video sessions.

2.3.4. Adverse Events

Throughout the experiment, adverse events will be monitored for each procedure. Any adverse events or reactions that are thought to be causally associated with the intervention will be recorded and managed.

2.4. Outcome Measures

2.4.1. Primary Outcomes

The primary outcome of this study will be the change in daily steps from baseline to three months and six months. Daily steps will be assessed using the Xiaomi Mi Band 5 smartwatch, a reliable wearable activity tracker that has been validated in previous research [35–37]. Days with missing data, defined as days with fewer than 1000 steps, based on previous literature [38,39], will be excluded from the analysis [40,41] to ensure accurate and comprehensive data representation.

2.4.2. Secondary Outcomes

The secondary outcomes include changes in: (1) physical activity; (2) anthropometric measurements and body composition; (3) physical capacities; (4) well-being. In addition, psychological mediators will also be examined.

The following criteria will be used in order to measure the WISE program adherence [42]:

- Retention (completion): the participants following the WISE exercise videos and showing up to the follow-up measurements.
- Attendance: percentage of videos completed of the total of 50 videos
- Duration: adherence to a minimum of 20 min of exercise two times a week.
- Intensity: intensity levels of the sessions will be assessed at the end of each session using the modified Borg Scale, which measures perceived exertion [43].

These items will be measured through a weekly online exercise diary, where the participants will be asked if they completed both video sessions. They will also be asked about their perceived exertion after each session. This will allow us to assess whether they are following the 20 min of exercise two times a week rule, completing the video sessions, so we can see the number of videos completed at the end of the program.

Table 1 presents the assessment schedule, following the Standard Protocol Items: Recommendations for Interventional Trials (SPIRIT) schedule template. Table 5 provides an overview of all the outcome measures.

Table 5. Outcomes measures of the WISE RCT.

Outcome	Assessment Method
	Primary outcome
Daily step count over 6 months	The Xiaomi Mi Band 5 smartwatch will be used to assess the daily step count. Days with missing data and/or days with fewer than 1000 steps will be excluded from the analysis.
	Secondary outcomes
	Physical activity
Self-reported physical activity and sedentary behaviors	Self-reported behaviors will be gathered using the International Physical Activity Questionnaire (IPAQ). The IPAQ assesses walking and activities of a moderate and vigorous intensity that are performed continuously for at least 10 min in all domains of everyday life (i.e., leisure, occupational, household and transport) in the last 7 days. The IPAQ demonstrates acceptable levels of test-retest reliability and fair to moderate associations with accelerometer measures [40,41].
	Anthropometric data and body composition
BMI, Body mass and height	Using a calibrated digital scale, body mass is measured to the nearest 0.1 kg; height is measured to the nearest 0.1 cm using a wall-mounted stadiometer. The BMI is automatically calculated by the body composition analyzer as the ratio of body mass (kg) to height squared (m^2).
Body composition	Assessment of body composition is conducted using bioelectrical impedance analysis with the multi frequency segmented body composition analyzer InBody 230 (InBody, Cerritos, CA, USA) [44].
	Physical capacities
Muscle strength	The plank test protocol requires participants to maintain a static prone position with only forearms and toes touching the ground. Proper form requires feet together with toes curled under the feet, elbows forearm distance apart, and hands clasped together against the floor mat. Participants maintain eye contact with their hands, a neutral spine, and a straight line from head to ankles. The test begins when the participant demonstrates the correct position. Participants are allowed to deviate from the correct position once and can continue the test if they immediately resume the correct starting position. The test is terminated on the second deviation from the correct position or if the participant does not return to the correct position after the first warning [45]. The plank test protocol is a reliable test [46] and has been used with children [45] and young adults [46].
Endurance	The 6 Minute Walk Test measures aerobic capacity and endurance through sub-maximal exercise. The outcome by which to compare changes in performance capacity is the distance traveled during a period of 6 min. Reference equations have been developed for healthy young adults [47], and have proved Excellent test-retest reliability, interrater reliability and intrarater reliability for different populations [48–52].
	Well-being
Quality of life	The Healthy Lifestyle and Personal Control Questionnaire (HLPCQ) will be used to examine several dimensions of daily living. The HLPCQ has been described as a good tool for assessing the efficacy of future health-promoting interventions to improve individuals' lifestyle and well-being. This questionnaire is a 26-item tool in which the respondent is asked to indicate the frequency of adopting 26 positively stated lifestyle habits using a Likert-type scale (1 = Never or rarely, 2 = Sometimes, 3 = Often and 4 = Always) [53].

Table 5. *Cont.*

Outcome	Assessment Method
Quality of sleep	Quality of sleep will be assessed with The Pittsburgh Sleep Quality Index (PSQI), a self-administered questionnaire used to evaluate sleep quality during the past month. The validity of the PSQI has been confirmed by several studies in different patient populations and languages [54–57]. The PSQI consists of seven clinically derived components that assess sleep difficulty, and the sum of these seven component scores yields a global score of subjective sleep quality [57]. The PSQI demonstrates moderate convergent validity compared to measures of insomnia and fatigue and good divergent validity with measures of daytime sleepiness and circadian phase preference in young adults [58].
Period pain	Pain intensity will be measured using the visual analogue scale (VAS). A 100-mm line bounded by "no pain" on the left (0) and "worst pain possible" (100) on the right will be used to indicate the average pain during the period. The minimum clinically detectable difference (MCID) of this scale has been set at 30 mm [59]. It is also a scale that has been validated [60].
Psychological mediators	
Perceived enjoyment	The Physical Activity Enjoyment Scale (PACES) is used to evaluate perceived enjoyment of physical activity during the intervention. The questionnaire comprises 16 items, where participants rate their feelings about the physical activity they have been engaging in using a 7-point Likert scale, ranging between 1 (not at all) and 7 (very much) [61]. PACES shows a very high internal consistency (Cronbach's alpha = 0.908) and the test-retest reliability indicates a good temporary agreement (Spearman rho = 0.815, $p < 0.001$) in adolescents with overweight and obesity [62].
Program adherence	
Retention, Attendance, Duration and Intensity	These four variables will be used to measure program adherence, as described before.

WISE, Women's Involvement in Steady Exercise; RCT, randomized controlled trial; BMI, body mass index; PPT, pressure pain threshold.

3. Statistical Analyses Plan

All the statistical analysis will be performed using the SPSS version 24 Software (SPSS Inc., Chicago, IL, USA). If adherence issues occur, all analyses will be evaluated by intention-to-treat principles, with a level of significance of 0.05. Continuous outcomes will be presented using mean and standard deviation if they follow a normal distribution. On the other hand, count outcomes, such as WISE sessions completed, will be reported through the median and percentage of achievement. At the basal assessment, an independent-samples Student's *t*-test will be performed to rule out differences between groups of age and height of participants.

In order to answer the first aim of this study, the changes in the adherence variables, such as daily steps, the Borg scale and the IPA questionnaire, will be analyzed by 1-within-subject factor multivariate analysis of variance (MANOVA). The Bonferroni adjustment will be used for post-hoc comparisons between the T0, T1 and T2 times assessments. On the other hand, the secondary outcomes of the study will be tested by 2-factor mixed MANOVA to analyze the effects of a within-subject factor (time assessment from T0 to T4), the between-subject factor (group) and their interaction. As mentioned above, Bonferroni adjustment will be used for times and groups comparisons. In addition, the same statistical test will be carried out considering the country of origin of the data as a covariate, and the findings will be reported according to their statistical significance. For all the statistical analyses, differences will be declared statistically significant if the *p*-value is less than 0.05. The exact *p*-value and 95% confidence interval will be reported.

4. Ethics and Dissemination

The WISE RCT adheres to the Helsinki declaration principles. The research protocol has been reviewed and approved by the Human Research Ethics Committee of the University of Valencia (protocol number 1944476). The study is registered in the ClinicalTrials.gov (NCT05467280). Written informed consent will be obtained from each participant. All of the investigators, the ethics committees and the trial registry will be informed of any modifications that must be made to the study protocol. The results will be disseminated through international conference presentations and in relevant scientific journals.

5. Conclusions

This study aims to investigate the effectiveness of an online HIIT program with nutritional plans and other health-related advice in improving exercise adherence and various health outcomes among sedentary young women. A significant contribution of this study is the development of guidelines to promote exercise engagement in this population.

We expect that the WISE program will improve the participants' knowledge of physical activity, proper nutrition and healthy habits. The program's physical activity component will involve monitoring the participants' heart rate, number of steps and body composition, including fat percentage, visceral fat, water and muscle mass. The diet and healthy habits component will offer basic guidelines and advice on healthy eating, quality sleep and adequate water intake, all delivered via a mobile application.

However, the study design has some limitations, including the program's duration of six consecutive months, which includes vacations and local holidays, as well as the challenge of ensuring that the participants adhere to the program, such as wearing the smartwatches. Depending on the results obtained, the WISE program may consider adjusting the sessions to be more individualized. Additionally, as noted by DiPietro et al. [63], it is important to acknowledge that the WHO recommendations used as a filter in this study are general guidelines for the population as a whole and may not be specifically tailored to certain subpopulations, such as sedentary young women. Finally, increasing the sample size will enable the development of a specification equation (via regression analysis) to determine the general health status of the women in each age group.

Supplementary Materials: The following supporting information can be downloaded at: https://www.mdpi.com/article/10.3390/healthcare11091279/s1, Table S1: WISE exercise protocol with links to the videos of the sessions; Table S2: Messages sent to the participants as readings about education in healthy habits.

Author Contributions: I.F.-T., Y.R.-C., I.P., N.Ć. and S.P. conceptualized the project and obtained the funding. Methodology, G.B. and F.P. All authors provided input into the study design. C.S.M.-V. and M.G. designed the data analysis plan. The first draft of the manuscript was written by I.F.-T. and L.D.; review and editing Y.R.-C. and S.M.; Supervision, P.P. and L.D. All authors have read and agreed to the published version of the manuscript.

Funding: This research was co-funded by the European Union through the Erasmus+ programme, grant number 622485-EPP-1-2020-1-BE-SPO-SCP.

Institutional Review Board Statement: The study was conducted in accordance with the Declaration of Helsinki and approved by the Ethics Committee of University of Valencia (1944476, 5 May 2022). The trial protocol has been registered on the Clinical Trials Registry (NCT05467280).

Informed Consent Statement: Informed consent was obtained from all subjects involved in the study.

Acknowledgments: The authors wanted to thank the European Union for their grant and all members of the WISE team for their contribution to the project.

Conflicts of Interest: The authors declare no conflict of interest.

References

1. Tremblay, M.S.; Aubert, S.; Barnes, J.D.; Saunders, T.J.; Carson, V.; Latimer-Cheung, A.E.; Chastin, S.F.M.; Altenburg, T.M.; Chinapaw, M.J.M.; Altenburg, T.M.; et al. Sedentary Behavior Research Network (SBRN)–Terminology Consensus Project Process and Outcome. *Int. J. Behav. Nutr. Phys. Act.* **2017**, *14*, 75. [CrossRef]
2. De Rezende, L.F.M.; Rodrigues Lopes, M.; Rey-López, J.P.; Matsudo, V.K.R.; Luiz, O.d.C. Sedentary Behavior and Health Outcomes: An Overview of Systematic Reviews. *PLoS ONE* **2014**, *9*, e105620. [CrossRef]

3. Wu, X.Y.; Han, L.H.; Zhang, J.H.; Luo, S.; Hu, J.W.; Sun, K. The Influence of Physical Activity, Sedentary Behavior on Health-Related Quality of Life among the General Population of Children and Adolescents: A Systematic Review. *PLoS ONE* **2017**, *12*, e0187668. [CrossRef]
4. Martins, L.C.G.; Lopes, M.V.d.O.; Diniz, C.M.; Guedes, N.G. The Factors Related to a Sedentary Lifestyle: A Meta-Analysis Review. *J. Adv. Nurs.* **2021**, *77*, 1188–1205. [CrossRef] [PubMed]
5. Bull, F.C.; Al-Ansari, S.S.; Biddle, S.; Borodulin, K.; Buman, M.P.; Cardon, G.; Carty, C.; Chaput, J.-P.; Chastin, S.; Chou, R.; et al. World Health Organization 2020 Guidelines on Physical Activity and Sedentary Behaviour. *Br. J. Sport. Med.* **2020**, *54*, 1451–1462. [CrossRef]
6. European Commission: Directorate-General for Education, Youth, Sport and Culture. *Sport and Physical Activity: Report*; European Commission: Brussels, Belgium, 2014.
7. Hallal, P.C.; Andersen, L.B.; Bull, F.C.; Guthold, R.; Haskell, W.; Ekelund, U. Lancet Physical Activity Series Working Group Global Physical Activity Levels: Surveillance Progress, Pitfalls, and Prospects. *Lancet* **2012**, *380*, 247–257. [CrossRef] [PubMed]
8. Encuesta de Hábitos Deportivos en España. Available online: https://www.culturaydeporte.gob.es/servicios-al-ciudadano/estadisticas/deportes/encuesta-habitos-deportivos-en-espana.html (accessed on 21 December 2021).
9. Theofilou, P.; Saborit, A.R. Adherence and Physical Activity. *Health Psychol. Res.* **2013**, *1*, e6. [CrossRef] [PubMed]
10. Ferreira Silva, R.M.; Mendonça, C.R.; Azevedo, V.D.; Raoof Memon, A.; Noll, P.R.E.S.; Noll, M. Barriers to High School and University Students' Physical Activity: A Systematic Review. *PLoS ONE* **2022**, *17*, e0265913. [CrossRef] [PubMed]
11. Baillot, A.; Chenail, S.; Polita, N.B.; Simoneau, M.; Libourel, M.; Nazon, E.; Riesco, E.; Bond, D.S.; Romain, A.J. Physical Activity Motives, Barriers, and Preferences in People with Obesity: A Systematic Review. *PLoS ONE* **2021**, *16*, e0253114. [CrossRef]
12. Portela-Pino, I.; López-Castedo, A.; Martínez-Patiño, M.J.; Valverde-Esteve, T.; Domínguez-Alonso, J. Gender Differences in Motivation and Barriers for The Practice of Physical Exercise in Adolescence. *Int. J. Environ. Res. Public Health* **2020**, *17*, 168. [CrossRef]
13. Moreno, J.P.; Johnston, C.A. Barriers to Physical Activity in Women. *Am. J. Lifestyle Med.* **2014**, *8*, 164–166. [CrossRef]
14. Tiggemann, M.; Slater, A. NetGirls: The Internet, Facebook, and Body Image Concern in Adolescent Girls. *Int. J. Eat. Disord.* **2013**, *46*, 630–633. [CrossRef]
15. Heinrich, K.M.; Patel, P.M.; O'Neal, J.L.; Heinrich, B.S. High-Intensity Compared to Moderate-Intensity Training for Exercise Initiation, Enjoyment, Adherence, and Intentions: An Intervention Study. *BMC Public Health* **2014**, *14*, 789. [CrossRef]
16. Rodríguez-Torres, Á.F.; Arias-Moreno, E.; Espinosa-Quishpe, A.; Yanchapaxi-Iza, K. Método HITT: Una Herramienta Para el Fortalecimiento de la Condición Física en Adolescentes. Available online: http://www.dspace.uce.edu.ec/handle/25000/24269 (accessed on 13 July 2022).
17. Gaesser, G.A.; Angadi, S.S. High-Intensity Interval Training for Health and Fitness: Can Less Be More? *J. Appl. Physiol.* **2011**, *111*, 1540–1541. [CrossRef]
18. Laursen, P.B.; Jenkins, D.G. The Scientific Basic for High Intensity Interval Training: Optimising Training Programmes and Maximising Performance in Highly Trained Endurance Athletes. *Sport. Med.* **2002**, *32*, 53–73. [CrossRef]
19. Alonso-Fernández, D.; Fernández-Rodríguez, R.; Taboada-Iglesias, Y.; Gutiérrez-Sánchez, Á. Impact of a HIIT Protocol on Body Composition and VO2max in Adolescents. *Sci. Sport.* **2019**, *34*, 341–347. [CrossRef]
20. Ouerghi, N.; Fradj, M.K.B.; Bezrati, I.; Khammassi, M.; Feki, M.; Kaabachi, N.; Bouassida, A. Effects of High-Intensity Interval Training on Body Composition, Aerobic and Anaerobic Performance and Plasma Lipids in Overweight/Obese and Normal-Weight Young Men. *Biol. Sport.* **2017**, *34*, 385–392. [CrossRef]
21. Bonato, M.; Turrini, F.; DE Zan, V.; Meloni, A.; Plebani, M.; Brambilla, E.; Giordani, A.; Vitobello, C.; Caccia, R.; Piacentini, M.F.; et al. A Mobile Application for Exercise Intervention in People Living with HIV. *Med. Sci. Sport. Exerc.* **2020**, *52*, 425–433. [CrossRef]
22. Shcherbina, A.; Hershman, S.G.; Lazzeroni, L.; King, A.C.; O'Sullivan, J.W.; Hekler, E.; Moayedi, Y.; Pavlovic, A.; Waggott, D.; Sharma, A.; et al. The Effect of Digital Physical Activity Interventions on Daily Step Count: A Randomised Controlled Crossover Substudy of the MyHeart Counts Cardiovascular Health Study. *Lancet Digit. Health* **2019**, *1*, e344–e352. [CrossRef]
23. Bellini, A.; Nicolò, A.; Bustos, A.S.O.; Sacchetti, M. Step Count Accuracy and Precision of the Xiaomi Mi Smart Band 5 in Healthy Young Individuals. In Proceedings of the 2021 IEEE International Workshop on Metrology for Industry 4.0 & IoT (MetroInd4.0 & IoT), Roma, Italy, 7–9 June 2021; pp. 198–202.
24. Pino-Ortega, J.; Gómez-Carmona, C.D.; Rico-González, M. Accuracy of Xiaomi Mi Band 2.0, 3.0 and 4.0 to Measure Step Count and Distance for Physical Activity and Healthcare in Adults over 65 Years. *Gait Posture* **2021**, *87*, 6–10. [CrossRef]
25. Chase, J.-A.D. Interventions to Increase Physical Activity Among Older Adults: A Meta-Analysis. *Gerontologist* **2015**, *55*, 706–718. [CrossRef] [PubMed]
26. Eysenbach, G.; Group, C.-E. CONSORT-EHEALTH: Improving and Standardizing Evaluation Reports of Web-Based and Mobile Health Interventions. *J. Med. Internet Res.* **2011**, *13*, e1923. [CrossRef] [PubMed]
27. Agarwal, S.; LeFevre, A.E.; Lee, J.; L'Engle, K.; Mehl, G.; Sinha, C.; Labrique, A. Guidelines for Reporting of Health Interventions Using Mobile Phones: Mobile Health (MHealth) Evidence Reporting and Assessment (MERA) Checklist. *BMJ* **2016**, *352*, i1174. [CrossRef] [PubMed]
28. Dick, B.; Ferguson, B.J. Health for the World's Adolescents: A Second Chance in the Second Decade. *J. Adolesc. Health* **2015**, *56*, 3–6. [CrossRef] [PubMed]

29. Niemeier, H.M.; Raynor, H.A.; Lloyd-Richardson, E.E.; Rogers, M.L.; Wing, R.R. Fast Food Consumption and Breakfast Skipping: Predictors of Weight Gain from Adolescence to Adulthood in a Nationally Representative Sample. *J. Adolesc. Health* **2006**, *39*, 842–849. [CrossRef]
30. Chan, A.-W.; Tetzlaff, J.M.; Altman, D.G.; Laupacis, A.; Gøtzsche, P.C.; Krleža-Jerić, K.; Hróbjartsson, A.; Mann, H.; Dickersin, K.; Berlin, J.A.; et al. SPIRIT 2013 Statement: Defining Standard Protocol Items for Clinical Trials. *Ann. Intern. Med.* **2013**, *158*, 200–207. [CrossRef]
31. Schulz, K.F.; Altman, D.G.; Moher, D. The CONSORT Group CONSORT 2010 Statement: Updated Guidelines for Reporting Parallel Group Randomised Trials. *Trials* **2010**, *11*, 32. [CrossRef]
32. Berlin, D.A.; Gulick, R.M.; Martinez, F.J. Severe COVID-19. *N. Engl. J. Med.* **2020**, *383*, 2451–2460. [CrossRef]
33. Gal, R.; May, A.M.; van Overmeeren, E.J.; Simons, M.; Monninkhof, E.M. The Effect of Physical Activity Interventions Comprising Wearables and Smartphone Applications on Physical Activity: A Systematic Review and Meta-Analysis. *Sport. Med. Open* **2018**, *4*, 42. [CrossRef]
34. Michie, S.; Richardson, M.; Johnston, M.; Abraham, C.; Francis, J.; Hardeman, W.; Eccles, M.P.; Cane, J.; Wood, C.E. The Behavior Change Technique Taxonomy (v1) of 93 Hierarchically Clustered Techniques: Building an International Consensus for the Reporting of Behavior Change Interventions. *Ann. Behav. Med.* **2013**, *46*, 81–95. [CrossRef]
35. Viciana, J.; Casado-Robles, C.; Guijarro-Romero, S.; Mayorga-Vega, D. Are Wrist-Worn Activity Trackers and Mobile Applications Valid for Assessing Physical Activity in High School Students? Wearfit Study. *J. Sport. Sci. Med.* **2022**, *21*, 356–375. [CrossRef]
36. Kim, C.; Kim, S.H.; Suh, M.R. Accuracy and Validity of Commercial Smart Bands for Heart Rate Measurements During Cardiopulmonary Exercise Test. *Ann. Rehabil. Med.* **2022**, *46*, 209–218. [CrossRef]
37. Degroote, L.; Hamerlinck, G.; Poels, K.; Maher, C.; Crombez, G.; Bourdeaudhuij, I.D.; Vandendriessche, A.; Curtis, R.G.; DeSmet, A. Low-Cost Consumer-Based Trackers to Measure Physical Activity and Sleep Duration Among Adults in Free-Living Conditions: Validation Study. *JMIR Mhealth Uhealth* **2020**, *8*, e16674. [CrossRef]
38. Bassett, D.R.; Wyatt, H.R.; Thompson, H.; Peters, J.C.; Hill, J.O. Pedometer-Measured Physical Activity and Health Behaviors in U.S. Adults. *Med. Sci. Sport. Exerc.* **2010**, *42*, 1819–1825. [CrossRef]
39. Kang, M.; Rowe, D.A.; Barreira, T.V.; Robinson, T.S.; Mahar, M.T. Individual Information-Centered Approach for Handling Physical Activity Missing Data. *Res. Q. Exerc. Sport* **2009**, *80*, 131–137. [CrossRef]
40. Tehard, B.; Saris, W.H.M.; Astrup, A.; Martinez, J.A.; Taylor, M.A.; Barbe, P.; Richterova, B.; Guy-Grand, B.; Sørensen, T.I.A.; Oppert, J.-M. Comparison of Two Physical Activity Questionnaires in Obese Subjects: The NUGENOB Study. *Med. Sci. Sport. Exerc.* **2005**, *37*, 1535–1541. [CrossRef]
41. Craig, C.L.; Marshall, A.L.; Sjöström, M.; Bauman, A.E.; Booth, M.L.; Ainsworth, B.E.; Pratt, M.; Ekelund, U.; Yngve, A.; Sallis, J.F.; et al. International Physical Activity Questionnaire: 12-Country Reliability and Validity. *Med. Sci. Sport. Exerc.* **2003**, *35*, 1381–1395. [CrossRef]
42. Hawley-Hague, H.; Horne, M.; Skelton, D.A.; Todd, C. Review of How We Should Define (and Measure) Adherence in Studies Examining Older Adults' Participation in Exercise Classes. *BMJ Open* **2016**, *6*, e011560. [CrossRef]
43. Borg, G. Psychophysical Scaling with Applications in Physical Work and the Perception of Exertion. *Scand. J. Work. Environ. Health* **1990**, *16* (Suppl. S1), 55–58. [CrossRef]
44. McLester, C.N.; Nickerson, B.S.; Kliszczewicz, B.M.; McLester, J.R. Reliability and Agreement of Various InBody Body Composition Analyzers as Compared to Dual-Energy X-Ray Absorptiometry in Healthy Men and Women. *J. Clin. Densitom.* **2020**, *23*, 443–450. [CrossRef]
45. Boyer, C.; Tremblay, M.; Saunders, T.; McFarlane, A.; Borghese, M.; Lloyd, M.; Longmuir, P. Feasibility, Validity, and Reliability of the Plank Isometric Hold as a Field-Based Assessment of Torso Muscular Endurance for Children 8–12 Years of Age. *Pediatr. Exerc. Sci.* **2013**, *25*, 407–422. [CrossRef] [PubMed]
46. Saporito, G.; Jernstedt, G.; Miller, H. Test-Retest Reliability and Validity of the Plank Exercise. Available online: https://www.semanticscholar.org/paper/Test-Retest-Reliability-and-Validity-of-the-Plank-Saporito-Jernstedt/8a8a3aa04a479ddb7a0e2649419e0fd64c7f7ac6 (accessed on 4 November 2022).
47. Zou, H.; Zhang, J.; Chen, X.; Wang, Y.; Lin, W.; Lin, J.; Chen, H.; Pan, J. Reference Equations for the Six-Minute Walk Distance in the Healthy Chinese Han Population, Aged 18–30 Years. *BMC Pulm. Med.* **2017**, *17*, 119. [CrossRef] [PubMed]
48. Scivoletto, G.; Tamburella, F.; Laurenza, L.; Foti, C.; Ditunno, J.F.; Molinari, M. Validity and Reliability of the 10-m Walk Test and the 6-Min Walk Test in Spinal Cord Injury Patients. *Spinal Cord* **2011**, *49*, 736–740. [CrossRef] [PubMed]
49. Ries, J.D.; Echternach, J.L.; Nof, L.; Gagnon Blodgett, M. Test-Retest Reliability and Minimal Detectable Change Scores for the Timed "up & Go" Test, the Six-Minute Walk Test, and Gait Speed in People with Alzheimer Disease. *Phys. Ther.* **2009**, *89*, 569–579. [CrossRef] [PubMed]
50. Kennedy, D.M.; Stratford, P.W.; Wessel, J.; Gollish, J.D.; Penney, D. Assessing Stability and Change of Four Performance Measures: A Longitudinal Study Evaluating Outcome Following Total Hip and Knee Arthroplasty. *BMC Musculoskelet. Disord.* **2005**, *6*, 3. [CrossRef]
51. Harada, N.D.; Chiu, V.; Stewart, A.L. Mobility-Related Function in Older Adults: Assessment with a 6-Minute Walk Test. *Arch. Phys. Med. Rehabil.* **1999**, *80*, 837–841. [CrossRef]
52. Tappen, R.M.; Roach, K.E.; Buchner, D.; Barry, C.; Edelstein, J. Reliability of Physical Performance Measures in Nursing Home Residents With Alzheimer's Disease. *J. Gerontol. A Biol. Sci. Med. Sci.* **1997**, *52*, M52–M55. [CrossRef]

53. Darviri, C.; Alexopoulos, E.C.; Artemiadis, A.K.; Tigani, X.; Kraniotou, C.; Darvyri, P.; Chrousos, G.P. The Healthy Lifestyle and Personal Control Questionnaire (HLPCQ): A Novel Tool for Assessing Self-Empowerment through a Constellation of Daily Activities. *BMC Public Health* **2014**, *14*, 995. [CrossRef]

54. Mollayeva, T.; Thurairajah, P.; Burton, K.; Mollayeva, S.; Shapiro, C.M.; Colantonio, A. The Pittsburgh Sleep Quality Index as a Screening Tool for Sleep Dysfunction in Clinical and Non-Clinical Samples: A Systematic Review and Meta-Analysis. *Sleep Med. Rev.* **2016**, *25*, 52–73. [CrossRef]

55. De la Vega, R.; Tomé-Pires, C.; Solé, E.; Racine, M.; Castarlenas, E.; Jensen, M.P.; Miró, J. The Pittsburgh Sleep Quality Index: Validity and Factor Structure in Young People. *Psychol. Assess.* **2015**, *27*, e22–e27. [CrossRef]

56. Curcio, G.; Tempesta, D.; Scarlata, S.; Marzano, C.; Moroni, F.; Rossini, P.M.; Ferrara, M.; De Gennaro, L. Validity of the Italian Version of the Pittsburgh Sleep Quality Index (PSQI). *Neurol. Sci.* **2013**, *34*, 511–519. [CrossRef]

57. Farrahi Moghaddam, J.; Nakhaee, N.; Sheibani, V.; Garrusi, B.; Amirkafi, A. Reliability and Validity of the Persian Version of the Pittsburgh Sleep Quality Index (PSQI-P). *Sleep Breath.* **2012**, *16*, 79–82. [CrossRef]

58. Dietch, J.R.; Taylor, D.J.; Sethi, K.; Kelly, K.; Bramoweth, A.D.; Roane, B.M. Psychometric Evaluation of the PSQI in U.S. College Students. *J. Clin. Sleep Med.* **2016**, *12*, 1121–1129. [CrossRef]

59. Lee, J.S.; Hobden, E.; Stiell, I.G.; Wells, G.A. Clinically Important Change in the Visual Analog Scale after Adequate Pain Control. *Acad. Emerg. Med.* **2003**, *10*, 1128–1130. [CrossRef]

60. Karcioglu, O.; Topacoglu, H.; Dikme, O.; Dikme, O. A Systematic Review of the Pain Scales in Adults: Which to Use? *Am. J. Emerg. Med.* **2018**, *36*, 707–714. [CrossRef]

61. Kendzierski, D.; DeCarlo, K.J. Physical Activity Enjoyment Scale: Two Validation Studies. *J. Sport Exerc. Psychol.* **1991**, *13*, 50–64. [CrossRef]

62. Latorre-Román, P.Á.; Martínez-López, E.J.; Ruiz-Ariza, A.; Izquierdo-Rus, T.; Salas-Sánchez, J.; García-Pinillos, F. Validez y Fiabilidad Del Cuestionario de Disfrute Por El Ejercicio Físico (PACES) En Adolescentes Con Sobrepeso y Obesidad. *Nutr. Hosp.* **2016**, *33*, 595–601. [CrossRef]

63. DiPietro, L.; Al-Ansari, S.S.; Biddle, S.J.H.; Borodulin, K.; Bull, F.C.; Buman, M.P.; Cardon, G.; Carty, C.; Chaput, J.-P.; Chastin, S.; et al. Advancing the Global Physical Activity Agenda: Recommendations for Future Research by the 2020 WHO Physical Activity and Sedentary Behavior Guidelines Development Group. *Int. J. Behav. Nutr. Phys. Act.* **2020**, *17*, 143. [CrossRef]

Protocol

An Active Retirement Programme, a Randomized Controlled Trial of a Sensorimotor Training Programme for Older Adults: A Study Protocol

Carolina Alexandra Cabo [1,*], Orlando Fernandes [1,2], María Mendoza-Muñoz [3,*], Sabina Barrios-Fernandez [4], Laura Muñoz-Bermejo [5], Rafael Gómez-Galán [6] and Jose A. Parraca [1,2]

[1] Departamento de Desporto e Saúde, Escola de Saúde e Desenvolvimento Humano, Universidade de Évora, Largo dos Colegiais 2, 7000-645 Évora, Portugal
[2] Comprehensive Health Research Centre (CHRC), University of Évora, Largo dos Colegiais 2, 7000-645 Évora, Portugal
[3] Research Group on Physical and Health Literacy and Health-Related Quality of Life (PHYQOL), Faculty of Sport Sciences, University of Extremadura, 10003 Caceres, Spain
[4] Occupation, Participation, Sustainability and Quality of Life (Ability Research Group), Nursing and Occupational Therapy College, University of Extremadura, 10003 Caceres, Spain
[5] Social Impact and Innovation in Health (InHEALTH), University Centre of Mérida, University of Extrema dura, 06800 Mérida, Spain
[6] Research Group on Physical and Health Literacy and Health-Related Quality of Life (PHYQOL), University Centre of Mérida, University of Extremadura, 06800 Mérida, Spain
* Correspondence: carolinaacabo@gmail.com (C.A.C.); mamendozam@unex.es (M.M.-M.)

Abstract: Research shows that exercise training programmes lead to several improvements in older adults' health-related quality of life (HRQoL) and well-being. This study will examine the effects of an active retirement programme on Portuguese older adults, investigating its effects on body composition, physical fitness, HRQoL, and physical activity level (PAL). Therefore, a parallel-group randomised controlled trial will be conducted, including body composition (height and body weight), physical fitness (strength, flexibility, agility, postural control, and gait), HRQoL, and PAL assessments before and after the application of the programme. The programme will be carried out for six months, two days per week (45 min), plus a year of follow-up. The programme will consist of six circuits with eight physical exercises each. The circuits will change at the end of the four weeks (one monthly circuit). The exercises' difficulty will increase throughout the programme, with alternatives for all the participants. If the effectiveness of the programme is demonstrated, implementation in different services and municipalities could be advised, as the actors involved in health and social services should promote the well-being of their citizens through, among others, health-related physical activity and the prevention of diseases associated with inactivity.

Keywords: elderly; falls; gait; postural control; exercise; quality of life

Citation: Cabo, C.A.; Fernandes, O.; Mendoza-Muñoz, M.; Barrios-Fernandez, S.; Muñoz-Bermejo, L.; Gómez-Galán, R.; Parraca, J.A. An Active Retirement Programme, a Randomized Controlled Trial of a Sensorimotor Training Programme for Older Adults: A Study Protocol. *Healthcare* **2023**, *11*, 86. https://doi.org/10.3390/healthcare11010086

Academic Editor: Filipe Manuel Clemente

Received: 17 November 2022
Revised: 19 December 2022
Accepted: 25 December 2022
Published: 28 December 2022

1. Introduction

Demographic ageing is the paradigm of the 21st century, and the older adult population is trending upwards. According to the central forecast scenario of the Office of Statistics National Institute, the number of older adults will increase from 2.1 to 2.8 million between 2015 and 2080 [1]. Thus, understanding and preventing disability in older adults is an important public health challenge, as ageing is a risk factor contributing to functional decline [2]. Ageing is continuous and irreversible and can be related to physical and cognitive function decline [3].

Physical activity (PA) plays an important role in healthy ageing, extending life, and quality of life, reducing chronic the incidence of physical and mental diseases. Moreover, every additional 15 min of daily PA (up to 100 min/day) reduces the mortality rate from any cause 4%. Although mechanisms underlying PA's positive effects on health have not been fully revealed, evidence points to inflammatory markers, cellular oxidative stress regulation, and the impact on telomere length that could delay the ageing process [4]. According to the World Health Organization (WHO)'s recommendations [5], more than a quarter of the worldwide population (27.5%) does not perform at least 150 min of moderate or 75 min of vigorous PA per week [6]. Older adults should conduct, at least 150 min of moderate-intensity aerobics, at least 75 min of vigorous-intensity aerobics, or an equivalent combination of moderate- and vigorous-intensity PA throughout the week for substantial health benefits. Older adults should also perform moderate or greater-intensity muscle-strengthening activities that involve all major muscle groups two or more days a week, as these provide additional health benefits. As part of their weekly PA, older adults should practise multicomponent PA, emphasising functional balance and strength training at a moderate or high intensity three or more days a week to prevent falls [5]. In this sense, Gopinath et al. [4] carried out a study for 10 years in Australia, following up with adults over 50 years involved in moderate-vigorous PA ($\geq$ 5000 MET minutes/week), which found that participants doubled their survival probability without chronic diseases, cognitive deficits, or functional incapacity. Moreover, PA levels (PAL) in older people should be higher than those recommended by the WHO (600 MET minutes/week) [5] to reach healthy ageing. There are 2.4 million individuals aged 65 and over in Portugal [7], with only 16.6% following the WHO's recommendations [8]. A lack of PA contributes to the loss of functional fitness in the elderly [9]. Aerobic exercise is important for optimal musculoskeletal performance; however, aerobic capacity gradually declines with advancing age. Likewise, exercised muscles are more sensitive to anabolic stimuli, such as protein ingestion, which allow muscle protein synthesis [10]. As a result, sedentary elderly people have lower muscle mass and reduced bone mass, and, consequently, their ability to perform PA is affected.

Psychosocial factors also impact the functional status of the elderly. Self-efficacy and fear of falling are important factors [11]. Self-efficacy is not only associated with gait speed and physical-function limitations but also with participation in physical exercise programmes. In turn, fear of falling negatively influences sedentary behaviour time. Other psychosocial factors associated with reduced functional fitness are loneliness, depression, and exhaustion [10]. Moreover, physical factors impact the ageing process. A decrease in strength leads to a lower functional capacity, loss of balance, and decreased gait speed [12]: older adults generally have a wider gait, postural sway, and greater gait variability when walking [13]. Moreover, the sensorimotor system incorporates all processing and the afferent, efferent, and centrally integrating components involved in joints' functional stability. As people age, this ability becomes limited, leading to important biomechanical changes in gait [14]. Loss of mobility, decreased muscle strength, and a balance deficit all contribute to dependence [15]. Hence, sensorimotor training improves older adults' balance and provides gait confidence, although it does not affect the biomechanical parameters [16]. Several studies on sensorimotor training reported improvement in performance in clinical balance and mobility [17]: sensorimotor training-induced shortened onset latency in postural muscles [18], decreased onset latency, and enhanced reflex activity [19]. Sensorimotor training also induces an improvement in maximal and explosive-force production of leg extensors' capacity [20] and improves lower-limb strength and power, static and dynamic balance, and mobility [21].

Research on how PA affects health-related quality of life (HRQoL) has shown that PA's benefits regarding quality of life were limited [22,23]. Several studies have evaluated the sensorimotor system in the elderly and obtained strength, mobility, and balance improvements, but these results were not consistent due to the reduced sample [20,21], being short-term (12 weeks or less), or only assessing some abilities [17]. This study will last 24 weeks, and its effects will be measured through gait, postural control, strength, flexibility, balance, and agility. Hence, this study will aim to (1) assess the effectiveness of an active retirement programme on elderly body composition, physical fitness, HRQol, and PAL and (2) assess the programme's effectiveness throughout the ageing process (follow-up).

2. Materials and Methods

2.1. Design

A parallel-group randomised controlled trial will be conducted, including a 6-month intervention phase and a 1-year follow-up period. For both groups (control and experimental), assessments will be performed at baseline (before starting the intervention), and after the intervention ends. The study will follow the Consolidated Standards of Reporting Trials Statement (CONSORT) [24].

2.2. Ethics

The Ethics Committee of the University of Évora approved this project (approval number: 21040). The study was registered with the Clinical Tri-als.gov PRS Protocol Registration and Results System (Registration Number: NCT05398354; https://www.clinicaltrials.gov/ct2/show/NCT05398354?term=NCT05398354&draw=2&rank=1 (accessed 12 June 2022).

2.3. Sample Size

Sample size calculations were performed using the G*Power 3.1.9.4 software (Kiel University, Kiel, Germany), selecting the statistical test to compare the difference between two independent means (two groups). Thus, accepting an alpha risk of 0.05 and a beta risk of 0.2 in a bilateral contrast and assuming a moderate effect size of 0.5, a total of 160 participants (80 subjects in the experimental group and 80 in the control group) were sufficient to reach a minimum potency of 90%.

2.4. Randomisation and Blinding

Participants will be randomly assigned to the experimental (active retirement programme) or control groups. To assign participants to each group (1:1), Research Randomizer software (version 4.0, Geof-frey C. Urbaniak and Scott Plous, Middletown, CT, USA; http://www.randomizer.org, accessed 12 June 2022) will be used to create a randomisation sequence. A member of the research team, who will not participate in the intervention, will carry out this process. Group assignment will be hidden in a password-protected computer file. Participants will know their group assignment, but outcome assessors and data analysts will not know the participants' group assignment.

2.5. Participants

Participants must meet the following inclusion criteria: (1) be retired; (2) be aged between 55 and 80; (3) show their agreement to participate in the study by providing a signed consent form; (4) be individuals without dentures (except dental prosthesis); and (4) not have undergone surgery for less than six months. Exclusion criteria will be individuals with (1) musculoskeletal diseases; (2) locomotion issues; (3) psychiatric disorders; (4) neurological disorders; and (5) cardiovascular diseases.

2.6. Intervention

Experimental group: This sensorimotor training programme will be carried out for six months, twice per week. As shown in Figure 1, as the programme progresses the load will be progressively increased.

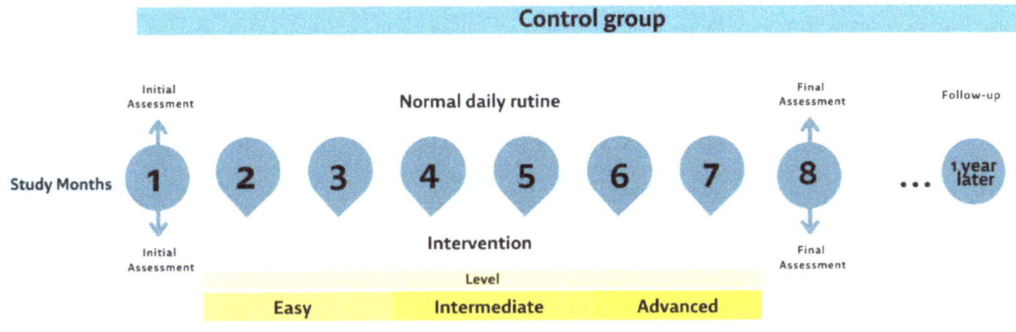

Figure 1. Study timeline graph.

Sessions will be divided into three intensity levels: easy (no external load during the first eight weeks), intermediate (application of external load: elastic bands, shin guards, and free weights, from the 9th week to the 16th week) and advanced (increase in external load for the previous level, from the 17th week to the 24th week). Each month, a different type of session will be developed (Figure 2). Each session will last 45 min and will be divided into three phases: an initial one, consisting of a 5 min walk followed by a joint warm-up (10 min); a fundamental phase (25 min), working on exercises circuits consisting of four cycles, with eight exercises each (50 s on and 15 s off); and a return to calm with muscle stretching (10 min). Additionally, at the end of every session, the intensity will be assessed using the Borg Rating of Perceived Exertion Scale, and the level of satisfaction of the participants will be assessed employing the Physical Activity Enjoyment Scale (PACES) [24].

Control group: Individuals will continue with their normal daily routine, only participating in the assessments. They will be offered to perform the exercise programme when the study ends.

Moreover, to guarantee participants' safety, different strategies will be applied:

- Anamnesis of all participants;
- Hygiene of the material to be used during the sessions;
- Breaks between exercises;
- More than one researcher assisting with the sessions;
- Application of the effort scale.

2.7. Measures

A variety of tools will be used. All measures will be undertaken at baseline, at the end of the intervention, and one year after the end of the intervention (follow-up). Before the first measurement, all participants will go through a familiarisation phase to familiarise themselves with the different instruments and assessments included in this project.

a. Anthropometrics and body composition. Bodyweight and height will be assessed. Before the measurements, participants will be asked to remove their shoes, socks, and heavy clothing (coats, sweaters, coats, etc.). They will also be asked to empty their pockets and remove belts and other accessories (bands, pendants, etc.). Height will be measured using a stadiometer (Seca 22, Hamburg, Germany). This instrument must be placed on a vertical surface with the measuring scale perpendicular to the ground. Participants will be positioned in a standing position, with their shoulders balanced, and their arms relaxed along their body. Height will be taken in cm and rounded to the nearest mm. Body weight will be measured using a scale. Body weight will be recorded in kg. and the body mass index (BMI) will be determined using the formula: $weight \times height^2$.

b. Physical fitness. Participants will wear tracksuit bottoms and will be asked to remove accessories and any objects in their pockets. The following measures will be carried out (Figure 3): (I) agility and execution speed will be assessed through the Timed Up and Go (TUG) test, which consists of getting up from a chair, walking in a straight line three meters away, and walking back and sitting down again [25]; (II and III) postural control will be tested through a force platform (Ber-tec4060-Columbus; USA). The assessment will consist of measuring the oscillations in a static bipedal position, with eyes open (two minutes) and with eyes closed (two minutes) [26]; (IV) gait will be assessed using the mobile application "Phyphox" on the surface of the skin, at the inner edge of the tibia, to quantify the number of steps and time. Participants will be asked to walk a pre-established route, without slopes or obstacles, for 10 min, at their natural cadence and, later, walk the same route at a pace determined by complex stimuli (auditory metronome—loudspeakers, which allow for hearing beats that correspond to steps) [27]; (V) muscular endurance will be assessed by rising up from the chair or bending and straightening for 30 s, during which the strength of the lower limbs involving the vastus medialis obliquus (VMO) and the vastus lateralis (VL) will also be calculated [28]; (VI) upper limb strength will be determined by the number of times that a weight can be lifted by performing a flexion–extension of the arms for 30 s [28]; (VII and VIII) flexibility will be evaluated by two tests: "the sit and reach" for lower limb flexibility, where participants, from a seated position with one leg extended, will slowly bend down, sliding their hands down the extended leg until they touch (or pass) their toes [28]; and "the behind the back reach" test for upper limbs, assessing the shoulder's full range of motion, which will consist of measuring with a ruler the distance between (or the overlap of) the middle fingers behind the back [28].

c. HRQoL. This will be assessed using the 36-Item Short-Form Survey (SF-36) [29] in its Portuguese version [30,31], a 36-question tool, which results in 8 dimensions of health status (physical function, physical role, bodily pain, general health, vitality, social function, emotional role, and mental health) and 2 summary components (physical and mental). Dimensions and components are scored from 0 to 100, where 0 is the worst state, and 100 is the best.

d. PAL. This will be assessed using the International Physical Activity Questionnaire Short Form (IPAQ-SF) [32]. This instrument consists of four questions informing on the frequency (days/week) and duration (minutes/day) of daily walks and activities requiring moderate to vigorous physical exertion, as well as the time (minutes/day) spent on sitting activities on weekdays and weekends. PA will be classified into three categories according to the IPAQ consensus group: sedentary (<600 Met-minutes/week), active ($\geq$600 Met-minutes/week), and very active ($\geq$3000 Met-minutes/week). This instrument will be completed by the participants in its Portuguese version [33].

e. Subjective Perception of Effort. This will be assessed through the Borg Rating of Perceived Exertion Scale [34] during the sessions, consisting of 10 items for which participants rate their effort from "not at all" (1) to "maximum" (10).

f. Adherence rate. This will be controlled. For this purpose, following previous studies [35,36] the following parameters will be monitored: proportion of participants completing exercise programmes, proportion of exercise sessions attended, average number of exercise sessions completed per week, class attendance expressed as a proportion of participants reaching certain cutoffs, number of weeks in which exercise was undertaken, and proportion of days on which exercise was undertaken. In addition, recommendations from previous studies will be followed to encourage adherence to the programme, including making instructions to subjects simpler and less demanding, addressing cognitive motivational factors such as self-efficacy and health beliefs, offering social support and reinforcement, and providing reminders [37,38].

g. Level of satisfaction. This will be evaluated through the Physical Activity Enjoyment Scale (PACES) survey, which consists of 8 items that are scored from 1 to 7 points (where 1 = "I enjoy it", 7 = "I hate it", and 4 = "neutral"). The total score will be calculated from the sum of the items, with a maximum possible score of 56 points and a minimum of 8; the higher the score is, the greater the enjoyment is. The Portuguese version will be completed by the participants [24].

2.8. Statistical Analysis

Descriptive statistics and computations will be performed with SPSS (version 25.0; IBM SPSS Inc., Armonk, IL, USA). Personal data will be kept anonymous.

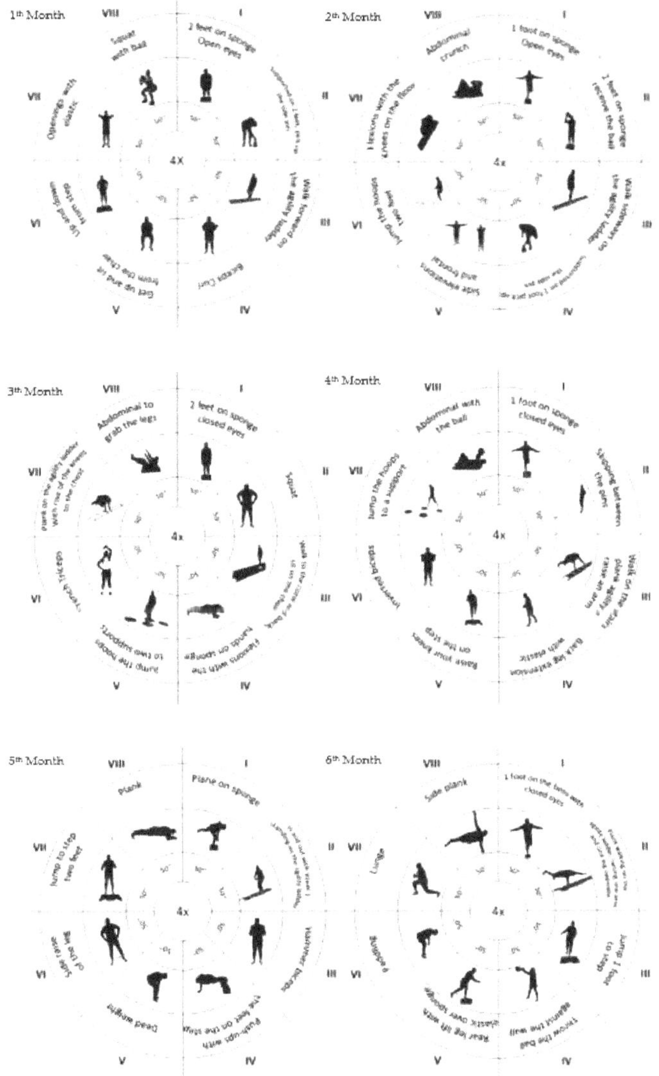

Figure 2. Sessions' main part planning.

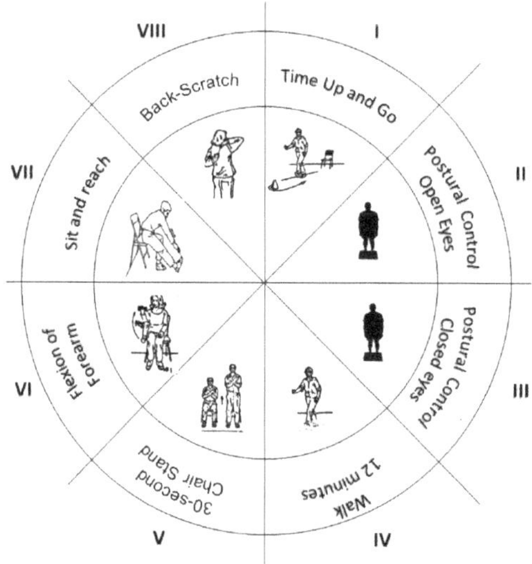

Figure 3. Assessments for physical fitness measurements.

The normality and homogeneity of data will be checked by applying Kolmogorov–Smirnov and Levene's tests, respectively. Data will be presented as means and standard deviation (SD) (parametric variables) or median and interquartile range (IR) (nonparametric variables). The independent samples t-test (parametric variables) or the Mann–Whitney U test (non-parametric variables) will be used to determine whether the experimental and control groups were comparable at baseline in terms of participant characteristics. Then, repeated measures of ANCOVA will be applied to analyse the intervention effects on the different dependent variables, adjusted by age and baseline outcomes. Cohen's d (with a 95% confidence interval) will also be included in the results as the effect size. Effect size thresholds will be interpreted as follows: >0.2, small; >0.5, moderate; >0.8, large [36]. Statistical significance will be computed for the effect of time and the interaction group × time. The alpha level will be fixed at $p \leq 0.05$.

3. Discussion

This project will be the first conducted developing a sensorimotor training programme for 24 weeks using different sensorimotor skills in Portuguese older adults, to test this programme's effects on HRQoL and fall prevention compared to usual care. Different difficulty levels will be included, facilitating individualisation. This training programme can be applied by multiple agents who might be interested in reaping the benefits of these types of activities, since this training method does not require any specific installation and can be completed indoors or outdoors.

In this sense, Avelar et al. [21] carried out a similar quasi-experimental study, with a similar training period composed of exercise circuits to improve strength, gait, functional reach, and static and dynamic balance, among others. Positive responses in lower limb strength, power, mobility, and static and dynamic balance were found, although the sample was small [20,21,27], compared to the 160 participants expected for our study. Other research [17,18,39] used sensorimotor training programmes on functional capacity and balance for a short time, from two to six weeks. Despite the short period, positive results were obtained concerning balance, mobility, activities of daily living performance, and fall risk. Thus, by improving physical capabilities, fall risk could be lower. Other studies [19–21] have applied 12–13 weeks of sensorimotor training, half of what is proposed in this protocol. There were positive results for mobility and balance and an improved ability to produce the maximum and explosive force of leg extensors. Several of these studies developed a short-time practical application [39], had a small size [21], or assessed only one or two sensorimotor abilities [17–20]. For all these reasons, no studies with consistent results have produced an advance on this topic, bearing in mind that the percentage of old people practising PA is very low [40], so it is essential to promote PA practise, as sensorimotor skills may be deficient.

The application of this study is expected to obtain an increase in physical capacities such as strength, flexibility, gait, postural control, balance, and agility. In addition, we are also aiming to improve the participants' HRQoL and reduce their fall risk. The programme could be applied to public and private entities. In the public sector, the application of active retirement programmes within the services offered by public health programmes exists at a regional level, as is the case of the "Alma Senior" programme, developed in the Municipality of Almada, Portugal, or the "The Exercise Looks After You" programme, developed in Extremadura, Spain, which increases its cost–benefit ratio for all aspects related to HRQoL and fall prevention in this sector. Likewise, the possibility of implementing this training system in different associations can be studied. In the private sector, a focus group study can include the heads of health and sports centres, focusing on the applicability of sensorimotor training in their centres, with the results of the study able to highlight the potential advantages of private sector application.

The Portuguese population has experienced a substantial increase in average life expectancy in the last four decades. During the same period, there has been a progressive increase in health expenditure [41]. Regarding the relationship between ageing and health expenditure, there does not seem to be a consensus in the literature on the effects of ageing on health expenditure. The literature refutes that population age is a major determinant of health expenditure and other factors, thus considering a strong positive correlation between the two variables. This correlation is not confirmed in the majority of econometric analyses that have been developed in the last two decades in the area of health economics [42]. Thus, PA should be considered an essential item for public health, since, if performed regularly, it is considered an important condition for health promotion and the prevention/remediation of chronic diseases, for different age groups [43].

Regarding limitations, one of the limitations that will exist for the application of this study is being after/during the COVID-19 pandemic. We must take extra care, by adding space and material for the hygienists, to give greater security to the programme. Travel to the space where the programme will be carried out may also be a limitation, considering that the population under study involves people who no longer travel by car; it may be too far away for them to travel on foot, so there will be increased costs due to the necessity of using public transport.

In the future, it would be interesting to evaluate all age groups, in addition to this age group, such as those who are still actively employed, to have a better perception of what happens throughout life in terms of sensorimotor behaviour, not just during the ageing process. This could result in people being able to act from an early age, to improve the capacities analysed in this study. In addition, it would be interesting to add other clinical variables more related to health, such as blood pressure and cholesterol, because, as reported in other studies [44], there were improvements in these variables in the physical condition of participants. We can also analyse the effects of the programme on the number of medical appointments. Finally, applying a pre-participation physical assessment of the participants by a sports physician would be beneficial.

The added value of this study will be the acquired knowledge about sensorimotor training for the elderly, which will present a scientific advancement for PA, HRQoL, and sedentary behaviour prevention. Finally, it will be innovative in creating an intervention manual and a digital library with specific exercises for sensorimotor training, which allow for better adaptation and motor coordination.

4. Conclusions

This study will investigate the effectiveness of an active retirement programme in the elderly on their body composition, physical condition, HRQoL, and fall risk, advancing the knowledge of PA, health, and wellness.

Author Contributions: Formal analysis, J.A.P. and O.F.; investigation, C.A.C. and M.M.-M.; supervision, O.F.; writing—original draft, C.A.C. and S.B.-F.; writing—review and editing, J.A.P., R.G.-G. and M.M.-M., visualisation, S.B.-F. and L.M.-B., funding acquisition, L.M.-B. and R.G.-G. All authors have read and agreed to the published version of the manuscript.

Institutional Review Board Statement: The study was approved by the Ethics Committee of the University of Évora (approval number: 21040).

Informed Consent Statement: Not applicable.

Data Availability Statement: Not applicable.

Acknowledgments: M.M.-M. was supported by a grant from the Universities Ministry and the European Union (NextGenerationUE) (MS-12).

Conflicts of Interest: The authors declare no conflict of interest.

References

1. Instituto Nacional de Estadística. Available online: https://www.ine.pt/xportal/xmain?xpgid=ine_main&xpid=INE (accessed on 4 July 2022).
2. Anton, S.D.; Cruz-Almeida, Y.; Singh, A.; Alpert, J.; Bensadon, B.; Cabrera, M.; Clark, D.J.; Ebner, N.C.; Esser, K.A.; Fillingim, R.B.; et al. Innovations in Geroscience to Enhance Mobility in Older Adults. *Exp. Gerontol.* **2020**, *142*, 111123. [CrossRef] [PubMed]
3. Wong, T.K.K.; Ma, A.W.W.; Liu, K.P.Y.; Chung, L.M.Y.; Bae, Y.-H.; Fong, S.S.M.; Ganesan, B.; Wang, H.-K. Balance Control, Agility, Eye–Hand Coordination, and Sport Performance of Amateur Badminton Players: A Cross-Sectional Study. *Medicine* **2019**, *98*, e14134. [CrossRef] [PubMed]
4. Gopinath, B.; Kifley, A.; Flood, V.M.; Mitchell, P. Physical Activity as a Determinant of Successful Aging over Ten Years. *Sci. Rep.* **2018**, *8*, 1–5. [CrossRef]
5. Bull, F.C.; Al-Ansari, S.S.; Biddle, S.; Borodulin, K.; Buman, M.P.; Cardon, G.; Carty, C.; Chaput, J.-P.; Chastin, S.; Chou, R. World Health Organization 2020 guidelines on physical activity and sedentary behaviour. *Br. J. Sport. Med.* **2020**, *54*, 1451–1462. [CrossRef] [PubMed]
6. Guthold, R.; Stevens, G.A.; Riley, L.M.; Bull, F.C. Worldwide trends in insufficient physical activity from 2001 to 2016: A pooled analysis of 358 population-based surveys with 1.9 million participants. *Lancet Glob. Health* **2018**, *6*, e1077–e1086. [CrossRef]
7. PORDATA. *População Residente Com 65 E Mais Anos, Estimativas a 31 de Dezembro: Total E Por Grupo Etário*; PORDATA: Lisboa, Portugal, 2021.
8. Portugal Physical Activity Factsheet. 2021. Available online: https://cdn.who.int/media/docs/librariesprovider2/country-sites/physical-activity-factsheet---portugal-2021.pdf?sfvrsn=d8fac7be_1&download=true (accessed on 15 July 2022).
9. Ghram, A.; Briki, W.; Mansoor, H.; Al-Mohannadi, A.S.; Lavie, C.J.; Chamari, K. Home-Based Exercise Can Be Beneficial for Counteracting Sedentary Behavior and Physical Inactivity during the COVID-19 Pandemic in Older Adults. *Postgrad. Med.* **2021**, *133*, 469–480. [CrossRef]
10. Tieland, M.; Trouwborst, I.; Clark, B.C. Skeletal Muscle Performance and Ageing: Skeletal Muscle Performance and Ageing. *J. Cachexia Sarcopenia Muscle* **2018**, *9*, 3–19. [CrossRef]
11. Garcia Meneguci, C.A.; Meneguci, J.; Sasaki, J.E.; Tribess, S.; Júnior, J.S.V. Physical Activity, Sedentary Behavior and Functionality in Older Adults: A Cross-Sectional Path Analysis. *PLoS ONE* **2021**, *16*, e0246275. [CrossRef]
12. Brief, A. Envelhecimento, força muscular e atividade física: Uma breve revisão bibliográfica. *Rev. Científica FacMais.* **2012**, *2*, 1.
13. Ready, E.A. *Optimizing Gait Outcomes in Parkinson's Disease with Auditory Cues: The Effects of Synchronization, Groove, and Beat Perception Ability*; The University of Western Ontario: London, ON, Canada, 2019.
14. Suzuki, K.; Niitsu, M.; Kamo, T.; Otake, S.; Nishida, Y. Effect of Exercise with Rhythmic Auditory Stimulation on Muscle Coordination and Gait Stability in Patients with Diabetic Peripheral Neuropathy: A Randomized Controlled Trial. *OJTR* **2019**, *7*, 79–91. [CrossRef]
15. Bacha, J.M.R.; Cordeiro, L.R.; Alvisi, T.C.; Bonfim, T.R. Impacto Do Treinamento Sensório-Motor Com Plataforma Vibratória No Equilíbrio e Na Mobilidade Funcional de Um Indivíduo Idoso Com Sequela de Acidente Vascular Encefálico: Relato de Caso. *Fisioter. Pesqui.* **2016**, *23*, 111–116. [CrossRef]
16. Rezende, A.A.B.; Silva, I.L.; Beresford, H.; Batista, L.A. Avaliação Dos Efeitos de Um Programa Sensório-Motor No Padrão Da Marcha de Idosas. *Fisioter. mov.* **2012**, *25*, 317–324. [CrossRef]
17. Steadman, J.; Donaldson, N.; Kalra, L. A randomized controlled trial of an enhanced balance training program to improve mobility and reduce falls in elderly patients. *J. Am. Geriatr. Soc.* **2003**, *51*, 847–852. [CrossRef] [PubMed]

18. Hu, M.-H.; Woollacott, M.H. Multisensory Training of Standing Balance in Older Adults: I. Postural Stability and One-Leg Stance Balance. *J. Gerontol.* **1994**, *49*, M52–M61. [CrossRef] [PubMed]
19. Granacher, U.; Gollhofer, A.; Strass, D. Training induced adaptations in characteristics of postural reflexes in elderly men. *Gait Posture* **2006**, *24*, 459–466. [CrossRef]
20. Granacher, U.; Gruber, M.; Gollhofer, A. Resistance Training and Neuromuscular Performance in Seniors. *Int. J. Sports Med.* **2009**, *30*, 652–657. [CrossRef]
21. Avelar, B.P.; Costa, J.N.D.A.; Safons, M.P.; Dutra, M.T.; Bottaro, M.; Gobbi, S.; Tiedemann, A.; de David, A.C.; Lima, R.M. Balance Exercises Circuit Improves Muscle Strength, Balance, and Functional Performance in Older Women. *AGE* **2016**, *38*, 14. [CrossRef]
22. De Oliveira, A.C.; Oliveira, N.M.D.; Arantes, P.M.M.; Alencar, M.A. Qualidade de Vida Em Idosos Que Praticam Atividade Física—Uma Revisão Sistemática. *Rev. Bras. De Geriatr. E Gerontol.* **2010**, *13*, 301–312. [CrossRef]
23. Marquez, D.X.; Aguiñaga, S.; Vásquez, P.M.; Conroy, D.E.; Erickson, K.I.; Hillman, C.; Stillman, C.M.; Ballard, R.M.; Sheppard, B.B.; Petruzzello, S.J.; et al. A Systematic Review of Physical Activity and Quality of Life and Well-Being. *Transl. Behav. Med.* **2020**, *10*, 1098–1109. [CrossRef]
24. Teques, P.; Calmeiro, L.; Silva, C.; Borrego, C. Validation and Adaptation of the Physical Activity Enjoyment Scale (PACES) in Fitness Group Exercisers. *J. Sport Health Sci.* **2020**, *9*, 352–357. [CrossRef]
25. Gerhardy, T.; Gordt, K.; Jansen, C.-P.; Schwenk, M. Towards Using the Instrumented Timed Up-and-Go Test for Screening of Sensory System Performance for Balance Control in Older Adults. *Sensors* **2019**, *19*, 622. [CrossRef] [PubMed]
26. Salehi, R.; Ebrahimi-Takamjani, I.; Esteki, A.; Maroufi, N.; Parnianpour, M. Test-Retest Reliability and Minimal Detectable Change for Center of Pressure Measures of Postural Stability in Elderly Subjects. *Med. J. Islam. Repub. Iran* **2010**, *23*, 224–232.
27. Malacrida, J.P. *Aplicativos Em Smartphones: O Despertar Científico No Estudo de Energia. Dissertação de Mestrado*; Universidade Estadual de Maringá: Maringá, Brazil, 2021.
28. Rikli, R.E.; Jones, C.J. Development and Validation of a Functional Fitness Test for Community-Residing Older Adults. *J. Aging Phys. Act.* **1999**, *7*, 129–161. [CrossRef]
29. Ware, J.E. *SF-36 Health Survey: Manual and Interpretation Guide*; Health Institute: Boston, MA, USA, 1993.
30. Ferreira, P.L. Development of the Portuguese version of MOS SF-36. Part II—Validation tests. *Acta Med. Port.* **2000**, *13*, 119–127.
31. Ferreira, P.L. Development of the Portuguese version of MOS SF-36. Part I. Cultural and linguistic adaptation. *Acta Med. Port.* **2000**, *13*, 55–66. [PubMed]
32. Lee, P.H.; Macfarlane, D.J.; Lam, T.H.; Stewart, S.M. Validity of the international physical activity questionnaire short form (IPAQ-SF): A systematic review. *Int. J. Behav. Nutr. Phys. Act.* **2011**, *8*, 1–11. [CrossRef]
33. Cruz, J.; Jácome, C.; Morais, N.; Oliveira, A.; Marques, A. Concurrent validity of the Portuguese version of the Brief physical activity assessment tool. *BMC Health Serv. Res.* **2018**, *18*, P150.
34. Borg, G. *Borg´s Perceived Exertion and Pain Scales*; Human Kinetics: Champaign, IL, USA, 1998.
35. Pavey, T.; Taylor, A.; Hillsdon, M.; Fox, K.; Campbell, J.; Foster, C.; Moxham, T.; Mutrie, N.; Searle, J.; Taylor, R. Levels and Predictors of Exercise Referral Scheme Uptake and Adherence: A Systematic Review. *J. Epidemiol. Community Health* **2012**, *66*, 737–744. [CrossRef]
36. Picorelli, A.M.A.; Pereira, L.S.M.; Pereira, D.S.; Felício, D.; Sherrington, C. Adherence to Exercise Programs for Older People Is Influenced by Program Characteristics and Personal Factors: A Systematic Review. *J. Physiother.* **2014**, *60*, 151–156. [CrossRef]
37. Flegal, K.; Kishiyama, S.; Zajdel, D.; Haas, M.; Oken, B. Adherence to Yoga and Exercise Interventions in a 6-Month Clinical Trial. *BMC Complement. Altern. Med.* **2007**, *7*, 37. [CrossRef]
38. Rivera-Torres, S.; Fahey, T.D.; Rivera, M.A. Adherence to Exercise Programs in Older Adults: Informative Report. *Gerontol. Geriatr. Med.* **2019**, *5*, 233372141882360. [CrossRef] [PubMed]
39. Gomes, C.S.; Rangel, G.M.B.; Sant'Ana, M.E.G.D.S.; Dos Santos, M.A.T.; Fraga, W.L.D.A.; Soares, E.V.; Ribeiro Junior, S.M.S. Efeitos do treinamento sensório motor por meio de dispositivos ecoeficientes sobre a capacidade funcional e equilibrio em idosos: Ensaio clinico controlado. *BS* **2018**, *8*, 28. [CrossRef]
40. World Health Organization. *2021 Physical Activity Factsheets for the European Union Member States in the WHO European Region*; World Health Organization: Copenhagen, Denmark, 2021.
41. Santana, P.; Almendra, R. The Health of the Portuguese over the Last Four Decades: Comments on the Course Taken. *Méditerranée. Rev. Géographique Des Pays Méditerranéens J. Mediterr. Geogr.* **2018**, *130*. [CrossRef]
42. Estevens, J.P.G. Envelhecimento E Despesa Em SaúDe: O Caso Português. Ph.D. Thesis, Universidade Nova de Lisboa, Lisbon, Portugal, 2015.
43. Elesbão, H.; Ramos, E.R.; Da Silva, J.O.; Borfe, L. A Influência Da Atividade Física Na Promoção Da Saúde Em Tempos de Pandemia de Covid-19: Uma Revisão Narrativa. *RIPS* **2021**, *3*, 158–164. [CrossRef]
44. Britto, L.S.; Vasconcelos Filho, F.S.L. Análise da relação de parâmetros imunológicos com o desenpenho de idoso na corrida. In *Educação Física Para Grupos Especiais: Exercício Físico Como Terapia Alternativa Para Doenças Crônicas*; Editora Científica Digital: São Paulo, Brazil, 2021; pp. 12–29. ISBN 9786589826859.

MDPI AG
Grosspeteranlage 5
4052 Basel
Switzerland
Tel.: +41 61 683 77 34

Healthcare Editorial Office
E-mail: healthcare@mdpi.com
www.mdpi.com/journal/healthcare